Also in the Variorum Collected Studies Series

ROGER LING
Stuccowork and Painting in Roman Italy

HANS BUCHWALD
Form, Style and Meaning in Byzantine Church Architecture

JACQUES HEYMAN
Arches, Vaults and Buttresses: Masonry Stuctures and their Engineering

ANTHONY CUTLER
Late Antique and Byzantine Ivory Carving

RICHARD GORDON
Image and Value in the Graeco-Roman World: Studies in Mithraism and Religious Art

ANTHONY CUTLER
Imagery and Ideology in Byzantine Art

PAULO M. COSTA
Studies in Arabian Architecture

MACHIEL KIEL
Studies on the Ottoman Architecture of the Balkans: A Legacy in Stone

THOMAS F. MATHEWS
Art and Architecture in Byzantium and Armenia: Liturgical and Exegetical Approaches

CHRISTOPHER WALTER
Prayer and Power in Byzantine and Papal Imagery

CYRIL MANGO
Studies in Constantinople

JOSEPH GUTMANN
Sacred Images: Studies in Jewish Art from Antiquity to the Middle Ages

JULIAN GARDNER
Patrons, Painters and Saints: Studies in Medieval Italian Painting

HENRY MAGUIRE
Rhetoric, Nature and Magic in Byzantine Art

ROBIN CORMACK
The Byzantine Eye: Studies in Art and Patronage

VARIORUM COLLECTED STUDIES SERIES

Structure in Architecture

Rowland J. Mainstone

Structure in Architecture

History, Design and Innovation

Routledge
Taylor & Francis Group
LONDON AND NEW YORK

First published 1999 in the Variorum Collected Studies Series by Ashgate Publishing

Published 2017 by Routledge
2 Park Square, Milton Park, Abingdon, Oxon, OX14 4RN
52 Vanderbilt Avenue, New York, NY 10017

Routledge is an imprint of the Taylor & Francis Group, an informa business

British Library Cataloguing-in-Publication Data
Mainstone, Rowland J.
Structure in Architecture: History, Design and Innovation.
(Variorum Collected Studies Series: CS659)
1. Structural Design. 2. Architectural Design. 3. Structural Design–History.
4. Architectural Design–History.
I. Title.
721

US Library of Congress Cataloging-in-Publication Data
Mainstone, R. J. (Rowland J.), 1923–
Structure in Architecture: History, Design and Innovation/Rowland J. Mainstone.
p. cm. – (Variorum Collected Studies Series: 659).
Includes indexes. (alk. paper).
1. Structural design.
I. Title. II. Series: Variorum Collected Studies: CS659.
TA658. M35 1999 99–26181
721–dc21 CIP

ISBN 13: 978-0-86078-763-1 (hbk)

VARIORUM COLLECTED STUDIES SERIES CS659

Contents

Acknowledgements

Grateful acknowledgement is made to the following persons, journals, institutions and publishers for their permission to reproduce the papers included in this volume: the Graduate School of Fine Arts, University of Pennsylvania and the editors of *VIA* 2 (for chapter I); the Institution of Structural Engineers, London (II, V, X); the *Architectural Review*, London (IV, XVII, XX); the Society of Architectural Historians, Chicago, IL (VI, XI); the College Art Association of America, New York (VII); the Newcomen Society, London (VIII, XVI); Dumbarton Oaks, Washington, DC (IX); the Society of Antiquaries, London (XIII); Dalu Jones and George Michell (XIV); the Institution of Civil Engineers, London (XV); Robert W. Berger and Pennsylvania State University Press (XXI); Alec W. Skempton and Thomas Telford Publishing, London (XXII).

Acknowledgements for illustrations other than the author's own photographs and drawings are made in the captions.

Preface

This volume brings together a selection of writings on the history of structure and structural design in architecture. They include both general studies of the genesis of new ideas and the overall pattern of development and more detailed studies of particular structures and buildings. Two appear in print here for the first time. With one exception, studies dealing with related topics such as structural conservation have been omitted, as have purely architectural studies.

The interest that gave rise to them began as a purely visual response to architectural form, space, light, and surface. But it soon grew to embrace also a fascination with the ways in which some buildings, such as the great medieval cathedrals, had been able to stand for centuries against gravity and the wind, and even against earthquakes and occasional misuse. Then, when I became professionally engaged in developing current approaches to structural design, questions arose about how the designers of the earlier structures could have conceived ways of giving their creations this ability without the help of our present calculation procedures, and about the relevance of their methods to present practice.

These were difficult questions to answer without any accounts by the designers themselves of how they proceeded, little other relevant documentation, and ample evidence of changes undergone by many surviving structures in the centuries since they were built. These changes, and the materials used, even make analysis of present structural behaviour difficult. Certainly such analysis could throw no direct light on approaches to design. Nevertheless a pattern did emerge from closer acquaintance with the structures themselves and from a study of such early writings and drawings as had survived. This pattern became clearer when I had the good fortune to be invited to participate in detailed studies of two of the greatest and best documented achievements of the pre-scientific age. Parallel studies of significant later achievements then illumined the changes that occurred when the scientific revolution began to have an impact.

In the following chapters, the papers that stemmed from these enquiries have been divided into two groups. The first group consists of five general studies. The other covers a larger number of more detailed studies of individual achievements, prefaced by a general exploration of the relevance of our present structural insights to such detailed studies.

Chapters I to V

Chapters I and II focus directly on the sources of new ideas, especially (but not exclusively) in the pre-scientific past, chapter I being a slightly revised version of a first attempt that I still find relevant, and chapter II a complementary more recent assessment. Chapter III is a shorter companion piece concentrating on the less easily identified role of tradition. Chapters IV and V then look at the continuities and changes in the whole structural design process that have been associated with developing concepts of structural behaviour and the changing backgrounds and responsibilities of designers.

Chapters VI to XXII

The studies of particular structures in Chapters VII to XXII are arranged in chronological order in relation to their subject matter and, where several chapters are devoted to a single structure, in the order of publication of the original studies. They mostly seek answers to such questions as when, in what sequence, and in what way was this structure built, and how was it designed and thought to stand by its designer? Usually this entailed asking also how it had actually withstood the loads acting upon it, although this was not the primary objective. And sometimes it was necessary to reconstruct the original form by identifying and discounting subsequent changes. To reach the answers, these studies drew chiefly on evidence such as the present form (including its precise geometry), visible clues of.past and present structural behaviour, and surviving documentary and epigraphic evidence – evidence which is rarely as complete and unequivocal as we would wish and which invariably calls for interpretation consistent with the practicalities and structural ideas of the time.

The two structures that went furthest beyond the bounds of current practice before modern theory came to the designer's aid – Hagia Sophia and the dome of Florence Cathedral – receive most attention, especially Hagia Sophia on account also of the complexity of the problems it presents. When the studies of these two structures were embarked upon it was expected that both would finally be published in book form, and the earlier papers were written as preliminary accounts in this expectation. But the book on Hagia Sophia that was planned to accompany the superb plates of the Van Nice measured survey never appeared, and the book on the dome of Florence Cathedral had to appear under the sole authorship of Howard Saalman after irreconcilable differences over the interpretation of some crucial documents left me no choice but to withdraw from our collaboration. The second paper on this dome, reproduced as chapter XVII, presents my main conclusions about Brunelleschi's solutions to the challenges he faced in building it. For Hagia Sophia, my conclusions were presented, after considerable further research, in another book, the main content of which is summarised in the paper reproduced in chapter X. Chapters IX and XI amplify some of the conclusions therein and the investigations that led to them, while chapter XII (the only one to deal with structural conservation) concentrates on the later history as basis for looking ahead.

Chapters XVIII to XXII consider significant later structures. They are mostly briefer because the survival, relatively unchanged, of the structures themselves or the availability of fuller and less ambiguous documentation usually meant that investigation was more straightforward. The particular interest of the first and fullest of them, dealing with the dome of St Peter's, lies partly in the relationship between this dome and the dome of Florence Cathedral but also in the amply documented (but hitherto little studied) inquiries into its stability almost two centuries after its completion – the first of which was chosen as the terminus for the paper reproduced as chapter IV. These inquiries made amply clear that the adoption of a new approach was far from simple, so that it could take place only gradually.

Chapters VII and XIII to XV were all written in response to earlier papers written by others on the same subjects. To make them fully intelligible, most of them are prefaced by introductory notes on these origins, identified (like a similar note preceding chapter XXI) by enclosure in square brackets. Chapter XIII attempts to elucidate the construction sequence of the present San Marco. Two others, XIV and XV, deal only with structural behaviour, while chapter VII deals with the wider historical significance of a particular series of structural analyses.

Structural analysis is, otherwise, kept mostly in the background. Chapter VI explains why. It stresses that no quantitative analysis can do more than explore the potential behaviour of a necessarily greatly simplified model of the real structure. It then emphasises the importance of a deeper, more intuitive understanding, both as an essential basis for valid modelling and as a source of more direct insights. The further importance of such understanding for the correct interpretation of documentary records is mentioned, but is perhaps not sufficiently stressed. These views differ from those of authors who have sought in the calculation procedures typically used in design in recent decades a more direct and less demanding route to historical understanding.

ROWLAND MAINSTONE

St Albans
December 1998

1. Bridge near Tiryns

I

Intuition and the springs of structural invention

'. . . all Arts were begot by Chance and Observation, and nursed by Use and Experience, and improved and perfected by Reason and Study.'
– Leone Battista Alberti

'The Imagination . . . I consider either as primary, or secondary. The primary Imagination I hold to be the living power and prime agent of all human perception, and as a repetition in the finite mind of the eternal act of creation in the infinite I Am. The secondary Imagination I consider as an echo of the former, co-existing with the conscious will, yet still as identical with the primary in the kind *of its agency, and differing only in* degree, *and in the* mode *of its operation. It dissolves, diffuses, dissipates, in order to re-create: or where this process is rendered impossible, yet still at all events it struggles to idealize and to unify.' – Samuel Taylor Coleridge*

This paper poses and attempts to answer certain questions about the art of structural design or, more particularly, about the genesis of structural ideas within the wider context of architectural design. How do these ideas arise? What are the springs of structural invention? How are they related, for instance, to the theoretical bases of structural analysis? These questions will be approached from the standpoint of present practice; but earlier practice, where it can throw a useful light on this, will be considered in some detail.

First, though, in order that these questions may be precisely understood, it is necessary to state what is meant here by invention in general and by structural invention in particular.

Inventions may come about in different ways. Chance and the unpredictable never cease to play their part in human creative endeavor.

2. Pseudo-arch at Segni

But, as Alberti clearly saw, the process does change when the simple recognition of the usefulness of some accidental discovery gives way to more purposeful and thoughtful innovation.

Primitive man, when he built his first bridge or shelter, no doubt stumbled on the simple arch or beam with little deliberate design or forethought. And initially, we may assume, it was largely through further chance discovery and observation that he slowly learned to improve on this simple beginning.

At this level his own contribution was merely in perceiving the usefulness of that which he chanced upon, the superior merit of each improvement, and enough of its essential characteristics to enable him to reproduce it successfully.

Suppose, however, that the building of archlike structures had not progressed beyond the rudimentary form represented by the rough, cyclopean work of a bridge on the old road between Tiryns and Epidaurus [*1*] by the time that skill in dressing and fitting stones together had made possible walls of fine polygonal masonry. Occasionally, in such walls, the inclusion of a stone considerably larger than average [*2*] or the incorporation in the lower part of the wall of a

 3

3. Pseudo-arch at Cerveteri

humped outcrop of rock [*3*] gave rise, quite accidentally, to 'pseudo-arches'. If then – as is possible, though we might equally well suggest other origins – some builder discerned in one of these pseudo-arches the possibility of a true load-bearing arch of cut voussoirs and recognized, in this, a promise of superior strength and stability, he would, albeit still at a simple level, have conceived something as yet unrealized but potentially feasible and useful. For in the wall there was, structurally speaking, no such arch. Only if we removed the larger stone or outcrop, which in fact largely carries the stones above, and then allowed the stones of the pseudo-arch to settle enough to close the gaps between them would these latter stones become voussoirs of a true arch bridging the void.

Invention such as this would call not merely for perceptive observation but also for the prior recognition of a need and for some conception, based on the observation, of a possible way of fulfilling it.

In what follows, invention will be considered as a conceptual process similar to that here involved, though usually a more complex one that is quantitative as well as qualitative. This conceptual process, rather than any measure of possibly adventitious novelty, will be taken as its distinguishing characteristic. It has, therefore, a part in every design that is not mere pastiche.

However, since architectural design usually involves much more than the devising of simple structural elements or expedients, it is still necessary to ask ourselves what part structural invention plays in this larger context.

It is not to be expected that this part will always be the same. The architect's primary concern is usually to enclose space and provide shelter and, in so doing, to create appropriately expressive and perhaps also symbolic relationships of forms. At the same time he also creates structures in that these enclosures, these forms, must stand and endure. They must be so proportioned and so built that the natural tendency of most things to give way eventually under the constant action of gravity and the varying assaults of other disruptive agencies is forestalled. But this necessary end may be achieved in different ways. Contemporary practice exhibits many variations, and these can easily be paralleled, less invidiously, in architectural history.

If, for example, it is true that most basic architectural forms had a purely functional origin in which structural necessity played a large part, it appears also that, through long use, many came to be valued purely as forms and were, as such, often transposed to new materials and new kinds of construction. Initially form followed function. But in its new transpositions the form came first and the structural means were considered later, if at all. The most striking examples of this are the numerous rock-cut replicas (devoid of all structural logic) of hallowed forms originally constructed in timber, stone, or brick and detailed to exploit the capacities and respect the limitations of these materials [*4*, *5*]. Architecturally more important, and for the present purpose more relevant, are the transpositions of the domical form to different uses and methods of construction from the Roman period on.

Originally no more than the simplest form in which to construct a small shelter out of a wide range of naturally occurring materials [*6*], the roughly hemispherical dome or 'beehive' shape acquired over a long period, simply as a space-enclosing form, many layers of symbolic overtones – as ancestral home, mausoleum, martyrium, temple, and church [7].[1] So valued, more for the sake of its associations than for its structural merit, it became the central germinal idea of many architectural conceptions. In bringing this about, the architects of the Roman Empire from the time of Nero to that of Hadrian played a decisive, if

4. *Rock-cut tomb at Cerveteri*

5. *Rock-cut church in Cappadocia*

6. *Stone beehive dome near Gordes*

7. *Monastery church at Daphni*

8. Pantheon, Rome

9. Serapeum, Villa Adriana, Tivoli

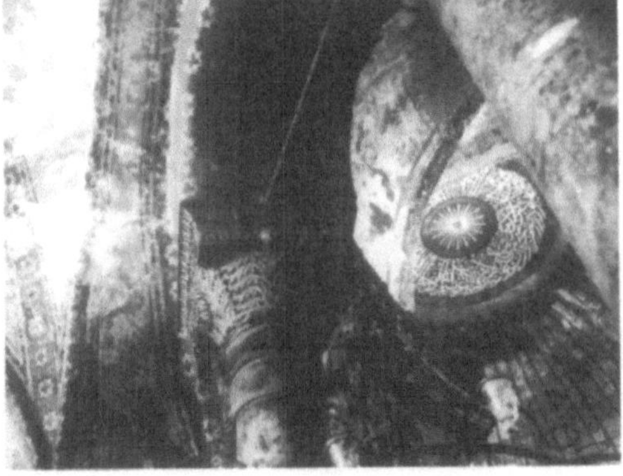

10. Hagia Sophia, Istanbul

somewhat fortuitous, role by demonstrating many new and exciting possibilities. Having chanced upon a concrete of unprecedented strength, they progressively exploited its ability to span wide spaces as a monolithic mass, proceeding tentatively at first, then with growing confidence in the Domus Aurea and the Domus Augustana, and finally with supreme assurance in the Pantheon [*8*] and the Villa Adriana [*9*]. Yet, apart from some constructional details, the domes of all these buildings appear to have been conceived wholly as forms and as interior spaces rather than as structures. It is as if the alternately projecting and receding surfaces of many of the domes had simply been carved out of the concrete mass: there is no sign of them externally. Where, indeed, something more than a simple modeling of the interior surface of a solidly based mass was attempted, as it may have been over the octagon at the south side of the Piazza d'Oro in the Villa Adriana, assurance possibly overreached itself.[2] Purely structural invention in these Roman domes is, therefore, to be seen less in their form or their scale than in constructional details like the selection and placing of the *caementa*, the incorporation of *amphorae* in the concrete to lighten it, and the use of brick ribs, revetments, and relieving arches.[3] When, however, these Roman achievements are transposed into Byzantine brick [*10*] or the brick and stone of medieval and Renaissance Italy or the more sparingly used concrete of today, the modeling of the whole form does again become a matter of structural concern and is often largely a structural invention.[4]

The development of the class of ribbed domes best known through the examples in the mosque at Córdoba [*11*] and in San Lorenzo in Turin [*12*] presents an interesting variant on this pattern of invention.[5]

11. Mosque at Córdoba

12. San Lorenzo, Turin

8

In Córdoba, and in the much smaller vaults of El Cristo de la Luz in Toledo, the ribs appear to be largely decorative and to have been conceived as little more than a surface pattern of a kind of which the Islamic world was particularly fond. This patterning originally had nothing to do with structure or architecture, having possibly a dual ancestry, both in the functional patterns produced in weaving and basketry and in the more playful superimposing or interlacing of simple geometric figures as in some Roman floor mosaics [*13*] and ceiling decorations. Yet it had a certain structural connotation when applied to a vault, and, at a later date, by analogy with load-bearing ribs of other kinds – or perhaps, if we may take a hint from the detailing of some early examples in the Friday mosque in Isfahan [*14*], by analogy with a family of intersecting squinches – a potential structural function was read into an existing form in much the same way as may have happened with the pseudo-arch; and similar ribs were conceived structurally. Finally, in the work of Guarini [*12*] and Vittone and, possibly, in a number of earlier exploratory sketches by Leonardo,[6] the ribs were given both an architectural and a structural autonomy and become the basis of new conceptions in which structure and form are indistinguishable. Conversely, in Islamic stalactite vaults, the basic forms of squinch and corbel, both originally purely structural elements, were elaborated for purely decorative ends [*15–17*].

13. 'House of Livia', Rome

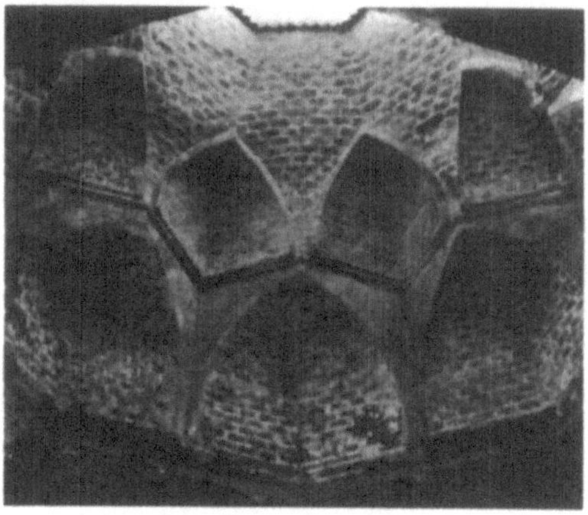

14, 15. Friday mosque in Isfahan

16. 'Green Mosque', Bursa

17. Alhambra, Granada

18. Priory church in Vignory

19. Saint-Philibert, Tournus

In one case, however, structural preoccupations interacted continuously with others of a purely architectural kind in determining the total conception. This was the development of the Gothic cathedral, as seen at Chartres, Reims, and Amiens, from the timber-roofed, aisled basilica of early Christian Rome and its Romanesque successors like the church at Vignory [*18*]. The aim was both to give each building the durability of stone without sacrifice of light or stability, and to make of it a new and more glorious *domus ecclesiae* as that was conceived by men like Abbot Suger and the theologians of Chartres.[7]

In the course of its fulfillment many diverse attempts were made to balance and concentrate the thrusts of high masonry vaults. In the upper narthex at Tournus [*19*], for instance, and in the pilgrimage churches on the route to Santiago de Compostela and the Auvergnat churches around Clermont-Ferrand, the thrust of the main vault is taken by flanking quadrant vaults, usually in a gallery over the aisles. At Fontenay, ranges of slightly pointed transverse barrel vaults over the aisles perform the same function as these quadrant vaults [*20*]. And over the nave at Tournus a similar succession of transverse barrel vaults, here semicircular, largely obviates the need for peripheral buttressing [*21*]. Gradually these and other experiments converge toward the final

 11

20. Abbey of Fontenay

21. Saint-Philibert, Tournus

solution. The buttressing quadrant vaults of the upper narthex at Tournus, the triforia of the pilgrimage and Auvergnat churches, and the transverse aisle vaults at Fontenay give way to the flying buttress; the largely decorative vault ribs of the ambulatory at Morienval and of Durham Cathedral become part of the main structural articulation; and a new unified structural system is evolved, which, through its very coherence and expressive clarity, becomes one of the most powerful and moving symbols that architecture has to show [*22*]. Here structural invention is to be seen both in the total concept and in nearly all its details. But in neither case is the invention solely structural.

We might conclude that, where there is a deliberate concern with the structural integrity of a projected building, that concern may extend no further than the choice of materials and method of construction or than certain detailed modulations of the forms, or it may, at the other extreme, strongly influence the entire conception. But, in all cases, it will do so within the framework of other concerns and of inventions of other kinds, conditioning these and being conditioned by them and, ideally, stimulating them and being stimulated in return.

We will now distinguish structural invention as that part of the total conception marked by any such concern with structural integrity.

22. Cathedral of Bourges

Creative knowledge: Intuition and insight

The process of conception is a complex one typically entailing much painstaking study of requirements and of potential ways of meeting them as well as repeated shifts of the focus of attention from the total concept of the evolving design to critical details and back again. Central to it though, at the levels of both the total concept and its details, must always be a sequence of imaginative apprehensions of possible solutions to the problems posed. These imaginative apprehensions must be refined, developed, and welded together by subsequent analysis and synthesis, but without them there could be no new conception, no invention.[8] And each of them implies, in turn, a prior awareness of the constellation of possibilities out of which it arises, just as Coleridge's creative 'secondary' imagination echoes the 'primary' one.

We will look, therefore, for the springs of structural invention in some kind of prior knowledge of the structural forms, materials, techniques, or modes of behavior that the new structure exploits. The emphasis will be on what might be called *structural* knowledge, but it should not be forgotten that, since every structural form is also a form

13

or an enclosure of space which invites consideration from other points of view as well, knowledge of forms and spaces and patterns which is not strictly structural may also be expected to play an important role.

First we must consider what kind of knowledge is requisite. All our knowledge is born of experience, and it differs in its efficacy for each one of us according to the nature, for him, of that experience. If it is to serve as a springboard for imaginative creative thought, then at least part of the experience must be truly personal, like that through which we learn to speak our native tongue. Through listening to others use it and through using it ourselves, in imitation of them at first and then more independently, we gain a mastery which can make it a powerful vehicle for the formulation of ideas. In contrast, a study of the vocabulary, grammar, and syntax of a foreign language – the distilled experience of others – gives only a very different kind of knowledge, of which, by itself, we can make no such creative use. If we want the fuller and potentially more creative mastery, such study will, at best, help us to gain it more rapidly through direct experience. What distinguishes the kind of knowledge that is potentially creative is our ability to bring it effectively to bear without the effort of explicit formulation in the mind and indeed without any conscious concentration on it. The conscious mind must be free to focus on the creative end. For the structural designer this means turning over in his conscious mind the different aspects of the structural requirements and concentrating, through these, on finding an as yet unknown solution.

When creative knowledge stems, like that we have of the use of words in our native tongue, directly from personal experience, we may call it *intuitive*. When, like that which more commonly constitutes the mastery of a foreign language, it is guided by explicit codifications of the experience of others but deepened through personal experience, the word *intuitive* is less appropriate, and we may call it *insight*. Intuitive knowledge (or intuition) and insight, so defined, differ little, however, in terms of their potential usefulness for creative ends. Essentially, they differ only in that the former may remain tacit throughout whereas the latter, by definition, has involved some degree of explicit formulation. Since this distinction is not a sharp one – the insights with which we will here be concerned usually going beyond the initial explicit formulations and tending to acquire a more intuitive character – the word *intuition* will generally be used in the discussion that follows and should be read, where appropriate, as also denoting insight.

14

Experience

Structural intuitions stem both from experience related to building and from experience gained in a variety of other fields. Most obviously they stem from looking at buildings under construction, in use, or, most revealingly, after some failure, or from handling and working the constituent materials and elements and helping to put them together during the process of construction. They may similarly stem from looking at and handling certain types of natural structure or from experience of man-made structures other than buildings. But they also stem, in a very fundamental sense, from some of our most deep-rooted experiences of our own muscular actions and of the nature of the world around us as that is apparent to the senses of sight and touch. In addition they may draw upon a further range of what might be called contrived experience, of the kind that the scientist gains in his laboratory.

The range, depth, value, and relative importance of these different kinds of experience does, of course, change from one period, and even from one designer, to another. Even the consciousness of muscular actions is not the same for everyone. For the artist of the Mausoleum frieze [*23*] and for Michelangelo [*24*] it must have had a clarity that is perhaps given only to the sculptor of the human form and to the athlete. Moreover, when construction was largely a matter of manually placing stone on stone and of other similar manual operations, this

23. Frieze from the Mausoleum, Halicarnassus

15

direct muscular consciousness could play a much larger part than it usually can today. On the other hand, our previously unaided senses can now be assisted in many ways by modern instruments, and the range of experience that we can fruitfully contrive is vastly enlarged by modern testing equipment.

With these changes in experience the intuitions also change. Yet the

24. Michelangelo's Moses

intuitions are always distinct from the experience: they are something that our minds take from it and by means of which transcend it, usually by relating it to a wider frame of reference.

Even the simplest intuitions are mostly found, on analysis, to involve correlations between experiences of different kinds. Our intuition of our muscular actions is intimately linked not only with our awareness of muscular efforts and of external pushes and pulls which may call for them but also with our senses of balance and movement and our awareness of spatial relationships. And our awareness of spatial relationships is itself compounded of both visual and tactile experience, visual data being interpreted in what seems to be the most plausible way in terms of schemata derived from earlier tactile explorations.

Intuitions of pushes and pulls which we do not ourselves exert or directly feel involve a further association of experiences. Out in the wind we feel buffeted and we keep our balance by leaning against it. Then seeing the grass and the trees bending with the wind, we 'feel' similar forces acting on them and their development of a resistance that we would directly feel if we took the wind's place. Or, lifting and easing a stone into place on a building or straining a tie to correct a distortion or forestall a collapse, the builder feels the effort that this demands and subsequently 'feels' a corresponding effort on the part of the structure. Indeed our whole intuitive tendency to consider inanimate structures as being subjected, as we are, to external forces like gravity and the wind and as reacting to them, as we do, by developing internal resistances is a complex outcome of experiences of many different kinds.

We can, however, penetrate such complexities to some extent by distinguishing broadly between groups of intuitions stemming primarily from each of the kinds of experience mentioned above. Looking at buildings and other man-made or natural structures and observing that certain forms constructed in certain ways are adequate for some purposes and others for different purposes leads to what we will call *intuitions of structural adequacy*. The basic muscular and sensory experiences lead to the simplest *intuitions of structural behavior*. And deliberately contrived experience leads to, and is usually an essential foundation for, more abstract intuitions of behavior, or *intuitions of structural action*. The distinctions are not clear-cut, but they can help in understanding the intuitive springs of invention if it is remembered that the intuitions guiding designers in practice are generally composite ones.

 17

25. Shrine on Mount Cynthus, Delos

Intuitions of structural behavior

The simplest intuitions of the orderly pattern which underlies all structural behavior – and which makes it all potentially determinate, if only we knew enough of the relevant parameters – are probably possessed to some degree by all men. We can even see them guide an intelligent child in his early efforts to build on the nursery floor. One aspect of the pattern may be described as *spatial*, the other as *physical* or *muscular*.

Our basic intuitions of spatial order in the sense here envisaged are such an integral part of our knowledge of the world around us that they usually pass unrecognized except in the incredulity with which we react to any apparent violation of the order. We see this world as composed of objects, each of which occupies a certain space to the exclusion of all others and seeks to occupy the lowest position available to it insofar as this is consistent with other bodies' seeking to do the same. We expect a stone released in the air to fall to the ground unless something is interposed to arrest the fall. Equally we would expect each of the large roofing slabs of the shrine on Mount Cynthus [*25*] to swing downward and drop off the supporting ledge in the absence of the corresponding slab on the other side. With that slab also in place and undeformed, we recognize, however, that neither can fall without violating the presence and equal tendency to fall of the other. We thus arrive at one kind of understanding – complete in itself without necessarily invoking any concept of the nature of the physical interaction between the slabs – of this simple structural system; and, by a simple extension, we can arrive at a similar understanding of the arch of several voussoirs, including the recognition of the possibility of collapse here through simultaneous compatible movements of the voussoirs. Similarly, also, we can see how a buttress can prevent a threatened movement by being placed as an obstacle in its way.

The corresponding basic intuitions of physical actions within and between structural members stem from our direct awareness of our own muscular actions and reactions. Through this awareness we come to expect or 'feel' analogous pushes and pulls to be acting on and through, and resisted by, inanimate bodies. In the simple cases just considered, such 'feelings', though entirely different from the spatial intuitions, are, in practice, closely associated with them for most men. Likewise some intuition of the thrust in a column is usually almost inseparable from our concept of it as a simple obstacle to the fall of the architrave above [*26*], and we can hardly visualize a domical structure as a purely spatial configuration (in the structural sense just considered) without also having some analogous intuition of radial thrusts [*27*]. Conversely we can hardly conceive a tie as an obstacle of another kind to threatened movement without also 'feeling' the tension within it. Coupled also with our recognitions that, in standing up to and transmitting pushes and pulls and bending actions, some materials and some forms are strong and others are weak, some almost rigid and others much less rigid, these spatial and physical (or muscular) intuitions lie at the root of all our understanding of structural behavior.

 19

26. 'Temple of Poseidon', Paestum

27. Chapel of Santo Zenone, Santa Prassede, Rome

The range of their usefulness is limited, though, by the lack of quantitative precision that is characteristic of the muscular component. We can directly perceive spatial magnitudes and directional relationships. Indeed these perceptions have long been given a precise theoretical statement in Euclidean geometry, and the spatial component has thereby become articulate in modern structural theory in the geometric conditions of compatibility of deformations and displacements. But we have no comparable sensory mechanism for perceiving quantitatively even the pushes and pulls we ourselves experience or exert. Still less can we perceive directly the magnitudes and directions of analogous physical actions in which we play no part. This means that, on the basis of the muscular kind of intuition alone, we can neither confidently assess the characteristic strengths and stiffnesses of the materials we use nor confidently predict the loads to which they will be subjected in particular structures. Since, moreover, the pattern of internal stresses is usually much more complex than in the simple cases just considered, it is often

20 INTUITION AND THE SPRINGS OF STRUCTURAL INVENTION

difficult even to be sure whether there will be primarily tension or compression in directions other than those in which the principal loads are applied or at appreciable distances from their points of application.

Intuitions of structural action

To overcome this limitation took many centuries. It was necessary to draw together many kinds of experience often, at first sight, not directly relevant to the behavior of static structures and, as has been said, deliberately contrived in the laboratory more frequently than gained in the course of everyday life or normal building operations. Clearly the highly imaginative synthesis that was called for was not reached on the nursery floor. It was very much the fruit of what Alberti called 'Reason and Study' on the part of men like Archimedes, Leonardo [*28*], Stevin, Galileo, Newton, and Hooke and their successors – assisted, since the pioneering work of Musschenbroek, by the development of accurate and versatile techniques for making structural actions (internal and external loads, stresses, strains, and small deformations) visibly manifest.[9]

The deeper understanding of structural actions which was thus progressively achieved was primarily the transformation of the basic muscular intuitions of pushes and pulls into a much more precise and

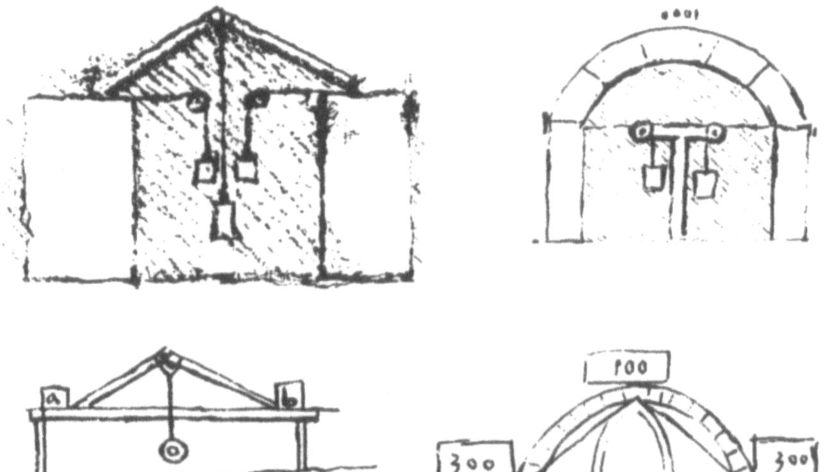

28. Studies of the statics of arches, by Leonardo da Vinci

powerful concept of forces and moments, coupled with a highly generalized understanding of the conditions for equilibrium. In turn, this new concept opened the door to a quantitative exploration and understanding of the response of different materials and structural elements to these actions.

The transformation became possible through learning to 'see' forces acting on material objects in the movements they induce and, conversely, to see the absence of movement as a state of equilibrium between the acting forces and between their moments. One of the forces most commonly manifest in this way is the force of gravity, and one of the simplest cases of equilibrium is that of two weights suspended from a balance arm. The Archimedean laws of the balance were largely empirical in origin, however, and could be extended to cover the equilibrium of forces acting in any direction only by a similar, almost ad hoc empiricism. The final clarification of the more Aristotelian concept of a force acting dynamically in any direction came only with Newton, and with it, as a simple corollary, an equally generalized statement of the conditions of equilibrium.

The subsequent quantitative understanding of responses was a much more straightforward extension of earlier recognitions of the different strengths and stiffnesses of different materials and different structural elements made from them. It was arrived at simply by applying known forces to specimens in the laboratory, measuring the response, and thereby building up a picture of the constancies underlying this response – of the ways in which distortions of different materials increase with increase in load and the ways in which different patterns of load lead to excessive distortion or complete disruption. Theories of elastic and plastic deformation, and related criteria of failure, are precise statements of simple idealizations of the pictures so obtained.

Today the structural engineer does not arrive at these fundamental insights unaided. Nor does he so arrive at much of the understanding derived from them, by his predecessors and contemporaries, of the more complex patterns of structural action associated with particular types of built form. He is guided to them by the theoretical statements and, so guided, draws from his own experience in both analysis and observation highly composite intuitions of the physical responses of the structures he is concerned with to imposed loads, imposed displacements, and other disturbances.

These responses can be considered in three ways. When considered

22

in terms of strains and resultant deformations and displacements (akin to consideration in terms of the basic spatial intuitions), the picture obtained will be a purely geometric one, though it may be difficult to visualize as a whole because of its three-dimensional complexity and its emphasis on small changes in the simpler geometry of the unstrained structure. When considered in terms of stresses and internal reactions (akin to consideration in terms of the basic muscular intuitions), the picture will be a complementary one, but drawn from the rather different vector geometry of the main tensions, compressions, shears, bending moments, and torsions for a structure composed largely of linear members, or of the principal tensile and compressive stresses or the principal bending moments for a surface structure like a slab or a shell; the decorative rib patterns used by Nervi on the soffits of some of his slabs [*29*] give visible expression to such 'isostatic networks'. Consideration of the stiffnesses and strengths of the structural elements and their interconnections will underlie both pictures and link both with the primary conception of any structure as something very real, to be built in a specific shape from specific materials used in a specific way.

These three aspects are always interrelated, but the extent to which they must be considered together varies with the type of structure and the importance of obtaining a full and accurate understanding of its

29. U.N.E.S.C.O. secretariat in Paris

action. When the deformations of each structural element can be accommodated by compatible deformations or displacements of other elements without inducing significant loads in these other elements, it is sufficient to consider each aspect in turn. However in most structures that are built today this is not so: all aspects must therefore be considered together if the action is to be seen correctly. If it is too difficult to do this from the start, a correct picture must be built up as far as possible, first by mentally breaking the structure down into simpler parts and visualizing the largely independent actions of these, and then by progressively correcting the pictures so obtained by allowing for the secondary actions necessary to reestablish the continuities that these simplified pictures ignore. For instance, in attempting to visualize the action of a shell structure with edge-stiffening ribs, we might first consider the shell and the ribs separately, taking into account the simple primary action in the shell itself and only the primary function of the ribs as collectors and transmitters of the primary stresses along the shell's boundaries. Considered in this way, the shell and the ribs would be seen to deform quite differently along their surfaces of contact because they would be subjected to different types of stress. We would then have to consider what changes would result from their actual monolithic interconnection. As in dealing with simpler structures, we would be helped in doing this by previous theoretical and experimental study but we would, ultimately, have to build up our own intuitive picture of the composite action.[10]

Intuitions of structural adequacy

These intuitions differ from those of behavior and action in focusing directly on the adequacy of a particular form for a particular structural purpose rather than on the precise manner in which this purpose is served. Some concept of the structural purpose is implicit, however, and, since it can be conceived in many different ways, these intuitions cover a wide spectrum. They range even more widely when they embrace, to a varying extent, some recognition of the effects of changes in scale and proportions.

The simplest of these intuitions remain largely at the level of perceiving the adequacy of an existing form for a purpose similar in nearly all respects to its present one. Perceptions such as this have already been shown to lie at the root of man's earliest and relatively unreflective acquisition of constructive skills, but they provide little guidance for innovative design. More developed intuitions both see more deeply into the essential character of the form and relate it to a more generalized (that is, less restrictive) concept of its purpose in terms of behavior or action.

As an example of the simplest type of intuition, the beehive dome in [6] may have been seen simply as a suitable form to use for other,

30. Forum, Pompeii

 25

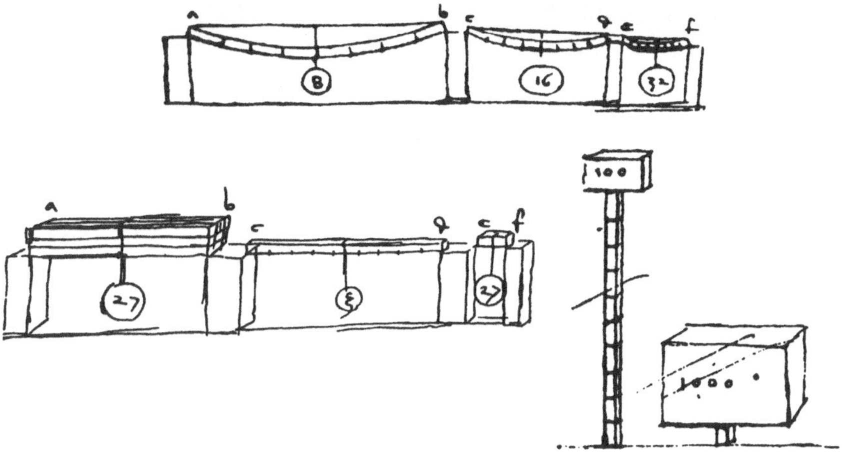

31. Studies of the statics of beams and columns, by Leonardo da Vinci

almost identical huts. A more developed intuition would both embrace some understanding of the structural action of the domical form and see particular proportions as being suitable for particular types of construction and particular conditions of loading and support. Similarly, the adequacy of particular materials or structural elements may, at the simplest level, be recognized only in terms of specific uses; or it may, with much greater generality, be seen in terms of characteristic strengths and stiffnesses.

Where such intuitions extend to a recognition of the ways in which adequacy varies with shape and size, they merge with intuitions of behavior and action, and the distinctions that have been made become blurred. The rules Vitruvius gives, and later architectural theorists quote, for the desirable variations in proportion of superimposed columns [*30*] may be considered as formal statements of intuitions of adequacy. Though they are seen as necessary in order to match the strength to the variations in load, they are derived from direct observation of practice in a particular type of construction and from observation of an analogous variation in girth of growing trees. On the other hand, Leonardo's explorations of the strengths and stiffnesses of beams and columns of different proportions [*31*] and the more general observations of Galileo set out in the opening pages of *Dialogues concerning Two New Sciences* drew, to a much greater extent, on intuitions of behavior and action and led to intuitions which are themselves probably better considered as intuitions of structural action.[11]

It is more important, however, to remember that there is a continual interplay between the two classes of intuition – for intuitions of behavior and action also merge into intuitions of adequacy when a new structural form, sought initially in terms of the abstractions of behavior and action, is conceived in more concrete terms as adequate for the specified purpose.

In focusing on concrete forms rather than abstractions, intuitions of adequacy, hybrid though they are, play an essential role in any conceptual process. Not only do they bring it to a conclusion: they also serve as the means whereby past experience in practical construction is most directly brought to bear as a starting point for fresh endeavor. None of us has enough imagination to think *ab origine* on more than a very limited front at any one time or to think far beyond what has been achieved already. Of this human limitation architectural history provides overwhelming evidence in the continuity and homogeneity of the pattern of development that it presents. This pattern is mostly one of slow evolution, to which, in the past, we have owed many of the subtlest refinements of style and most perfect marriages of form and function [*32*]. And even where it is punctuated by major innovations, familiarity with the milieus in which these arose often discloses a number of closely related antecedents. Rather than springing quite independently to life, the innovations bring together, or turn to a new purpose, existing possibilities.

32. Mykonos

Invention in the past

Most of the structural inventions described earlier may be seen as springing very largely from a combination of relatively simple intuitions of adequacy and simple intuitions of structural behavior.

These inventions were, as compared with much structural invention today, limited in scope. Most construction, other than the *opus caementicium* of imperial Rome, was essentially a piling up of units of masonry, balanced on one another so that they were able to resist the action of wind, subsidence, or earthquake. Timber or metal ties were introduced to assist in achieving and maintaining a state of balance by taking tensions where necessary within structural systems otherwise kept in equilibrium largely by the compressions resulting from dead weight of the masonry. But they did not modify its character radically, as do the reinforcing bars and cables of modern reinforced or prestressed concrete or the moment-resisting connections of modern steelwork. Also the range of materials used at any time and place was small. So there were fairly constant relationships between structural behavior and the simple geometry of structural form. Structure was, in this sense, an invariant of form.

Innovation was, moreover, more directly empirical than it is today, in that new structural forms and techniques of construction were mostly conceived to make good a shortcoming already observed in an existing structure or to overcome a difficulty already encountered in the course of construction. Either the innovation was a corrective measure, applied to a structure that had been found wanting, or an attempt, in a new design, to forestall shortcomings or difficulties that had previously been encountered elsewhere. Potential structural weaknesses rarely seem to have been anticipated and provided for until they had made themselves felt somewhere in undesirable movements, local failures, or more extensive collapses.

Where cracking or buckling or other excessive deformation was met solely by increasing girth or using a stronger material, little more than trial and error was involved, calling for only the very simplest intuitions of the ways in which strength and stiffness varied with shape, size, and material. But where they were met more radically – by changing the form, introducing new members, or varying the method of construction – the simple spatial and muscular intuitions of behavior played a larger and more creative role. Usually the new conception was then primarily

28

a geometric answer to the observed geometry of deformation or failure, whether it was arrived at during construction or subsequently. The spreading of the haunches of an arch or vault, encountered frequently when heights were increased and supports were made less massive, called for an obstacle to this movement in the form of a buttress or tie. The spatial intuition showed where it should be placed; the associated muscular intuition showed whether it would be subject to compression or tension and whether, therefore, it should be constructed of fairly heavy masonry [*33*] or could be a light section of timber [*34*] or metal. Open cracks, such as tended to appear around the bases of domes [*35*] or apsidal semidomes, always indicated excessive tensions either to be resisted by peripheral buttressing or encircling ties or to be reduced by a change of form. Usually the aim was to make a structure that was as rigid and immovable as possible. But the deliberately flexible structure of today, designed to remain strong and stable even when subjected to large settlements, also found its forerunner in the hinged

33. Cathedral of Avila

34. Saint-Philibert, Tournus

35. Baths of Trajan, Rome

36. Bridge in Venice

 29

37. Hagia Sophia, Istanbul

38, 39. Santa Maria del Fiore, Florence

masonry 'chain arch' of certain bridges built on shifting sands [*36*][12] and, to a more limited degree, in the relieving arch – an intentional pseudo-arch – of Augustan Rome.

During construction the stability of the incomplete structure had to be assured, and often, where timber for centering was scarce, the stability of each individual brick or stone as it was put in place was the main consideration determining both the bond of the masonry and the finished structural form. A preoccupation with the stability of the individual unit leads, through the simple spatial and muscular intuitions, to the characteristic profiles and the characteristic bonds of the soffits [*37*] of many arches and vaults, from those of ancient Mesopotamia to the brick domes built without centering in the Middle East today.[13] It leads, for instance, to the *a spinapesce* ('herringbone') brickwork [*38*] of Brunelleschi's dome for Santa Maria del Fiore (as well as, in all probability, to the pointed profile of this dome [*39*]) and to the slightly domical form of the webs of many ribbed masonry vaults.[14] A similar preoccupation with the practicalities of the process of construction, rather than a concern with the stability of the completed structure, is possibly the chief reason for the adoption of the pointed profile for arches and ribs in Gothic vaulting.

Some aspects of structural behavior and some of the choices that the designer would have had to make must, however, have remained puzzling on the basis of the simple intuitions of behavior alone.

How, for instance, was he to visualize the structural role of the stone ribs of a Gothic vault or the brick ribs, relieving arches, and bond courses typical of Roman concrete construction? Did the ribs carry the rest of the vault in the Gothic structure? Did the ribs and relieving arches direct the thrusts in the Roman one – or did they not? And how was he to choose between many alternative profiles for an arch if he was aiming at maximum strength of the completed structure rather than at ease and economy of construction?

In all such cases the evidence furnished by existing structures was, in itself, confusing. Arches of many different profiles remained standing, and arches of as many different profiles no doubt fell. In Gothic ribbed vaults, the webs often cracked away from the wall arches and stood independently of these. Sometimes they remained standing, like simple groined vaults, when all the ribs had fallen away [*40*]. But elsewhere, occasionally even in the same building, cracks extended continuously through web and rib, or the web fell away leaving the rib behind [*41*].[15] Cracks and failures which can now be seen in Roman concrete structures have usually paid little respect to the embedded brick ribs and arches, often passing through them as if they were just like any other part of the concrete. Where these ribs and arches have fallen or been quarried out, the concrete usually seems well able to take over their role [*42*]. But this may not have been so when the concrete was much younger.

Yet we are faced with the historical fact that structures like Hagia Sophia and the cathedrals of Chartres, Reims, and Amiens were actually built, and that they sometimes came closer to the limits of what is possible with the means available than we would care to approach today without many exploratory tests. Were they, as structural inventions, merely the fortunate outcome of intelligent trial and error guided only by the simpler intuitions of structural behavior and adequacy? Or

 31

40, 41. Abbey of Ourscamp

42. Accademia, Villa Adriana, Tivoli

were they also guided by some of the more powerful and accurate intuitions of structural action, on which designers today draw?

A definite answer to these questions would be of much more than merely historical interest. The theories which give articulate form to the essence of these latter intuitions are, with the sole exception of the laws of the balance codified in Archimedes' *On the Equilibrium of Planes*, all comparatively recent. Some are still being extended and perfected. So if these intuitions can be identified as having guided earlier work, they too must have been reached by designers independently of theory. An answer would, therefore, have an important bearing on the role of structural *theory*, as distinct from the intuitions behind it, in invention.

Unfortunately a definite answer is less easy to give than one would hope, and may well be unobtainable. It is always difficult, if not impossible, to relive the creative experiences of others – or at least to be certain that one is doing so. The effort of historical imagination here called for perhaps needs to be guided by more documentary evidence of past modes of thought and work than has yet come to light. Nevertheless a tentative answer is worth attempting.

Simple muscular intuitions of the forces acting in a column, arch, tie, or buttress are not now in question. They have already been discussed, and there is documentary evidence of them in the expressive terms used for such elements in some of the records of expert opinions called for, in the Gothic period, to resolve specific problems of design or construction.[16] Simple intuitions of the responses of different materials to load or deformation must also have been possessed by designers of all periods. The question to be investigated is whether these intuitions ever achieved the clarity and quantitative precision essential for an intelligent statical choice between alternative configurations of structural elements or for a proper understanding of the behavior of a composite structure like a ribbed vault.

Occasionally a design does exhibit such structural virtuosity that we are tempted to say that the guiding intuitions did have this clarity and

43. Saint-Urbain, Troyes

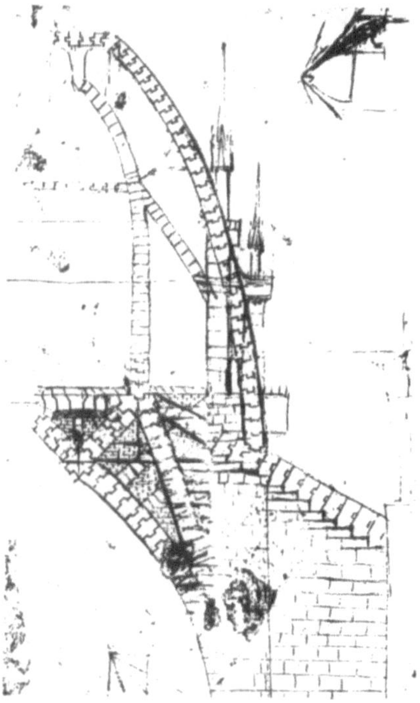

44. *Study for the cathedral of Milan, by Leonardo da Vinci*

45. *Study for Saint Paul's Cathedral, London, by Sir Christopher Wren*

precision. In the church of Hagia Sophia, especially after the rebuilding of its dome by Isidorus the Younger [*10*], in the church of Saint-Urbain at Troyes [*43*], in the dome of Santa Maria del Fiore [*39*], in Leonardo's studies for the *tiburio* of the cathedral of Milan [*44*],[17] and in the domes of San Lorenzo in Turin [*12*] and Saint Paul's in London [*45*], a remarkably clear understanding of static equilibrium does seem to be manifest. Yet these designs are exceptional; much more often there is evidence of indecision or lack of understanding.

Where, for instance, a composite structure like a ribbed vault presented alternative paths for transmission of loads to the supports, this transmission appears always to have been visualized directly in terms of the form rather than statically. Whatever the geometry and whatever the relative stiffnesses of ribs and webs, the loads were always seen as being concentrated as axial thrusts on the main geometric articulation of the ribs – on the rough, though incomplete, analogy with the concentration of our bodily actions in direct pushes and pulls along the axes of bone or muscle.[18] That Gothic ribs were, apart from their decorative and constructional roles, looked upon as actually collecting and bearing the weights of the vaults is suggested by a number of details, including the care always taken to meet the thrust assumed in any rib with an opposing thrust or thrusts from the other side [46]. And that the brick ribs of Roman concrete vaults were similarly conceived is suggested by details like the variation in the width of the ribs in the dome of 'Minerva Medica' [47]. It is a hemispherical dome with equal spaces between the ribs, but those answering to the piers below are more than twice the width of those over the arched openings between the piers.

Likewise, the line of thrust in an arch appears always to have been conceived as following the curve of the arch, whatever the curve and the loading; and the choice between curves, even when aiming at maximum strength, seems to have been made, with few exceptions, not statically but on the basis of purely geometric ideas. This falling back on pure geometry is well exemplified both by the widespread preference at many times for the semicircular arch and barrel vault and for the hemispherical dome, and by Alberti's arguments in their favor.[19] The geometrically perfect curve of the semicircle is preferred to a statically determined curve like the catenary, which is a configuration of minimum potential energy. And if Alberti's ideas may not here be representative of those of the best designers, we may note that a similar falling back on pure geometry runs through much of past design. Even Guarini's surviving writings treat of vaults and domes in wholly geometric terms, while Wren's, at best, consider the choice of form, with respect to minimizing the need for substantial abutment, in terms solely of the simple geometric statics of Archimedes; this involved making the fallacious assumption (for normal masonry construction) that each half of a vault together with its abutting pier or piers would

46. Cathedral of Noyon

47. 'Temple of Minerva Medica', Rome

 35

48. Different ways of envisaging the stability and the possible collapse of a vault or dome: left, the basis of Wren's static analysis of the size of pier needed to provide stable abutment for a vault; right, the mode of collapse of the dome of Saint Peter's Cathedral, Rome, as visualized in 1742-43

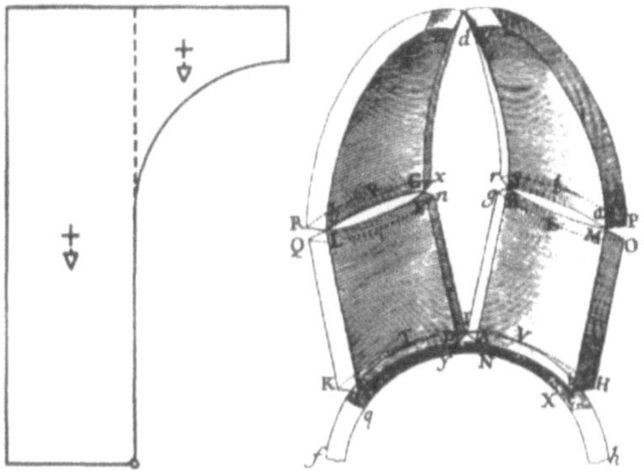

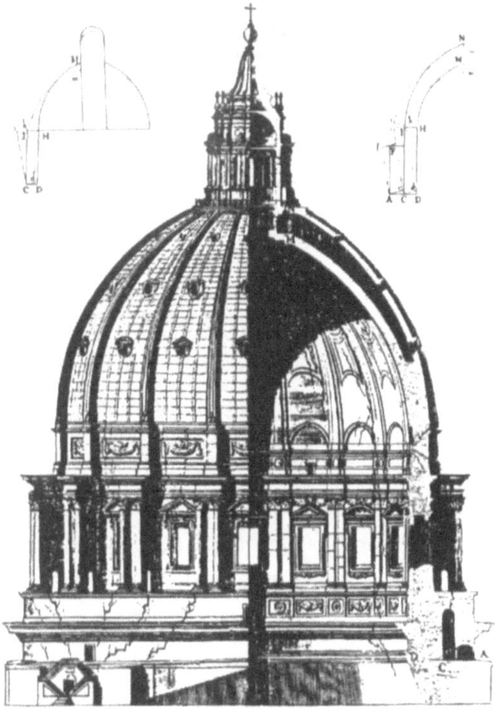

49. The dome and drum of Saint Peter's Cathedral as observed in 1742

constitute a single rigid body, the only threat to whose stability would be a tendency to overturn about the inner edge of the base [*48, left*].[20]

Only much later (in 1742–43, when precise abstract concepts of forces acting in any direction and the conditions of their equilibrium had already been arrived at theoretically) is the first unequivocal evidence found of a more valid approach. It was then, moreover, the work of professional mathematicians called in to advise on the stability of a dome constructed some two hundred years before and already clearly exhibiting by an extensive pattern of cracking [*49*] its likely ultimate mode of failure [*48, right*]; given a reasonably adequate picture of this, it was possible not only to see that the failure might be forestalled by encircling the dome with a circumferential tie at about the level of the top of the drum but also to calculate the strength that would be necessary in this tie.[21]

While we cannot on this limited basis, and certainly cannot a priori, deny to the designers of the past some intuitive knowledge of aspects of structural action that received no precise theoretical treatment until after their time, the evidence largely points against such anticipations. It appears that the breadth and depth and clarity of perception of an

Archimedes or a Newton are vouchsafed to one kind of mind (the mind which 'struggles to idealize and to unify'), that the power of invention is given to another (the mind which 'dissolves, diffuses, dissipates, in order to re-create'), and that this latter mind needs, especially in facing the more difficult structural problems, to be guided to some of the abstractions from which invention can spring. In so guiding it, the precise theoretical statement plays the role of intermediary. Experience, intuitive abstraction of the essence of the experience, and precise statement of the abstraction by someone of unusually acute perception thus enables others, less perceptive in this sense but more inventive, to proceed more readily from the primary experience to the related, potentially creative personal insight.

More recent invention

As soon as we turn to typical new structural forms of the past two hundred years or so [*50, 51*], we have to recognize that they have been far more complex both in their manner of construction and the variety of materials that they have used and in the nature of the structural actions that have been critical for their success or failure. A whole range of new materials of widely differing structural characteristics has been exploited – cast and wrought iron, steel, high-strength concrete,

50. T.W.A. terminal at John F. Kennedy International Airport

 37

51. Severn Road Bridge, near Chepstow, under construction

and intimate combinations of the last two in reinforced and prestressed concrete – and the forms in which they have been used and the manner in which elements of them have been put together have been such as to create increasingly complex internal interactions vital to the overall strength or stability of the form. Structure has certainly ceased to be an invariant of the simple visible geometry of the form. Strength, stiffness, and stability under externally imposed loads of magnitudes comparable with the weight of the structure itself – and often varying widely with time in their distribution and even in the directions in which they act – have, moreover, become generally more important, over all but some of the longest spans, than simple static stability or balance under the action of self-weight alone.

In these circumstances there are many critical aspects of behavior that would clearly have remained incomprehensible even qualitatively, and not merely puzzling, on the basis of the simpler intuitions alone. Innovation in the earlier, directly empirical way would, therefore, have been at best slower than it was before. In most cases it is almost inconceivable that innovation could ever have progressed as far as it has

38

today. Thanks to guidance from the newer insights into structural actions, there have not only been great advances, but at an ever increasing pace. Even if we aggregate the widely separated periods of the most intensive development of masonry construction, it took several hundred years to learn how to exploit its major structural possibilities; yet with cast iron, wrought iron, steel, and reinforced concrete, the possibilities were mastered to a comparable degree in around fifty years, and we have seen a similar mastery of prestressed concrete, only recently introduced, in a mere twenty years or so.

Indeed the unmistakable correlation between the rate at which these new potentialities have been exploited and the depth of understanding of the structural actions brought into play is perhaps the most powerful prima facie evidence of the importance of the more highly developed understanding behind recent invention. Alongside this evidence might be set the steadily increasing role, particularly over the past one hundred and fifty years, of explicit quantitative analysis of the likely actions of new forms before all the details of their designs are finalized and they are actually built.

Such analysis, supplemented where necessary by tests, has not only given desired assurances of safety before final commitment to a design. By disclosing shortcomings or unexploited possibilities in early, tentative proposals, it has also often stimulated further invention. But for the simple reason that no structure can be analyzed or tested until both it and the mode of behavior that one desires to analyze or test have been conceived, it is hardly possible to count the analysis or test per se among the primary springs of invention, which must, as before, be sought in the personal insights or intuitions of the designer.

The importance of structural theory in the development of these intuitions or insights has been sufficiently stressed already, but the recurrent need for tests, particularly when major innovation has been attempted, clearly demonstrates its shortcomings both in this role and as a basis for formal analysis. For, in spite of continual development, the theory has never quite kept pace with the full complexities of the actions that have become critical for the safety or serviceability of new forms (or even of forms based on extrapolations from experience gained with structures already built). Indeed it would have been a severe brake on invention to have insisted that it should. Maillart, for example, made full use of the possibilities of the theoretical analysis then open to him in developing the forms with which his name is

 39

52. Maillart's Schwandbach bridge, near Schwarzenburg
53. Maillart's bridge over the Thur at Henau

particularly associated – the mushroom slab with two-way reinforcement and smoothly flared column-heads, and two very elegant forms of bridge, each using the relatively thin slab as the basic element [*52, 53*]. But he saw very clearly and correctly that they mostly implied idealizations – of the basic deformational characteristics of concrete and of the behavior of a continuous form like a slab – that were grossly at variance with the true characteristics and behavior, and he therefore undertook numerous tests to clarify his personal understanding of the true behavior.[22]

A further and more fundamental shortcoming of the theory, and even of the directly related personal insight, is that it can at best do no more than indicate how a structure may in certain respects probably

40

act: by itself it can neither indicate whether it is in these respects that structural action will be critical nor give much guidance on what would be an acceptable level of performance – of stiffness under a given pattern of loading, for instance, or of ultimate strength under a given type of load or loads. Probably as many, if not more, collapses of recent structures that attempted some innovation in form or scale have been due to failures to ask a relevant question in the analysis, or to interpret correctly the results of the analysis, as to errors in the analysis itself. The collapse of the Tacoma Narrows Bridge [*54*] is a classic example, resulting from a failure to give proper attention to a mode of behavior that subsequently proved to be critical, and several serious collapses in tall buildings assembled from large precast-concrete elements – under circumstances which the usual alternative forms of construction would have survived with only local damage even without special provisions in design – are among more recent examples [*55*].[23] Only an imaginative intuitive visualization of the possible critical modes of action of a new structure can help to obviate this shortcoming.

Today, therefore, structural theory must be regarded as a necessary basis for structural invention, yet as a basis that is incomplete as well as essentially intermediary. If it is indeed to lead the designer to creative personal insights, its source in the kinds of experience from which it

54. Failure of the Tacoma Narrows Bridge

 41

55. Partial collapse of apartment building, Ronan Point, London

initially sprang must first be itself adequately perceived. Only thus will it come to life. Then it must be supplemented, through direct experience, with largely intuitive knowledge of some of the many differences between the idealized actions of which it necessarily treats and the more complex and variable actions of real structures and the real materials of which they are built. To this end, expectations must be continually checked against observation, making full use of the increasingly powerful new techniques of observation that have themselves been associated, to some extent, with advances in theory.

Modern structural theory thus serves the creative designer best as a guide, as a frame of reference and ready source of powerful interpretive schemata, rather than as a self-sufficient tool. Some practice in the discipline of 'exact' analysis gives him, moreover, a more quantitative feeling for structural behavior than he is likely otherwise to attain and discourages him from jumping to overly facile conclusions. Yet this discipline is perhaps best learned and then, temporarily at least, put aside. In the truly creative phase of invention, the precision of mathematics is more likely to be an impediment than an aid. For to create, one 'dissolves, diffuses, dissipates'; and the mathematical abstraction can, if dwelt in exclusively, 'shut out the real world and the sky'.[24]

NOTES TO THE TEXT

The introductory quotations are from Leone Battista Alberti, *Ten Books on Architecture*, trans. James Leoni (1755; reprint ed. [ed. Joseph Rykwert], London: Alec Tiranti, 1955), p. 113, and Samuel Taylor Coleridge, *Biographia Literaria* (New York: E. P. Dutton & Co., 1906), pp. 159–60.

1
See especially Karl Lehmann, 'The Dome of Heaven', *Art Bulletin* 27, no. 1 (March 1945): 1–27; Rudolf Wittkower, *Architectural Principles in the Age of Humanism*, 3d ed., rev. (New York: W. W. Norton & Co., 1971); and E. Baldwin Smith, *The Dome: A Study in the History of Ideas* (Princeton: Princeton University Press, 1950).

2
Erik Hansen, *La 'Piazza d'Oro' e la sua cupola* (Copenhagen: Ejnar Munksgaard, 1960), gives a careful reconstruction.

3
The most comprehensive surveys of Roman building are Marion E. Blake, *Ancient Roman Construction in Italy from the Prehistoric Period to Augustus* (Washington: Carnegie Institution, 1947); idem, *Roman Construction in Italy from Tiberius through the Flavians* (Washington: Carnegie Institution, 1959); and Giuseppe Lugli, *La tecnica edilizia romana, con particolare riguardo a Roma e Lazio*, 2 vols. (Rome: G. Bardi, 1957). For the Pantheon, Roberto Vighi, *Il Pantheon* (Rome: Tipografia Artistica, 1957), and Kjeld de Fine Licht, *The Rotunda in Rome: A Study of Hadrian's Pantheon* (Copenhagen: Gyldendal, 1968), summarize present knowledge. See also Giuseppe Cozzo, *Ingegneria romana* (Rome: Libreria editrice Mantegazza di P. Cremonese, 1928), and Norman Davey, *A History of Building Materials* (London: Phoenix House, 1961).

4
For Byzantine building methods, see J. B. Ward Perkins, 'Notes on the Structure and Building Methods of Early Byzantine Architecture', in *The Great Palace of the Byzantine Emperors, Second Report*, ed. D. T. Rice (Edinburgh: University Press, 1958), pp. 52–104. For Hagia Sophia, see William Emerson and Robert L. Van Nice, 'Haghia Sophia, Istanbul: Preliminary Report of a Recent Examination of the Structure'. *American Journal of Archaeology* 47 (1943): 403–36; Rowland

 43

J. Mainstone, 'The Structure of the Church of St. Sophia, Istanbul', *Transactions of the Newcomen Society* 38 (1965–66): 23–49; and idem, 'Justinian's Church of St. Sophia, Istanbul: Recent Studies of Its Construction and First Partial Reconstruction', *Architectural History* 12 (1969): 39–49.

5

This chapter of architectural and structural history is still largely to be written, and the present exegesis is only tentative. A full study would have to take into account, besides the works here cited, Mudejar work in Spain, pre-Seljuk work in Armenia, and central-European High Gothic work, plus a number of now isolated examples elsewhere. For later Spanish vaults, see *Ars Hispaniae: Historia universal del arte hispánico* (Madrid: Editorial Plus-Ultra, 1946–), vols. 4, 5, 7; for Armenia, Jurgis Baltrusaitis, *Le problème de l'ogive et l'Arménie* (Paris: E. Leroux, 1936); and for High Gothic, Walter Ueberwasser, 'Spätgotische Bau-Geometrie: Untersuchungen an den "Basler Goldschmiedrissen" ', *Jahresbericht, Oeffentliche Kunstsammlung, Basel*, n.s. 25–27 (1928–30): 79–122. See also Rowland J. Mainstone, 'Guarini and Leonardo', *Architectural Review* 147, no. 880 (June 1970): 454.

6

Mostly in MS B (Institut de France, Paris). The majority are projects for centralized churches, but a few small sketches have, possibly, a purely structural intention and a greater significance than has hitherto been recognized: MS B, fol. 10 verso and fol. 27 recto. A similar group of small sketches by Dürer – MS II (British Museum, London), fol. 167 verso – may, however, be little more than a record of contemporary Gothic practice.

7

The structural aspect of Gothic architecture has been discussed by many writers, from Viollet-le-Duc and Choisy to John Fitchen in his *The Construction of Gothic Cathedrals: A Study of Medieval Vault Erection* (New York: Oxford University Press, 1961). Paul Frankl, *The Gothic: Literary Sources and Interpretations through Eight Centuries* (Princeton: Princeton University Press, 1960), gives an extensive bibliography. For the architectural aspirations of the Gothic, see Suger, abbot of Saint-Denis, *Abbot Suger on the Abbey Church of St.-Denis and Its Art Treasures*, ed. and trans. Erwin Panofsky (Princeton: Princeton University Press, 1946); Erwin Panofsky, *Gothic Architecture and Scholasticism* (Latrobe, Pa.: Archabbey Press, 1951); and Otto von

44

Simson, *The Gothic Cathedral: Origins of Gothic Architecture and the Medieval Concept of Order*, 2d ed., rev. and enl. (New York: Pantheon Books, 1962).

8

The psychology of invention in different fields is discussed by, among others, Jacques Hadamard, *An Essay on the Psychology of Invention in the Mathematical Field* (Princeton: Princeton University Press, 1945); W. I. B. Beveridge, *The Art of Scientific Investigation*, 2d ed. (London: William Heinemann, 1953); Brewster Ghiselin, ed., *The Creative Process* (Berkeley and Los Angeles: University of California Press, 1952); and Michael Polanyi, *Personal Knowledge: Towards a Post-Critical Philosophy* (Chicago: University of Chicago Press, 1958).

9

For fuller accounts of this synthesis, see Rowland J. Mainstone, 'Structural Theory and Design', *Architecture and Building* 34 (1959): 106–13, 186–95, 214–21; idem, 'Structural Theory and Design before 1742', *Architectural Review* 143, no. 854 (April 1968): 303–10; and S. B. Hamilton, 'The Historical Development of Structural Theory', *Proceedings of the Institution of Civil Engineers*, n.s. 1, no. 1, pt. 3 (April 1952): 374–402. The last two give references to the principal sources from the earlier and later phases respectively.

10

Eduardo Torroja, *Philosophy of Structures*, trans. J. J. Polivka and Milos Polivka (Berkeley and Los Angeles: University of California Press, 1958), gives the best description of one designer's intuition. But, as Torroja himself knew it would, his intuition died with him, and no mere verbal description could convey it to others.

11

Both Leonardo and Galileo also made use of analogies from natural forms. Leonardo, for instance, likens the cracking of a dome or semidome to the splitting of the husk of a pomegranate in the Arundel MS (British Museum, London), fol. 141 verso. Much later, D'Arcy Wentworth Thompson, *On Growth and Form*, 2d ed., 2 vols. (Cambridge: At the University Press, 1942), draws analogies between natural and man-made structures chiefly in the reverse direction.

12

Such 'hinged' bridges are still to be found in the Yangtze delta; see H. Fugl-Meyer, *Chinese Bridges* (Shanghai: Kelly & Walsh, 1937). In cities like Venice and Amsterdam, the characteristic alternation of long

45

and short voussoirs seems today to be mostly decorative; but even this decorative use no doubt betrays a functional origin.

13

Other examples are illustrated in Perkins, op. cit. in note 3; Davey, op. cit. in note 3; Arthur Upham Pope, ed., *A Survey of Persian Art from Prehistoric Times to the Present*, 14 vols. to date (New York: Oxford University Press, 1938–), vols. 2, 4; and André Godard, 'Voûtes iraniennes', *Athar-é Irān: Annales du Service archéologique de l'Irān* 4, fasc. 2 (1949): 187–360.

14

For Santa Maria del Fiore, see Piero Sanpaolesi, *La cupola di Santa Maria del Fiore: Il progetto, la costruzione* (Rome: Istituto poligrafico dello Stato, 1941); Frank D. Prager, 'Brunelleschi's Inventions and the "Renewal of Roman Masonry Work" ', *Osiris* 9 (1950): 457–554; and Rowland J. Mainstone, 'Brunelleschi's Dome of S. Maria del Fiore and Some Related Structures', *Transactions of the Newcomen Society* 42 (1969–70): 107–26. For Gothic vaults, see Fitchen, op. cit. in note 7.

15

Pol Abraham, *Viollet-le-Duc et le rationalisme médiéval* (Paris: Vincent, Fréal & Cie, 1934), classifies different varieties of cracking, and Marcel Aubert, 'Les plus anciennes croisées d'ogive, leur rôle dans la construction', *Bulletin monumental* 93 (1934): 5–67, 137–237, illustrates many examples and summarizes earlier discussions of their significance.

16

Panofsky, in *Gothic Architecture and Scholasticism*, cited in note 7, page 55, gives Gothic trade expressions for buttresses, 'such as *contrefort*, *bouterec*, (hence our "buttress"), *arc-boutant*, or the German *strebe* (hence, interestingly enough, the Spanish *estribo*), all of which denote the function of a thrust or counterthrust'. Frankl, op. cit. in note 7, refers to and quotes much of the other evidence. Of greatest interest are the minutes of meetings held during the building of the cathedral of Milan, for which see especially James S. Ackerman, ' "Ars sine scientia nihil est": Gothic Theory of Architecture at the Cathedral of Milan', *Art Bulletin* 31, no. 2 (June 1949): 84–111.

17

The most detailed drawings of the *tiburio* project are in the Codex Atlanticus (Biblioteca Ambrosiana, Milan), fol. 310 recto and verso; the recto is here reproduced in part. A number of preparatory sketches are in the Codex Trivulzianus (Castello Sforzesco, Milan).

46 INTUITION AND THE SPRINGS OF STRUCTURAL INVENTION

18
The analogy is incomplete in that it takes no account of the fact that we continually make adaptive movements as we resist or exert different forces.

19
Leone Battista Alberti, *Ten Books on Architecture*, trans. James Leoni (1755; reprint ed. [ed. Joseph Rykwert], London: Alec Tiranti, 1955), pp. 57–59.

20
General accounts of this predilection for pure geometry are given in the works by Mainstone cited in note 9. For the Gothic period, see Maria Velte, *Die Anwendung der Quadratur und Triangulatur bei der Grund- und Aufrissgestaltung der gotischen Kirchen* (Basel: Verlag Birkhäuser, 1951), and Paul Booz, *Der Baumeister der Gotik* (Munich: Deutscher Kunstverlag, 1956). Most Renaissance architectural treatises follow the pattern of Vitruvius and Alberti or, like that of Philibert de l'Orme, concentrate on the geometry of stonecutting, on the pattern of Hero of Alexandria's *Stereometrica*. For Wren, see the *Parentalia* . . . (1750; reprint ed., London: Gregg Press, 1965), tract 2; for Guarini, the *Architettura civile* . . . (1737; reprint ed., London: Gregg Press, 1964).

21
The well-known analyses of 1742–43 of the stability of the dome of Saint Peter's are recorded in Tommaso Le Seur, Francesco Jacquier, and Ruggiero Giuseppe Boscovich, *Parere di tre mattematici sopra i danni, che si sono trovati nella cupola di S. Pietro sul fine dell'Anno MDCCXLII* (Rome, 1743), and Giovanni Poleni, *Memorie istoriche della Gran Cupola del Tempio Vaticano* . . . (Padua, 1748).

22
See Robert Maillart, 'Aktuelle Fragen des Eisenbetonbaues', *Schweizerische Bauzeitung* 111, no. 1 (1 January 1938): 1–5. For the development of Maillart's forms, and tests of many of them, see Mirko Ros, *Versuche und Erfahrungen an ausgeführten Eisenbeton-Bauwerken in der Schweiz, 1924–1937*, supp. to *Jahresbericht des Vereins schweizerischer Zement-, Kalk-, und Gips-Fabrikanten* 26 (1937), and Max Bill, *Robert Maillart: Bridges and Constructions*, trans. W. P. M. Keatinge Clay, 3d ed. (New York: Frederick A. Praeger, 1969).
In the early days of reinforced concrete, testing was usually carried out at or near full scale. Today full-scale tests are less common, but tests on reduced-scale models are frequently made, and we are even approaching

a situation in which techniques of analysis by computer and knowledge of the relevant characteristics of the materials will permit the behavior of the real structure to be simulated as accurately in a theoretical analysis as in a model test. The choice between the two will then be made primarily on the basis of relative cost and the different ways in which they allow the effects of changes in the design to be explored. But since such a theoretical analysis can still deal only with a mathematical abstraction of the real structure – and an analysis based on model testing, only with a simplified analogue of it – an essential part of the skill of a good designer relying on these aids will be an ability to make valid abstractions or simplifications and to make due allowances later for factors left out; this has been repeatedly stressed by men like Torroja and Nervi also.

23
On the failure of the Tacoma Narrows Bridge, in which torsional motion of the deck persisted for about an hour under a fairly moderate wind before the collapse, see U.S., Federal Works Agency, *The Failure of the Tacoma Narrows Bridge*, prepared by Board of Engineers, Othmar H. Ammann, Theodore von Kármán, and Glenn B. Woodruff (n.p., 1941). On the Ronan Point collapse, which occurred after local damage caused by a gas explosion, see Great Britain, Ministry of Housing and Local Government, *Report of the Inquiry into the Collapse of Flats at Ronan Point, Canning Town*, prepared by Hugh Griffiths, Sir Alfred Pugsley, and Sir Owen Saunders (London: Her Majesty's Stationery Office, 1968).

24
Slightly adapted from Joyce Cary, *Art and Reality: Ways of the Creative Process* (New York: Harper & Brothers, 1958), p. 165.

ACKNOWLEDGEMENTS

The author wishes to acknowledge his indebtedness to friends among architects and engineers for stimulus received while the essay was taking shape – in particular to his late colleagues at the Building Research Station, Norman Davey, William Allen, and Graham Denyer – and to critics who have subsequently drawn his attention to ambiguities in the earlier version of the present text which appeared, under the title 'The Springs of Structural Invention', in the *Journal of the Royal Institute of British Architects*, 3d ser. 70, no. 2 (February 1963): 57–71.

Although the argument remains substantially unchanged – 'I see the invention of new forms (indeed any sort of structural design that is not content merely to play variations on what has been done already according to the set routine of some Code of Practice) as stemming first from close familiarity with, and experience of, previous achievements and secondly, and very importantly, from a generalized understanding of structural actions which gives the necessary depth of meaning to the experience and makes it possible to move beyond it with some sense of direction and some assurance of success' – the author has revised and extended the text and illustrative material, drawing in the concluding section on his 'Scientific Analysis and the Springs of Structural Invention' (paper delivered at the First National Conference on Civil Engineering: History, Heritage, and the Humanities, Princeton University, 14–16 October 1970, and reproduced in the conference's 'Selected Papers', 2:105–33).

Illustrations: *1–27*, *29*, *30*, *32–43*, *46*, *47*, *50–53*, *55*, photographs by the author (all rights reserved by the author); *28*, redrawn by the author from the Forster MSS (Victoria and Albert Museum, London), MS II, fol. 47 recto (*top left*), MS II, fol. 92 recto (*top right*), MS III, fol. 4 verso (*bottom left*), and MS III, fol. 79 verso (*bottom right*); *31*, redrawn by the author from the Codex Atlanticus (Biblioteca Ambrosiana, Milan), fol. 332 recto b (*top*) and fol. 152 recto b (*bottom left*, *bottom right*); *44*, reproduced from the Codex Atlanticus, fol. 310 recto b; *45*, British Museum, London; *48 left*, redrawn by the author from the Wren *Parentalia*, cited in note 20, pages 356–58; *48 right*, reproduced from Poleni, op. cit. in note 21, pl. H; *49*, reproduced from Le Seur, Jacquier, and Boscovich, op. cit. in note 21; *54*, Washington State Historical Society, Tacoma.

II

The springs of invention revisited

Introduction

Last year, in the first of these lectures to bear his name, James himself returned to the subject of the relevance of history which Sir Alfred Pugsley and he and I introduced here in 1975. I shall not follow his example, partly because I have already presented my own further views elsewhere[1].

I have chosen instead to return to one of the interests that has stimulated my own historical studies – an interest in how it has repeatedly been possible in the past as well as today to move forward to fresh structural achievements with some sense of direction rather than by blind trial and error.

In doing so I shall look chiefly at structural invention as I did in a paper published many years ago[2]. But I have dropped the word structural from the title because it is often hardly possible to distinguish structural invention in past centuries from invention of the kind that we should now call primarily architectural, and it is unlikely that it would have been so distinguished then.

I start from the premise that the designing of buildings and bridges is not only an art, calling (like all creative arts) for imagination, but is also a practical art serving the needs of those who provide the resources for construction and constrained by the available means. In using his imagination to seek an appropriate form, the designer must focus on these needs while holding necessary knowledge of the means somewhere at the back of his mind so that it can be brought to bear with the minimum of conscious concentration on it (just as, when preparing this lecture, I had to concentrate on what I wished to say rather than on the words I could use and the ways in which I could use them).

This knowledge of possible means, knowledge of the end to be served, and the requisite imagination, should perhaps be regarded together as the springs of invention. But I have chosen to concentrate on the specifically structural one – the knowledge of possible means. For want of a better word, I previously called knowledge so held 'intuitive' and I shall do so again, though sometimes using the words 'insight' or 'understanding' where they seem more appropriate.

Structural intuitions

In the earlier paper I identified three fundamental types of structural intuition:

First, intuitions of the potential stability or instability of different arrangements of objects in space. These I called 'spatial' intuitions.

Secondly, intuitions of forces in and on a structure and their balance. These I called 'muscular' intuitions because of their relation to our bodily sensations of forces that we exert or that act upon us; and thirdly, apprehensions of the adequacy of certain materials and forms for particular roles and more general apprehensions of strengths and stiffnesses.

All these three are general in character, applicable to structures of all kinds.

There are also more complex derivative intuitions of the ways in which particular structures and classes of structural form bear their loads, deform under them, and may eventually collapse. These arise when the more fundamental intuitions are brought to bear on the particular structures and forms. They are more limited in their applicability.

Fig 1. Adam's first house according to Filarete

I intend to trace the roles of these intuitions at different times. I shall begin by looking back to the beginnings and moving forward as far as the creative period of early Gothic. I shall then look at two early achievements that went further than usual beyond what had been done before, and finally at more recent times[3].

The beginnings

There are considerable difficulties, of course, in understanding how designers worked in the earlier periods because of the dearth of evidence. There are few contemporary records and fewer structures have survived from among the great numbers that must have been built.

These difficulties did not deter the mid-15th century writer Filarete. He

Fig 2. Chephren's Valley Temple, Giza (26th century BC)

visualised Adam, when expelled from the Garden of Eden and caught in a shower of rain, placing his arms over his head as protection and then, on a somewhat similar principle, building a more permanent shelter (Fig 1). Without Divine guidance, even this simple form, and even more so the

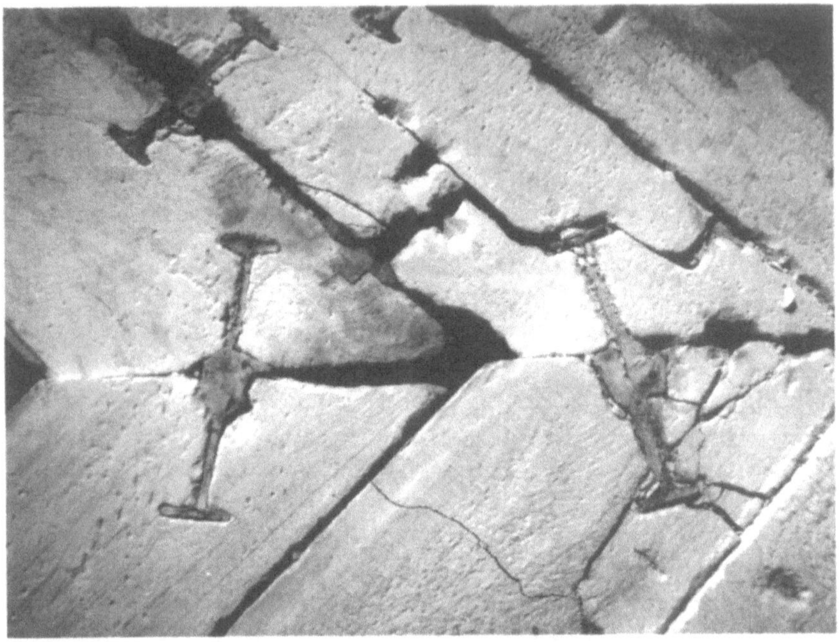

Fig 3. Iron cramps between blocks of the architrave in the southwest corner of the Parthenon (5th century BC) exposed during the current restoration (fractures of the marble blocks are due to fire and explosion in later centuries)

alternative in which the roof was raised on posts and beams, would have been an amazing first attempt starting from nothing.

It seems more likely that the earliest huts were simpler in concept and simpler to put together from available logs, sapplings or rocks. Logs or saplings could more easily be assembled by arranging their feet in a ring and joining their heads at a central meeting point. Small rocks, or mud bricks where there were no rocks, could be piled up to create walls which could then be spanned by logs laid horizontally. Imitation of similar forms occurring by chance cannot be ruled out as inspirations. But in any event the concepts would have been largely spatial ones with the main emphasis on the process of construction.

Whatever the actual sequence of their development, such forms provided the models for the earliest structures of durable stone masonry that have survived to our day. At this stage inventive skill would have been called for chiefly in devising new details. Thus in Chephren's Valley Temple at Giza (Fig 2) it was necessary to devise a way of dressing and fitting together the blocks for the walls with minimum labour to give a stable mass, and to provide adequate bearing on both walls and columns to receive the monolithic lintels. When, later, disruption occurred as a result of settlements or ground movements the response was to interlock the blocks or to hold them together with inset wooden dovetails or with bronze or iron cramps (Fig 3).

Clearly, there would have been some appreciation of the weight of the masonry, the downward forces engendered in it, and the tensions that had to be developed if the cramps were to serve their purpose. But even in the Parthenon, with all its geometric refinements, necessary cross-sections must have been determined empirically. The spanning capacity of the marble, for instance, would already have been tested to some extent in the processes of quarrying and would finally have been proven on assembly. If the structure stood when complete, the close fitting and cramping of the blocks would have prevented continuing deformations that could threaten stability.

Roman arched and vaulted structures

Two innovations of the 3rd and 2nd centuries BC made possible the great arched and vaulted structures that were the chief Roman achievement. These were the stone voussoir arch and a new concrete-like material.

Fig 4. Corbelled brick vaulting at Mari

Brick arches and barrel vaults with spans up to a few metres had long been in existence in those parts of the Middle East where timber was scarce and brick rather than stone was the usual building material. Small openings in walls were probably first spanned by simply jutting forwards successive courses until they met, as at Mari on the Euphrates (Fig 4). But with relatively thick joints it was almost as easy to start by laying a few bricks with thick mortar joints and without oversailing on each side of an opening and then simultaneously to tilt over both piles until the topmost bricks met face-

Fig 5. Defence wall at Segni (4th century BC)

to-face. This gave a true arch one brick thick on which other rings were then set as desired.

Nevertheless, the true voussoir arch seems to have been conceived quite independently. There would not, in any case, have been the same need for it in stone masonry. Monolithic lintels could span as far as early brick arches, and greater spans could easily be achieved by the same expedient as that used at Mari with a lintel as the final closure, as in a gateway in the defence wall at Segni (Fig 5).

Sometimes, however, where the wall masonry was dressed into polygonal blocks, arch-like forms would occur in it by chance, as indeed happened at Segni not far from the gate (Fig 6), and I like to think someone had enough imagination to see that, if the blocks were fully dressed on their meeting faces and the one around which the voussoir-like ones had been set were removed, there would be a self-sufficient stone arch. If so, we should have a perfect example of that mode of invention which sees and exploits a latent possibility in an existing form. But the concept would still have been a purely spatial one, as in all instances so far, and it could, with more imagination, have been arrived at directly.

The earliest voussoir arches were also over gateways in defence walls as at Falerii Novi, so that horizontal thrusts would have caused no trouble. But there would have been no doubt about their existence as soon as arches were set on isolated piers. Safe widths of piers could then have been established

Fig 6. Detail of the polygonal wall masonry at Segni

only by trial and error. Though we now know nothing of the errors, surviving examples – such as those in the market hall in Ferrentino and in the Pont du Gard – suggest that the usual approach was a fairly cautious one.

That it was primarily for its geometric simplicity and the consequent ease of cutting the voussoirs and setting out that the semicircular profile was usually preferred is suggested by the fact that the only other common Roman form was the flat arch which served as a direct substitute for the straight lintel. Interesting variants on the usual forms were arches with joggled voussoirs. Clearly, the joggling was intended to reduce the risk of slippages both during construction – allowing the use of lighter centering – and subsequently.

The basis of the new concrete-like material was the mortar produced by the natural pozzolanic earths of the Campagna and the region around Rome. There was nothing new in the way in which this mortar was made and first used. Its superiority over similar mixtures using other local earths was pure chance. So the innovation lay in the recognition of this superiority and its progressive exploitation.

Surviving structures suggest that up to and including the construction of the Pantheon in the early 2nd century AD the new concrete was simply substituted for cut stone in the less heavily loaded elements – notably the vaults – with progressive increases in span. There are early concrete conical domes in two bath buildings in Pompeii and by the mid 1st century AD wide-spanning domes like that in Nero's Golden House (Fig 7) were being built – all with open eyes at the top to evade the increased difficulty of a final closure. The great thickness given to all these domes up to about midheight and cer-

Fig 7. Domed octagonal audience hall in Nero's Golden House (AD 64-68)

tain other details of the Pantheon dome seem to indicate that they were conceived as acting only 2-dimensionally as rings of vertical arches. Despite its 3-dimensional form I suspect, also, that the complex arched construction of the Pantheon's cylindrical drum was also conceived 2-dimensionally with the arches now spanning around the circumference. The principle is little different from that of the circumferential arcading of the Colosseum except that the arches are of brick instead of cut stone and are filled in on the outside or both outside and inside. Concern for strength and stability is seen chiefly in the diminishing thickness of the dome above midheight and especially in the use of lighter materials as it closes in towards the central eye.

Fig 8. The so-called Temple of Minerva Medica, Rome (cAD 310) showing later buttresssing additions on the exterior

In some of the roughly contemporary structures of Hadrian's Villa near Tivoli we see (albeit on a smaller scale) much freer invention of an essentially formal kind. Domes with gored soffits are set over complex undulating plans. But their concrete is still so thick that the shaping of these soffits is of no more structural significance than the internal coffering of the Pantheon dome. Structural invention is to be seen rather in a further opening-out of the supporting structure. But there is nothing to indicate that this was more than an empirical exploration of limits. When it went too far it is likely that there were some collapses – perhaps, for instance, in the pavilion at one end of the Piazza d'Oro at Tivoli. Or collapse was forestalled by adding buttresses in the places where the spatial intuition showed them to be desirable as at the so-called Temple of Minerva Medica (Fig 8).

The Gothic achievement

I now jump almost 1000 years to what many would regard as the greatest structural achievement of past centuries – the Gothic cathedral. In seeking to understand how this achievement was possible we can draw on more evidence than for the Roman period and earlier. There are more surviving structures and, for the first time, there are contemporary drawings – though these drawings (by Villard de Honnecourt[4]) date only from about 1230, towards the end of the most inventive phase.

The characteristic structural elements – pointed arches, ribbed vaults, and flying buttresses – had already appeared in the preceeding Romanesque peri-

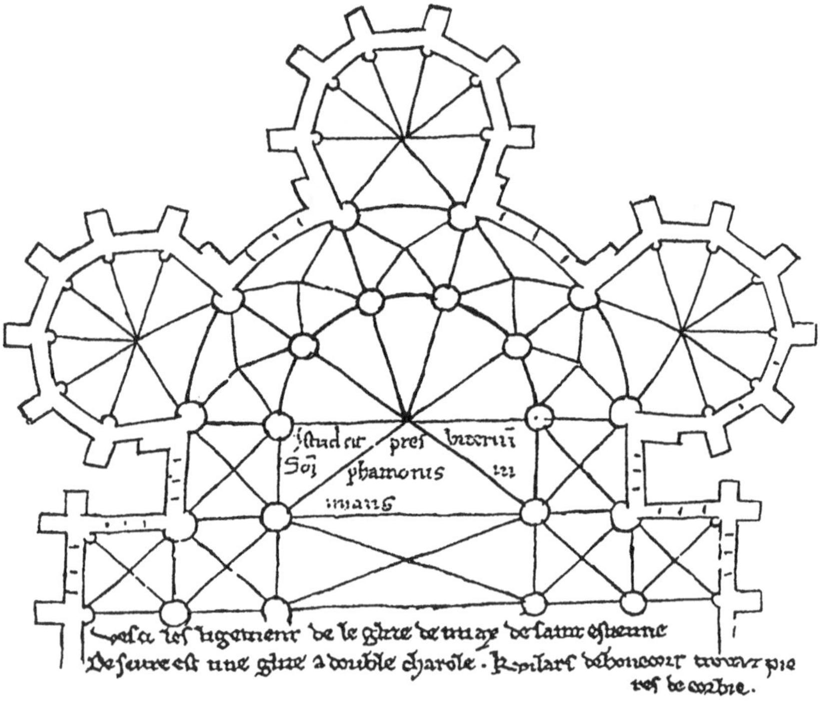

Fig 9. Plan of a church apse sketched by Villard de Honnecourt

od when renewed prosperity, coupled with religious fervour in NW Europe, led to the start of a great wave of new church building.

What was new in the rebuilding of the east end of the Abbey of St Denis around 1140 – which is usually seen as the first Gothic – was primarily an integrated spatial concept for chancel, ambulatory and radiating chapels to provide a more fitting setting for the high altar above the shrine of the saint in the existing crypt[5]. This called for vertical support to be concentrated on the outer walls and on two rings of relatively slender cylindrical columns which together carried a single integrated system of ribbed vaults over chapels and ambulatory. Ribbing of the vaults probably served primarily to simplify setting out and erection. With the chancel vaulted rather lower than we see it today, there was probably no need for flying buttresses, so that no major structural innovation would have been called for.

By the following century at least, this primarily architectural innovation at St Denis had led to a new integrated structural concept (Fig 9). This was an almost linear structure in which columnar supports and the arches and ribs they carried had wholly replaced the Romanesque concept in which walls, punctuated or not by arched openings, provided the vertical support and allowed the roofs and any vaults to be conceived almost independently.

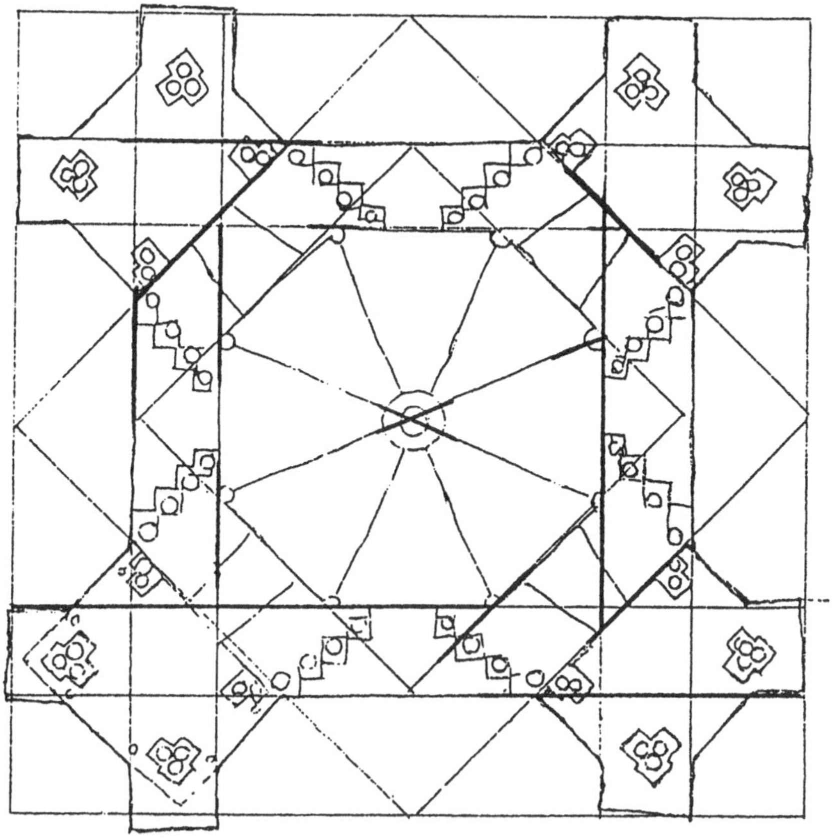

Fig 10. Villard's plan of one of the towers of Laon Cathedral with the scribed setting-out lines added in heavier and faint line (several different levels of the tower are superimposed)

Other sketches in Villard's notebook suggest, however, that designs were conceived primarily in terms of geometry. Since the stability of a masonry structure is largely a matter of achieving a compressive static balance between internal forces and self-weight loads which are themselves determined by the geometry and since (with certain qualifications) stresses are not critical, it can be argued that a purely geometric basis was all that was needed. But it must be remembered that an essential component of this basis, if it is to be adequate in itself, is that which was later exemplified by the 19th century graphic statics employed by Guadet in a suggested re-design of the buttresses of St Ouen in Rouen[6] and in Gaudi's catenary modelling. Moreover, the practicalities of serial construction in the Gothic period would have precluded the branching tree-like forms of the columns of Gaudi's Sagrada Familia.

Villard's geometry is far simpler than theirs. It is little more than the old

Fig 11. The 'crazy' vaults, Lincoln Cathedral

techniques handed down by Roman land-surveyors for halving a square by joining the mid points of its sides and for conversely doubling it – techniques that are shown in two small sketches added to his manuscript by a slightly later hand. Their use for developing a plan, and from it an elevation, is demonstrated in several 15th century mason's manuals of a more didactic character, and some 25 years ago I was able to find confirmation of its use for the former purpose by Villard himself by closely examining his plan of one of the towers of Laon Cathedral. The plan is inked-in like other drawings in the manuscript and this inked drawing is all that is seen in normal light. But in a strong, glancing light it was still possible to see some of the lines scribed into the parchment surface to guide it. They have been superimposed in heavier line in Fig 10 and extended in much thinner line to show what must have been a complete expanding set of squares, each twice the area of the previous one.

Such a procedure provides an ascending scale of dimensions derived from a chosen unit. It would also have provided both a simple way of indicating intentions with the minimum of precise drawing and a way of setting out a ground plan or marking a stone for cutting on any desired scale. But since it could be used in countless ways with complete freedom – and indeed was so used – it could hardly guarantee stability, whatever Platonic overtones of stability its geometric basis had for some clerical clients.

For the real springs of invention we must therefore look elsewhere. The simple awareness of weight pressing down and the basic spatial insight into the primary condition of equilibrium must have led to some relevant insights into the need, for instance, to plan vaults so that no rib could exert a thrust that was not countered by an opposing rib. This is seen not only in the typical apse vault (Fig 9) but even in the 'crazy' vaults at Lincoln (Fig 11).

Beyond this I see as most important the insights gained through observation of what was being done by others at the time and through hands-on experience on the building site, both observation and experience being interpreted in terms of the three fundamental intuitions to build up composite understandings of the strengths, potential weaknesses and modes of behaviour of the new forms.

Indeed I think that the most revealing characteristics of Villard's notebook are perhaps the emphasis he places on noting down the plans and elevations of other buildings that he saw and the freedom with which he drew the elevations. Not only are these not strictly to scale: they show significant variations from the buildings themselves. Most surprisingly he shows the flying buttresses at Reims abutting the piers that carry the high vaults lower down than they actually do. Since he took such pains with the tower plan (which he probably copied from a design drawing seen on the site) these variations seem to indicate that he saw what others were doing chiefly as possible starting points for his own designs rather than as precise guides for details or dimensions.

Continued observation as construction proceeded, and for some time afterwards, must have been the chief guide to the behaviour and safety of what was actually being built and to potential problems – though with construction using considerable quantities of slow-hardening mortar it would have been less safe to rely on the absence of immediate signs of distress than it would have been in the Doric temple.

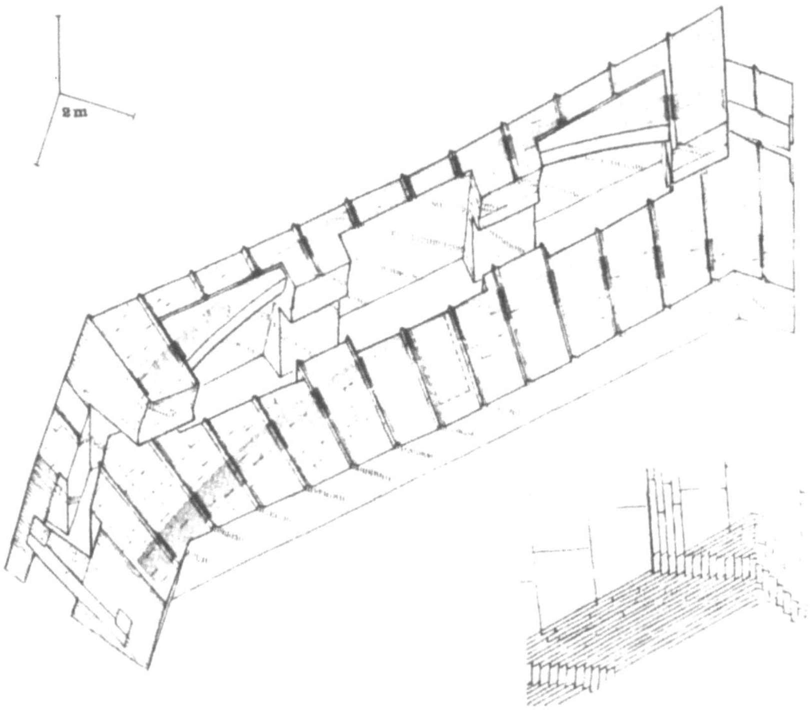

Fig 12. Cut-away isometric detail of the construction of the dome of Florence Cathedral (early 15th century)

Further evidence of the purely qualitative and intuitive nature of the more general insights and for the step-by-step manner in which designs were developed is provided by the changes that can often be seen to have been introduced as construction proceeded, by the common practice of dealing with doubts by calling together a group of experts, and by the ways in which they then justified their opinions. The changes most frequently introduced were the addition of external buttresses or internal ties to halt or reduce progressive settlements or progressive bowing of the piers under the thrusts of the high vaults. At Wells we see internal buttresses added to halt major differential settlements of the western crossing piers.

Changes made for other than structural reasons sometimes had serious consequences. At Beauvais, for instance, it is clear from the placing of two surviving original fliers on the north side that, after construction had reached triforium level, the springing height of the unusually wide main vaults of the choir was increased by some 5m without any strengthening of the relatively weak piers in critical locations in the north and south aisles. An early collapse of these vaults ensued and called for a major partial rebuilding[2].

Fig 13. A likely source of the final idea: a typical flat arch in the Baths of Trajan, Rome (2nd century AD)

The dome of Florence Cathedral

The early achievement that I want to look at first in slightly more detail is the construction of the dome of Florence Cathedral between 1420 and 1436. Here Brunelleschi did not merely take current practice a little further along roughly the same road, as had happened in the Romanesque and Gothic periods. Unusually also, the significant invention was purely structural. The basic form had already been decided upon 50 years ago when – in 1417 with the octagonal drum already under construction – it became necessary to decide how to built it with its most significant diameter slightly greater than that of the largest dome constructed hitherto, that of the Pantheon.

For various reasons, to build it on full centering would have presented insuperable problems. And the octagonal plan precluded straightforward adoption of the technique for centreless construction of a circular dome that relied on the fact that each successive horizontal ring could provide an effective key to the still incomplete vertical arches.

Though there was more to it than this, the essence of Brunelleschi's solution was three-fold[8]. He saw that the supporting drum was wide enough for there to be two shells of which the inner one was still thick enough to contain a circular ring at all levels. He then saw that, despite the octagonal surface geometry, construction could proceed as if the dome were a thicker circular one from which the unwanted parts on the intrados and extrados were simply omitted. Finally, having realised that this procedure could leave parts of the shell unsupported towards the centre of each side, he devised a bond that simultaneously avoided this risk and greatly assisted the laying of the bricks of each course. It did so by dividing the mass into voussoir-like blocks by means of bricks set vertically and radially to the dome centre at regular intervals (Fig 12).

It is notable that the procedure was not fully worked at the start so that there had to be two formal amendments to the initial specification during construction. Indeed, the initial specification explicitly stated that some details were to be decided later because 'in building only practical experience could teach what should be done'.

Unless he reached the solution completely *ab initio* (which I think is unlikely), I think that he must have seen a new possibility in the only analogous existing form – that of the Roman flat arch as constructed of brick faced concrete from the early 2nd century onwards (Fig13)[9]. If so, he was doing much as I suggested may have been done in arriving at the idea of the voussoir arch. In any case the basic intuition drawn upon was, as there, the spatial one.

Hagia Sophia and the Suleymaniye Mosque

The other achievement that I wish to look at in more detail is the 6th century church of Hagia Sophia in the New Rome of the eastern empire, Constantinople. I am dealing with it out of sequence because (as I explained when discussing it here 5 years ago[10]) its designers had a theoretic background closer to our own than any of those we have encountered hitherto. They were masters of all the statical and geometric knowledge of the ancient world.

Like others before and since, they certainly drew widely on past experience. But they went far beyond it in seeing new possible uses of existing part-spherical forms such as the semidome, and combining these forms in a new way and on the grandest scale to cover the huge central space in one continuous sweep. The basic insight here must have been the recognition that such a combination could not deform without considerable disruption of the surfaces and that the part-spherical geometry throughout would have the great practical merit of all surfaces meeting on vertical planes in circular arcs.

Appropriate support systems were also devised. These showed clear intuitive recognition of the different ways in which the longitudinal and transverse outward thrusts would have to be resisted. But with the knowledge of the time it was not possible to calculate the necessary dimensions. This inability led them to take a chance in solving a problem created by the requirement for continuous side galleries. The critical depth of the interconnection of main and outer buttressing piers at gallery floor level was excessively reduced and the resulting weakness was accentuated by the use of brickwork with thick mortar joints subjected too soon to large shearing

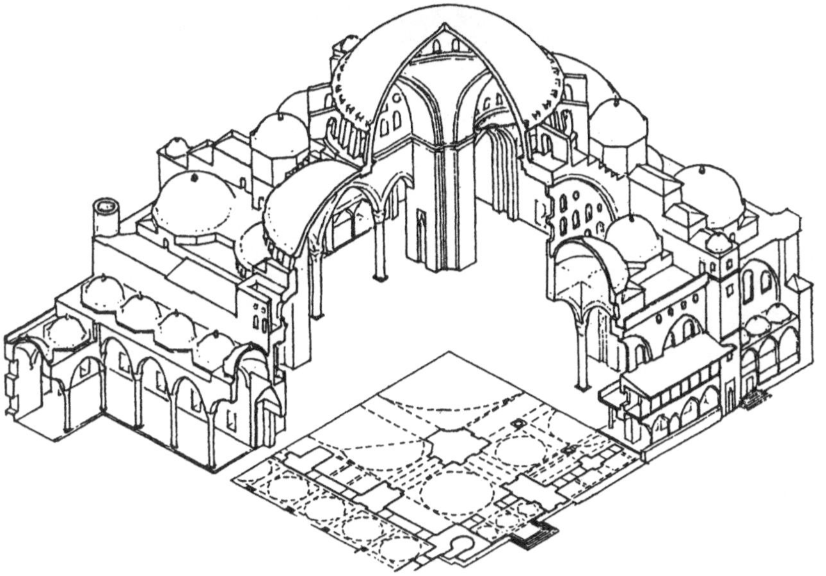

Fig 14. Partly cut-away isometric of the Suleymaniye Mosque, Istanbul (16th century)

forces on account of an excessive speed of construction that was possibly enforced by the client. Large early deformations did lead to a need for extra works which again demonstrated the empiricism of innovation when the only way to test the adequacy of a design was to build it.

It was only a thousand years later that the Ottoman architect/engineer Sinan, having closely examined Hagia Sophia and sought out its weaknesses, built the first structure to face up to its challenge on anything approaching its scale – the Suleymaniye Mosque (Fig 14). Taking advantage of the fact that, in the mosque, there was no need for wide side galleries, he greatly improved the lateral buttressing and, drawing on Ottoman construction techniques, he replaced its brickwork and that of all the main arches by cut stone and introduced additional ties to avoid the excessive deformations that had occurred in Hagia Sophia[11].

More recent developments

In turning finally to more recent developments to explore the role of the new insights into structural behaviour that have stemmed from developments in theory and from the experience gained through designing, analysing, building, observing and testing an ever-increasing range of structures, I shall have to be even more selective.

A century and a half before Sinan, Leonardo had already been seeking a more generalised understanding of the conditions of static balance and I think that we can see this quest reflected in some of his sketches of ideas for vault-

Fig 15. Detail of the iron reinforcement of the architraves of the east colonnade of the Louvre, Paris (17th century)

ing the crossing of Milan Cathedral. These sketches may, in the second half of the 17th century, have been one source for Guarini's design for the novel vaulting system of the church of San Lorenzo in Turin. Yet, though Guarini was among other things a mathematician, there is no evidence that he took even as much interest in the statics of his vaults as Leonardo had done. His most relevant writings deal only with their stereometry.

In France, at the same time, there was a different reaction against the massive arched forms typical of the Roman High Renaisance. This was a new classicism that took as its model the column-and-beam structures of most classical temples. It emerged in the east front of the new Louvre then under construction for Louis XIV. The design called for colonnades of paired Corinthian columns carrying flat architraves and standing in front of recessed wings between a central and two end pavilions. No available stone could span between the columns in single blocks, so it was necessary to construct the architraves as flat arches. It was well understood that these would exert large outward thrusts and that the unbalanced thrusts of the transverse architraves might push out the columns unless some means of resisting these thrusts was provided.

To serve this purpose a complex and ingenious system of iron ties was devised, anchored to other reinforcement embedded in the columns. It used to be known only from an engraving published 100 years later and therefore slightly suspect. But a few years ago a colleague and I were able to take advantage of a temporary closure of the top-floor gallery in the north pavilion to gain access to an inspection void at roof level and to confirm, at least in part, the accuracy of the engraving (Fig 15). Two things are of particular interest. First, it will be seen that wedged connectors were provided to permit adjustments to the tension in the ties if and when this seemed desirable.

Fig 16. Interior of the dome of the Paris Pantheon (18th century)

Secondly, it is recorded that a structural model was made to give confidence in the effectiveness of the ties. This tie system furnished the prototype for the reinforced masonry flat arch known as the platte bande which was extensively used in the 18th century, notably in the Paris Pantheon[12].

This is of particular interest for the unprecedentedly slender piers that carry the central dome and for the use of largely hidden flying buttresses to carry most of its outward thrusts to the external walls. What were probably the first such statical calculations were made to investigate the safety of the design[14]. But the source of the buttressing idea was a close study of the

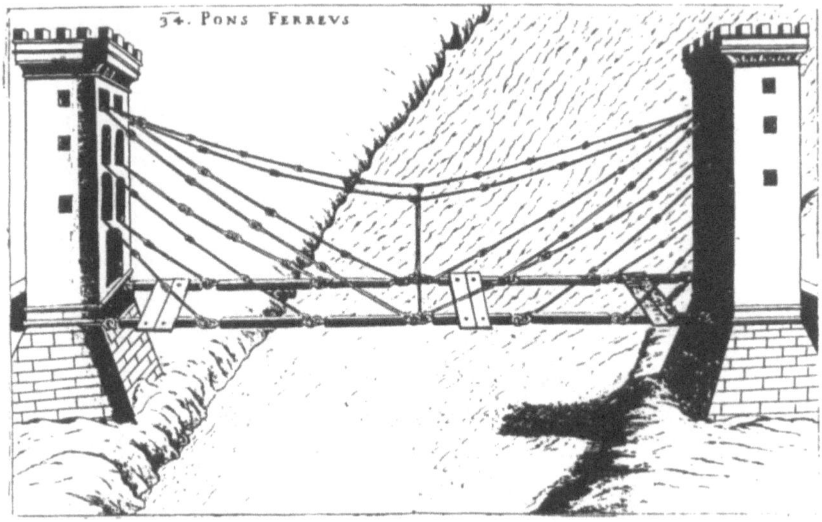

Fig 17. One of two suggested designs for suspension bridges by Faustus Verantius (from Machinae Novae, Florence and Venice, c1615)

Gothic structural system and the statical insights gained from this. These insights are also evident, I think, in the superb design of the triple-shell dome which, like several earlier domes, was tied circumferentially at several levels (Fig 16).

These structures, however, were all still masonry ones whose stability was determined largely by a geometry that permitted a safe balance between internal forces and self-weight loading. Where calculations or tests were made, they were not the prime sources of the designs.

As our present structural theory developed further, new materials became available and new requirements arose, especially for longer-span bridges on which loads other than self-weight were increasingly important.

The first projects to meet these new requirements, such as those of Verantius and Pope (Figs 17 and 18), were highly unrealistic as we can easily see today. They apparently could not. Pope claimed that his slender timber arch, constructed by cantilevering from each side, could span up to 2400ft. Real progress involved much practical experiment, initially much as before by simply building full-scale and proving the design in service.

The major difficulty with the suspension bridge was, of course, to make the decks stiff enough to resist wind and the effects of moving loads. They had frequently to be stiffened or totally rebuilt. Roebling (first at Niagara and Wheeling and then in the Brooklyn Bridge) added a multiplicity of ties and stays to the basic suspension system to restrain movements (Fig 19). In this country Stephenson and Fairbairn embarked on the classic series of model tests that led them to move away from early impractical reinterpretations of the suspension idea to the freely spanning tubes of the Conway

Fig 18. Timber bridge designed by Thomas Pope to span the Hudson River, New York (from A treatise on bridge architecture, *New York, 1811*)

and Britannia Bridges. And although Eaton Hodgkinson was simultaneously advancing the relevant theory on the basis of the tests, it was the tests themselves that played the main role in developing the designers' understanding and indicating desirable changes in the evolving design.

In the present century, theoretic advances have played a far greater and indispensable role. The superb designs of Robert Maillart (Fig 20) would not have been possible without calculations of reinforcing needs. But he clearly saw and insisted upon the limits of current calculation methods and the need to assess carefully the answers they gave on the basis of a more intuitive understanding which he developed partly by a long series of full-scale load tests[14].

The designs of more recent structures, whether in steel or prestressed concrete, have, of course, been far more dependent on calculation. Only with its help have the dreams of Verantius and Pope been realised. This calculation and the insights derived from it have increasingly served as the means of bringing the experience gained by previous tests to bear on new designs with less restriction to particular forms, and they have thereby reduced the need for full-scale testing.

Calculation yields quantitative precision. But the intuitive insights are needed to judge which calculations are desirable and what weight should be placed on the results. Even more importantly, calculation alone rarely generates new ideas. As in the past and for the same reasons, creative design springs best from the personal intuitive understanding and imagination of the individual designer.

Thus I see no fundamental change in the springs of invention. The change I see is rather a progressive coming together and deepening of the intuitions that I identified at the start. The greatest change has been in the understanding of the characteristics of materials and the ways in which these determine the strengths and stiffnesses of what we make from them. But part of this understanding has been a new insight into the internal balances of compressions and tensions and into the ways in which these stresses determine deformations. The first of these is, of course, linked to the basic understanding of the conditions of equilibrium furnished by the muscular

Fig 19. Ellet's Wheeling Suspension Bridge (1849/1860) with stays added by Roebling in 1872

Fig 20. Maillart's Valtschiel Bridge (1925)

intuition. And what I called the spatial intuition now embraces possibilities of exploiting tension and flexural and torsional restraints that did not previously exist. Drawing on these insights we can see much more in, and make wider use of, the experience gained through an ever-expanding range of successes and failures. This ability to move more freely and confidently beyond

Fig 21. TWA Terminal, Kennedy Airport (1962)

the limits of past achievement is perhaps what most distinguishes our position today from that of our forerunners.

Whether we call our present personal understanding intuitive or merely qualitative is immaterial, provided that we acknowledge its primacy. Sometimes we may visualise it in terms of trajectories of principal stresses or bending moments, sometimes in terms of load paths and relative stiffnesses, and sometimes in terms of deformations and possible modes of collapse. Always a good capacity for 3-dimensional visualisation is called for. And it can never, I feel, be replaced by computer-held expert systems because these must always be circumscribed and because they do call for some conscious concentration when we should be concentrating on the end to be served.

Structure and architecture

I conclude with a glance at the relationship between what we now see as the architectural and structural aspects of design. I said earlier that they could not really be distinguished in the past. This was especially true when new forms arose from a multitude of tiny changes in a process of almost pure trial and error, as possibly happened in, for instance, the evolution of better ways of making small boats than by hollowing out logs. The resulting artifacts then seemed to have an inevitable rightness. But so wide are the possibilities open to us today that the initial concept can even largely ignore structural feasibility. This can produce striking works when the architect has a fine imagination and the engineer has enough ingenuity to make his con-

Fig 22. Clifton Cathedral (1975)

cept work (Fig 21). But I hope you will agree that it is when the concept is, throughout, still both a structural and an architectural one that the good engineer can give of his best (Fig 22).

Notes and references

1. Mainstone, R J: 'The uses of history', *Architectural Science Review*, 20, 1977, 30-34
2. idem: 'The springs of structural invention', *Journal of the Royal Institute of British Architects*, 70, 1963, 57-71 (revised version 'Intuition and the springs of structural invention', VIA, 2, 1973, 42 67).
3. In delivering the lecture I used far more slides to supplement the spoken word than can be reproduced here. So what follows is necessarily more selective. Further relevant illustrations will be found in idem: *Developments in structural form*, London, 1975.
4. In the so-called *Sketchbook*, Ms.19093, Bibliotheque Nationale, Paris. Facsimile editions: Lassus, Paris, 1858 (reprinted 1976); Willis, London, 1859; Hahnloser, Vienna, 1935; Bowie, Bloomington, !957; Bucher (in *Architector* vol.l), New York, 1979. The folios reproduced and/or referred to are xxix, xxxix, xviii, and lxiv.
5. Radding, C M and W W Clark: *Medieval architecture, medieval learning*, New Haven, *1992*
6. For the former see Gaudet, *Éléments et théorie de l'architecture*, vol.III, Paris, 344, fig.1096

7. Murray, S: *Beauvais Cathedral,* Princeton, 1989
8. Mainstone, R J: 'Brunelleschi's dome', *Architectural Review,* CLXII, 1977, 156-166 and 320.
9. idem: 'Le origine della concezione strutturale della cupola di Santa Maria del Fiore' (The sources of structural ideas for the dome of S. Maria del Fiore), *Filippo Brunelleschi: la sua opera e il suo tempo,* Centro Di, Florence, 1980, 883-892.
10. idem: 'Justinian's Church of Hagia Sophia, later the Mosque of Ayasofya, in Istanbul', *The Structural Engineer,* 68, 1990, 65-71. (Note that, through editorial errors corrected on p.101, the scales of Figs 4,5,7 and 13 are not as stated in the captions, and the captions of Figs 4 and 7 are interchanged.) See also idem: *Hagia Sophia – architecture, structure and liturgy of Justinian's Great Church,* London, 1988.
11. idem: 'Sinan's Suleymaniye Mosque and Hagia Sophia', *Public Assembly structures from antiquity to the present,* Mimar Sinan University, Istanbul, 1993, 53-62
12. idem and Berger R W : 'Materials and structure' in Berger, R W: *The Palace of the Sun: the Louvre of Louis XIV,* University Park, 1993, 65-74
13. Gauthey, E M: *Mémoire sur l'application des principes de la méchanique à la construction des voûtes et des dômes,* Dijon, 1771
14. for instance in *Schweizerische Bauzeitung,* 111, 1938, 1-5. Partial translation in Max Bill, *Robert Maillart,* Erlenbach-Zurich, 1949, 15-16.

III

The role of tradition in the development of form in domes, shells, and space structures †

The word tradition can mean many different things. It can be taken to embrace all that is available to a designer at a given time and place as a result of previous endeavour and experience – everything that distinguishes his position from that of Adam at the gates of Eden. Or it can be used in much more restricted senses – as it usually is in the form 'traditional'.

I shall use it to denote primarily the operational lore, customary usages, and values of a particular society at a particular time. These will, in part, have been inherited. But a living tradition is continually open to criticism and development and will also, in part, be a product of its own time. Where the tradition is handed on and developed largely through direct example and practice, it will, at any time, be clearly defined and all-of-a-piece. Change will involve a partial forgetting of what went before and not just an addition to it. This is what can still be observed, to a limited extent, in surviving simple or 'primitive' societies. Typical outward signs are the characteristic rituals and taboos of the society and its typical physical artefacts such as the hut that serves as shelter or dwelling.

In more developed societies, the position is more complex. We are no longer so dependent on direct example and on our own direct experience in trying to do better. Much of the experience and knowledge, many of the insights, and even some of the wisdom of the past are accessible to us also through the written word and through surviving buildings and other artefacts. Thus embodied, traditions that have now, as living traditions, been changed out of all recognition are still, in part, accessible to us. Our contemporary traditions are thus much more complex and no longer all-of-a-piece. For they include, and have for a considerable time included, our attitudes to these earlier traditions and our ways of drawing on them.

Our present tradition of designing and building shell and space structures has roots at least as far back as the Roman phase of the original Classical tradition. The architects of Republican Rome were themselves heirs to much earlier traditions of arch and vault construction exemplified by partial survivals of great antiquity, by slightly later rock-cut replicas (fig. 1) and, later still, by the much larger vaulted tombs of Mycenae. But they and their successors completely transformed these earlier traditions by exploiting the special characteristics of the concrete made from their natural pozzolanic earths. The transformation began fairly slowly around the end of the 2nd century BC. It proceeded with greatly increased momentum in the

1. 'Holy of Holies', Saflieni hypogeum, Malta, c. 3000 BC.

latter half of the 1st and early 2nd century AD spurred on by Imperial patronage, notably by Hadrian. It continued to develop further for another two centuries. Structures like the Pantheon, the great Imperial Baths and the Basilica Nova, still bear eloquent witness to it and still, in some ways, challenge our understanding. But all the available evidence suggests that the process of development differed little in kind from that which led to the most primitive hut forms – an empirical process of successive trial and error. The significant differences – apart from the very different demands and resources available to meet them – lay most likely in the greater wealth of relevant experience on which to draw as the development proceeded and the keener critical intelligences of some of the designers. For all the variety of form of the surviving structures, the tradition remained a clearly defined and unified one except to the extent that it developed somewhat differently in its later phases in parts of the Empire where the resources, particularly of materials, were different.

Much the same was true when, after a long period of limited building activity, Roman techniques of vaulted construction were revived by early Romanesque architects and then much further developed in the Gothic period. Here also, particularly in the 13th century, there are strong suggestions of some powerful critical intelligences at work seeking to understand the structural principles inherent in the more daring structures. But the more plentiful surviving evidence shows even more clearly that development proceeded almost entirely by successive small steps beyond what had already been done – steps that not infrequently resulted in a partial failure and an enforced modification of the design.

Up to the Renaissance there was probably only one instance of a major, and ultimately supremely successful, attempt to look outside the immediate tradition for a new concept and for hints as to how it might be realized. This was the rebuilding, to the orders of Justinian in the early 6th century, of the church of Hagia Sophia in Constantinople (fig. 2). The two architects, and perhaps Justinian himself, must have looked beyond the main current tradition in church building to both the great groin-vaulted halls of the earlier Roman metropolitan tradition and the great domed rotunda of the Pantheon. They combined these forms in a wholly novel way and created thereby an interior of breathtaking splendour. Though precedents for all the individual elements can be found (except perhaps for the clearly differentiated pendentives), the basic idea for their combination must have come from outside the respective traditions and from the different discipline of geometry of which both architects were masters. The lack, at the time, of a similarly developed tradition of structural understanding was responsible for several mishaps during construction and a partial failure not much later.

2. Hagia Sophia with the dome of Hagia Irene to its left and a small dome of Topkapi Serai in the foreground. Just visible to the right of Hagia Sophia is the dome of the Mosque of Sultan Ahmet.

The Renaissance, of course, owes its very name to such a looking beyond the traditions of the previous few centuries to those of Classical Rome, in so far as they could be retrieved through the surviving literature, buildings, and other works. By far the most important single event in relation to the present theme, and one of outstanding importance also in its own time, was Brunelleschi's achievement in building the dome of Florence Cathedral – the first for 1300 years to match the Pantheon in span. Indeed, being carried at a considerably greater height on an octagonal instead of a circular base, it did more than match it in other respects.

For those who initially planned a dome of this size and octagonal form, the much earlier dome of the adjacent Baptistery (fig. 3) must have served as the model. Indeed it would have been a more relevant model, structurally, than the bulbous nomad's tent would have been for the slightly later bulbous Islamic dome. But it is unlikely that it was still known quite how the Baptistery dome had been built and it must have been a pure act of faith – if not of foolhardiness – to assume that a way would be found, when necessary, of almost doubling the scale. When detailed design began, the current tradition (exemplified in the construction of the part-octagonal vaults of the tribunes) was to use centring. All the masters who had been brought up in this tradition seem to have proposed the use of similar timber centres, notwithstanding the difficulties that would have arisen in constructing and using them over more than twice the span. Brunelleschi's solution was to build without centering, essentially by dividing the dome into inner and outer shells, giving the inner one enough thickness to accommodate, at all levels, a continuous circular ring, and relying on the self-sustaining capacity of each completed ring to ensure stability at all stages.

3. Baptistery and Cathedral, Florence. Beneath the pyramidal roof of the Baptistery is a pointed octagonal dome similar, in form, to the much larger Cathedral dome built three centuries later.

Today we can only imaginatively reconstruct the thought processes by which Brunelleschi arrived at the solution. It seems to me that he was led to it by considering the possibilities implicit in three traditional Roman practices still visible in numerous surviving structures. These were the leaving of an open central eye in a dome (as in the Pantheon), the frequent later adoption of deeply lobed or coffered inner forms for domes that nevertheless contained a circular ring at all levels within their thickness, and the practice, when constructing flat concrete arches over doors and similar openings, of having bricks penetrate the concrete mass in such a way as to divide it into voussoir-like sections. If so, Brunelleschi was drawing on earlier Roman traditions in much the same way as Anthemius and Isidorus, the architects of Hagia Sophia, had done. He was seeing in them possibilities that the Romans themselves never exploited in quite the same way and may not even have fully realized.

It seems significant that Brunelleschi, again like Anthemius and Isidorus, had not been brought up just within the circumscribed tradition of the master masons and carpenters of his day. (Indeed he was briefly imprisoned during the later stages of construction because he was not a member of their guild.) With a wider outlook and a more broadly trained mind, each was able to look beyond the contemporary tradition, aware nevertheless that it could be drawn upon for the purpose of putting new ideas into practice.

This kind of enrichment and rejuvenation of the contemporary tradition by men with a wider outlook and background became increasingly common from the late 15th century. Sometimes the enrichment was primarily structural, as in Leonardo's projects for ribbed domes – early masonry space structures stimulated initially by the search for a way of vaulting the crossing of Milan Cathedral and inspired by certain Gothic forms. It was structural also in Guarini's later practical realization of such structures – notably in San Lorenzo in Turin – and in Wren's adaptation of the double-dome prototype for the dome of St Paul's.

4. Looking westwards from the dome of the Pantheon. On the left is the dome of Sant'Ivo; in the centre that of Sant'Agnese; and to the right that of St Peter's.

More often at this time, the enrichment was seen chiefly in formal or spatial invention, keeping well within the limits of contemporary structural practice but responding to new aesthetic sensibilities, new objectives, and a wider scale of values. This was most typical of the High Renaissance and the Baroque. It is well illustrated in a single westward glance from the dome of the Roman Pantheon (fig. 4). This dome stands for the basic tradition and its development up to the contributions of Brunelleschi. In Sangallo's work on St Peter's, Brunelleschi's way of vaulting without centring was directly employed in the simpler context of the secondary circular domes around the future high central dome. In this dome we see chiefly Michelangelo's unsurpassed plastic skill. Technically the circular form made construction much easier, and it is always easier to do something that has already been demonstrated to be possible than something whose feasibility remains in doubt. In the relatively small dome of Borromini's Sant'Ivo plastic invention had free rein. The culmination of this phase of development was in the late Baroque of Southern Germany, particularly in Neumann's handling of interpenetrating vaults to weld a whole interior into one continuous movement.

By the time of Neumann's death in the mid 18th century, other momentous changes had begun or were about to begin. A major dome – that of St Peter's – had for the first time been subjected to a statical analysis of the kind we might perform today; understanding of structural behaviour and the ability to predict it at the time of design were soon to grow at an ever increasing rate; and major developments were to take place in materials and in construction and fabrication techniques. Closely associated with these changes was an increased commitment to paper of what had hitherto been largely unwritten operational lore in spite of treatises like those of Vitruvius and Alberti, plus a growth of specialised professions to take responsibility for different aspects of design. In consequence we are now heirs to the much more widely diversified body of tradition referred to earlier. Among other things, I would say that it includes traditions, sometimes in a state of fairly rapid evolution, of the ways in which analytic and predictive theories are used to guide design and new traditions of planning and formal or aesthetic choice. Among the latter, the Beaux Arts tradition was the most clearly formulated, and those associated with the Modern Movement of the 1920s and 1930s were the most influential in later decades.

Underlying these changes there has, however, been a different sort of change that is highly relevant to the development of form today. It stems from our greater understanding of structural behaviour. At the same time as permitting analyses of the likely behaviour and adequacy of a particular structure before it is built, this understanding has greatly expanded the range of proven practice on which the designer can draw. The feasibility of a whole family of structures of a certain type may now be demonstrated by the construction of just one or two – and demonstrated not just to an Anthemius or Brunelleschi, a Leonardo or Guarini, but to every fully competent engineer.

To say that everything is thereby now made possible would, of course, be a gross exaggeration, particularly in the case of slender wide-span forms. But the limits are usually so wide that a measure of choice nearly always exists where previously it existed only when working well within the confines of recent practice. In these circumstances the architect – or the engineer as architect – can either exploit the freedom of choice to primarily aesthetic or other non-structural ends as Michelangelo and Borromini and Neumann exploited their more limited freedoms. Or he can seek and submit to the discipline of a particular structural logic and allow the form to evolve through this discipline as Guarini did in San Lorenzo.

As fine examples of the first approach we might take Torroja's tongue-in-cheek design of the coal bunker at his institute outside Madrid, knowingly or unknowingly echoing Michelangelo (fig. 5), or almost of any of Saarinen's wide-span roofs. In our more puritanical moods we may be tempted to condemn the structural illogicicality of the Kresge and TWA roofs. But I think that the latter at least, and that of the Yale Hockey rink, display a mastery of the handling of surface forms to define space and create a memorable and life-enhancing presence equal to that of, and in the true tradition of, the best Baroque architects. We should be poorer without them. I do not know to what extent any of them was inspired by a conscious study of this tradition. But I do think that an understanding of it would save us from some of the banalities that stem from a mere seizing on forms that happen, at the time, to be fashionable.

5. Gilded polyhedron by Michelangelo above the dome of the Medici Chapel, San Lorenzo, Florence.

As examples of the second approach we might, at first sight, take many space frames and other doubly curved tension roofs, because in such forms the structural and fabrication and constructional requirements tend to impose their own logic more insistently than they do in shells – which can always be unobtrusively thickened where pure membrane action becomes impossible. But here also I think we must, on closer inspection, see a much freer initial choice of form, exercised particularly in the placing of supports and anchorages.

Typically, if we pass from early experimental trusses to recent space frames, we find that the overall form of the latter is a framed analogue of a slab, a vault, or a dome, and that the overall geometry has been chosen with as much freedom as if the slab or shell itself had been used. Thus, even if the triangular/hexagonal framing of the St Louis Climatron roof is the outcome of purely geometric, fabrication, and structural considerations, and unconnected with the very long tradition of earlier explorations of similar geometries for other purposes, the choice of the basic form of the analogous shell must have been made with much the same freedom as Saarinen's similar choice at MIT. Likewise the choice of a single horizontal slab widely cantilevered beyond its outer supports as the basic form of the McCormick Center roof (fig. 6) was made possible by the spanning capacity of the space frame used but not enforced by the inner logic of this frame. It was a choice wholly within the Constructivist/Elementarist tradition of some of the earliest architecture of the Modern Movement and, very significantly I think, of painters and sculptors of the same time like Malevich.

6. McCormick Center, Chicago.

I find the affinity with a major recent sculptural tradition even more striking in the case of the doubly-curved tension roof and related anticlastic shell forms. More striking because, as far as I am aware, it is not explicable here by direct involvement of people like Gabo, Hepworth, and Moore in the structural developments, as most similar affinities have been. Perhaps this is because the structural and constructional logic are so insistent once the basic decisions on supports, anchorages, and boundaries have been made, and differ too much from the disciplines imposed by the sculptor's media. No matter. The tradition must have played its part in awakening sensibilities.

Taking a narrow view of tradition, my reaction, when I first considered your invitation to speak on this subject, was that its role as a stimulus to new developments had been slight, even if its role as an essential foundation was unquestionable. I put it to you now that the broader view, including traditions in related arts and seeing tradition not as a dead legacy from the past but as something very much alive – a means whereby the vitality of the past enriches the life of the present – is more helpful, and that, in these terms, a proper understanding of our shared traditions will do as much as anything to further fruitful interprofessional interactions.

† This paper, unpublished at the time, was the invited opening address at the 1978 conference of the International Associate of Shell and Space Structures, held at the University of West Virginia.

I am particularly indebted to Dr Reuben Grima and Mr Joseph Farrugia of the Malta Museums Department for kindly allowing access to the Saflieni hypogeum during conservation works and for providing lighting for photography and other assistance.

IV

Structural Theory and Design Before 1742

'*Ars sine scientia nihil est: Scientia sine arte nihil est.*' This assertion by Jean Mignot, a French consultant, and counter-assertion by the Milanese masters were recorded in Milan in 1400 during one of the long succession of formal discussions or debates on the design of the new cathedral.[1] Though stated with unusual force in the heat of argument, the dependence of sound design on both theory and practical experience which each emphasized from a different point of view was not, even then, either new or previously unrecognized. Mignot's championship of theory under the provocation of Milanese intransigence, for instance, merely echoed an earlier emphasis on its importance which can be traced back through Islamic, Byzantine and late Hellenistic writers to the first surviving architectural treatise, that of Vitruvius, written in the first century B.C.[2] The changes that have taken place since Vitruvius have been changes in the theory itself, in the ways in which it has been used and supplemented by practical experience, and in the types of structure to which it has been applied.

Hitherto structural historians have largely confined their attention to the development of the present theories of equilibrium, deformation and strength. The first recorded application of these theories in structural practice in 1742–43[3] may be said to mark the birth of the present art of structural design. This art has, however, grown out of what preceded it and still in some ways reflects its origins. Our understanding of it will be deepened by a better knowledge of these origins, and some knowledge of the manner in which the surviving structures of the past were conceived seems essential if they are to be studied with profit today as part of the wider education of the young architect or engineer.

There are difficulties, of course, in any attempt to establish the working methods of designers more than 200 years ago. It is difficult enough to describe adequately the methods adopted today. As the enquiry is pressed further back the evidence becomes, in general, increasingly fragmentary. Different facets of the total changing picture are illuminated at different times, and through them its main

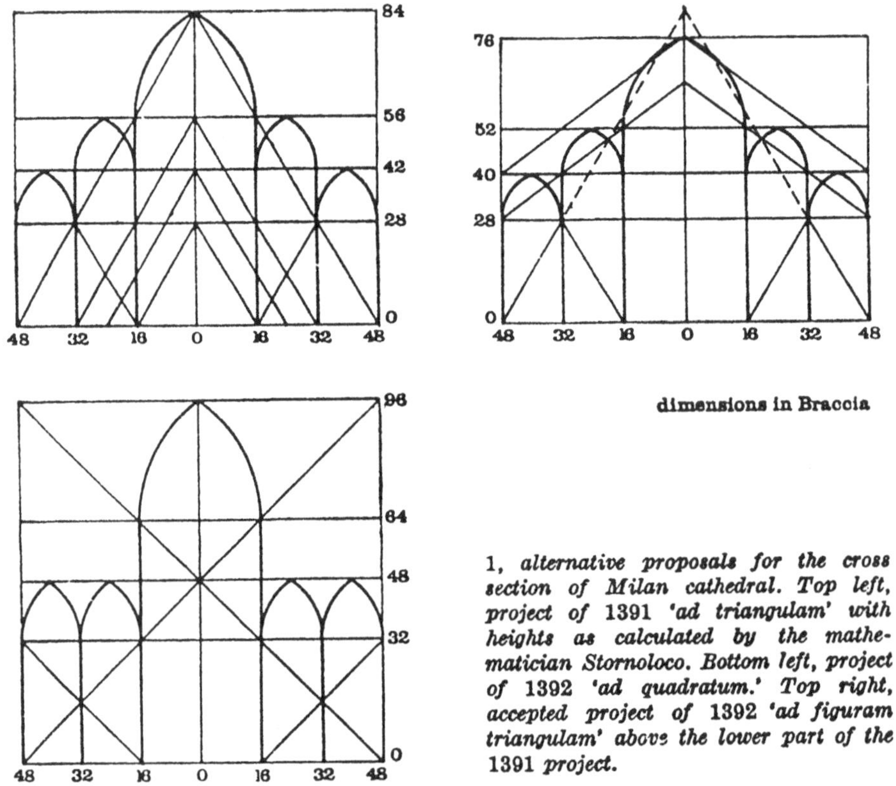

1, *alternative proposals for the cross section of Milan cathedral. Top left, project of* 1391 *'ad triangulam' with heights as calculated by the mathematician Stornoloco. Bottom left, project of* 1392 *'ad quadratum.' Top right, accepted project of* 1392 *'ad figuram triangulam' above the lower part of the* 1391 *project.*

outlines must be reconstructed and the significant underlying continuities discerned. The surviving structures themselves are a primary source of information, but they alone can tell us nothing of theory and may easily mislead us into projecting too much of our own approach to design into the past. Supporting evidence must be sought, particularly in treatises, in contemporary drawings and in other contemporary records.

Attention will be concentrated on the design of those structures which presented a real structural challenge on account of their form or scale or the adoption of new methods of construction. Whereas today such structures might be long-span bridges, they were in the past mostly large churches—large essentially unicellular structures which presented as the principal structural problem the need to span a high wide space with some kind of permanent vault.

THEORY (1)

The theory invoked at Milan will be considered first.

Mignot was quite explicit. The theory of which he spoke in the exchange quoted above was '*scientia geometriae*'.[4] Essentially this required the main lines and forms of a structure—such as the heights of vaults and the proportions of piers and buttresses—to conform to a certain predetermined geometric grid. In the earlier discussions between visiting experts and the local master more than one such grid had been proposed, 1, but the need for some control of this kind on the design was generally accepted and was strikingly acknowledged by the calling in of a professional mathematician in 1391 to compute the irrational height of the equilateral triangle then favoured as the control on the cross section.[5]

This geometric theory can be traced in considerable detail from the birth of Gothic to the early Renaissance and something closely analogous can be seen already in Vitruvius.[6] Evidence of it is found in surviving contemporary drawings of buildings,[7] in precise measurements of the buildings themselves,[8] in the records of other discussions like those at Milan (notably in Siena in 1321[9] and in Bologna in the late sixteenth century[10]) and in the architectural treatises of Filarete,[11] Francesco di Giorgio,[12] Serlio[13] and Philibert de l'Orme.[14] Most explicitly it appears in a number of short treatises from the end of the Gothic period, such as that by Roriczer published in 1486.[15] Roriczer demonstrates the theory by means of a simple example of its application to the design of a small pinnacle, 2. Starting from one dimension in the ground plan he derives all others by means of the simple geometric construction for doubling or halving the square. This construction, repeated often enough, provides all the dimensions needed for both plan and elevation. Variations in the method arise largely from the substitution of other basic figures for

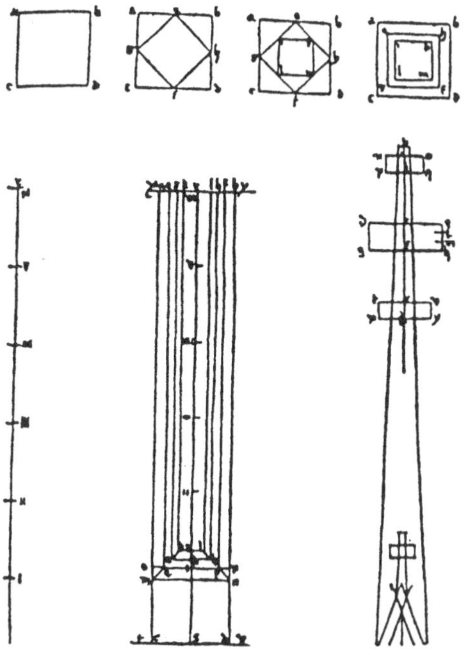

2, *stages in the design of a pinnacle according to Roriczer.*

the square. During the Gothic period the equilateral triangle and probably the regular pentagon (which leads to the 'golden section') were also used,[16] while in the early Renaissance there was an increasing use of the circle and rejection of constructions which led to irrational ratios between dimensions.[17]

Nowhere, however, are the basic premises of the theory stated in surviving documents, and it is unlikely that they were ever precisely formulated. It was not solely, and it is unlikely that at

3, thirteenth-century practical geometry in the pattern book of Villard de Honnecourt. 4, two thirteenth-century church plans.

first it was even primarily, a structural theory. Initially the immediate practical value of a simple definition of the form of any large structure and of the resultant simplification of the process of setting it out on the site would have been paramount. The preoccupation of designers with the practical problems of setting out is clearly illustrated in the group of small diagrams taken from a text in the Roman agrimensorial—or land surveying—tradition and inserted by a second hand in the mid-thirteenth century pattern book of Villard de Honnecourt,[18] 3. Surviving medieval architectural drawings, other than details and large presentation copies of elevations with more emphasis on the decoration of the facade than on the structure behind, are mostly plans, which are neither dimensioned nor drawn accurately to scale,[19] 4. Dimensions, in both plan and elevation, must, therefore, have been determined largely on the site in terms of the predetermined geometric system and by means of the simple constructions with square and compass or lengths of cord shown in the pattern book;[20] and this procedure alone would have placed a premium on the use of those simple regular geometric figures which readily form repetitive or expanding grids.

The structural connotations were more likely gradual and, in the circumstances of the times, almost inevitable accretions to the theory, which slowly transformed it. Practice in any lodge would tend to crystallize around a particular geometric system, and this system, in the absence of an adequate structural theory in the modern sense, would provide a ready framework for the assimilation of that kind of structural experience which teaches that certain forms are adequate and others are not. That a transformation of this kind had already occurred before the decision was taken, in the fourteenth century, to rebuild Milan Cathedral on an unprecedented scale, is shown time and again in the records of the discussions there of successive structural difficulties. For Mignot and his colleagues the structural relevance of the theory was not only unquestioned: it was their only recorded reason for invoking

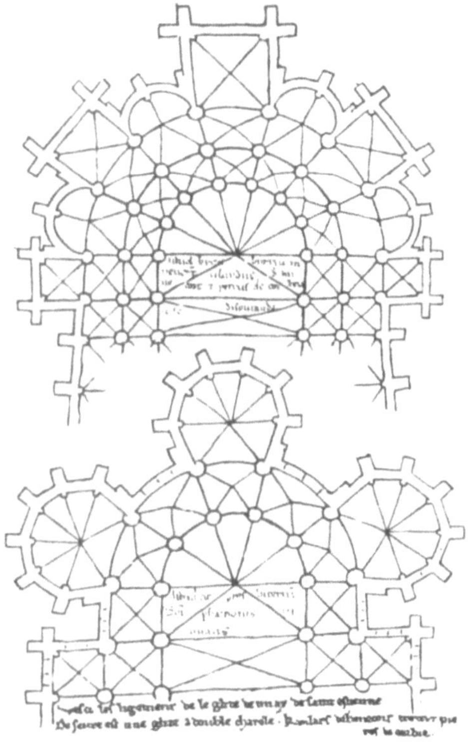

4

it.[21] Later similar discussions[22] and the early Italian and French architectural treatises[23] show that the structural connotation survived well into the Renaissance, while, at a much earlier date and in an earlier form, there are hints of it again in Vitruvius.[24]

As a structural theory lacking any conceptual distinction between structure and form, the *scientia geometriae* was, in fact, potentially relevant to the extent that there was a close identity of the two in actual construction. To a large extent this identity did exist.[25] The wide choice of materials and almost unlimited freedom in using them in combination which are enjoyed today were unknown in the past. Even structures which took generations to complete show a remarkable homogeneity of construction. Almost invariably, with the exception of ties inserted to control the spreading of arches, vaults and domes, their primary structural action is solely that of masonry in compression under its own weight. Mean stress levels due to this dead weight are low[26] and such other stresses and overturning moments as are present are chiefly those due to wind or earthquake and are also, therefore, direct functions of the form. The chief structural problem of design was to ensure the stability of this masonry, and it could indeed be solved largely by appropriate geometric rules.

This potential relevance was, nevertheless, limited in practice by the impossibility, without loss of generality, of removing the inherent ambiguities of the basic geometric symbolism for the conditions of structural integrity. In its basic form the theory called merely for conformity to one of several geometric disciplines. One could choose the discipline of the equilateral triangle, or the square, or one of the other ideal figures; and one could then conform to it in a virtually unlimited number of ways. To remove this ambiguity called for a number of written or unwritten operational rules. With these rules the adequacy and practical validity of the theory turned largely on their adequacy and validity and, in turn, on the extent to which past experience could be validly interpolated or extrapolated in terms of direct geometric proportionality. The theory was, in fact, proved wanting chiefly when, as at Milan, attempts were made to apply it,

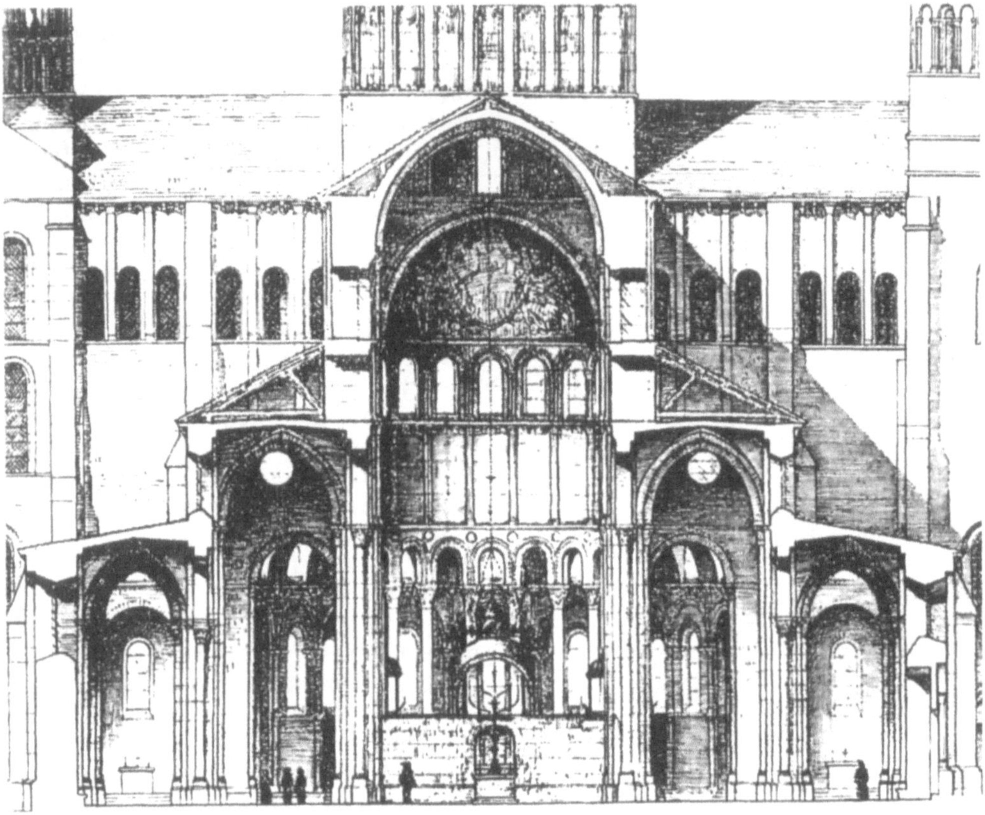

5, cross section of the nave of the third Abbey Church of Cluny as originally constructed without flying buttresses to the main vault. These were added after the vault collapsed in 1125.

as developed in one tradition of construction, to structures in a radically different tradition or of unprecedented form or scale.

DESIGN (1)

At Milan the cathedral had been started with great confidence on the pattern of much smaller brick-built Lombard Gothic churches like the Carmine in Pavia; but for the previous two hundred years Lombardy had been a back-water, and the provincial *ars* of the local masters—which went little beyond a knowledge of local materials and constructional techniques—soon proved unequal to the structural problems which faced them. Possessing little or no *scientia* of their own, they had to call on that developed in the different traditions of Northern Gothic. This, however, presupposed the different *ars* of these northern traditions as well as other characteristics, like the greater heights of northern cathedrals, which the Milanese had no desire to emulate. The essential partnership of northern theory and northern practice could not be broken, and the Milanese were thus thrown back on a highly

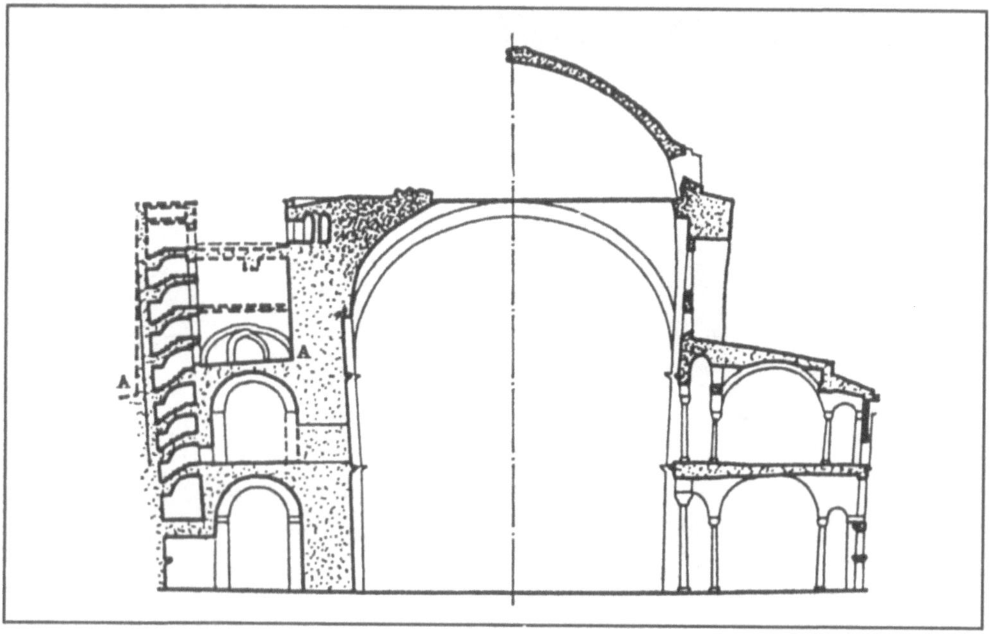

6, composite transverse cross section of St. Sophia, Istanbul, cut, at the left, on the centreline of the eastern piers and buttresses and, on the right, on the transverse centreline of the church. The buttressing masses above the level AA appear to have been added during the initial construction, but only when the action of the thrusts from the main arches had become apparent. They nevertheless proved inadequate and the dome and other parts of the structure indicated by the heavy stippling had to be rebuilt after a partial collapse about twenty years later. All inclinations from the vertical are exaggerated two times and broken lines signify later additions to the sixth-century structure or later rectifications of the facings.

empirical approach based largely on their own judgments and practical experience.

Radical innovation on this scale was always rare: in the past, even more than today, the norm was a much more gradual evolution which taxed any theory far less. When it was attempted elsewhere it usually called again, ideally, for a synthesis of earlier achievements in different traditions. But, as at Milan, an empirical approach was usually inescapable.

Empiricism betrays itself in various ways. Most commonly it is seen in continual changes in the design as construction proceeds, coupled with an apparent reluctance to face, or inability to make, decisions until they are inescapable. In no great building of the past for which adequate records remain was the form of the superstructure finally determined until its supports were complete or nearly so; and this is equally true whether the final choice was made by the original designer (as in the case of St. Paul's)[27] or left to his successors (as in most other cases).[28] Almost as frequently it is seen in the need to undertake remedial works or even partial reconstructions to forestall or make good the damage caused

7, *St. Sophia, Istanbul.*

8, *Florence cathedral dome.*

9, *the vault of San Lorenzo, Turin.*

by partial collapses. All flying buttresses in the great northern churches prior to the second half of the twelfth century seem, for instance, to have been added as casual expedients only after weaknesses had become apparent[29] or, as at Cluny III, 5, the vaults had already pushed the walls aside and collapsed.[30] At both St. Peter's[31] and St. Paul's[32] extensive consolidation or other remedial work on the main piers was necessary before the domes could be safely constructed, and at the former additional chains were necessary at a later date to contain the dome thrusts.[33] Most instructively, it is possible to read, in the distortions and lacks of bond of the brickwork of the great lateral buttresses of St. Sophia, 6, a whole sequence of modifications in their design in an attempt to halt the lateral spread of the dome base.[34] The attempt was only partly successful, and not much later the dome collapsed and had to be entirely reconstructed to a different design.

Occasionally, as in the development of Brunelleschi's design for the dome over the crossing of the cathedral at Florence, new structural ideas were deliberately tried out first on a smaller scale.[35] It is possible also that the oak chain which circles the base of the dome was intended to serve in part as a measuring device to give an early warning of any bursting tendency as construction proceeded.[36] But controlled experiments of this kind seem to have been very rare and their value will always have been limited by

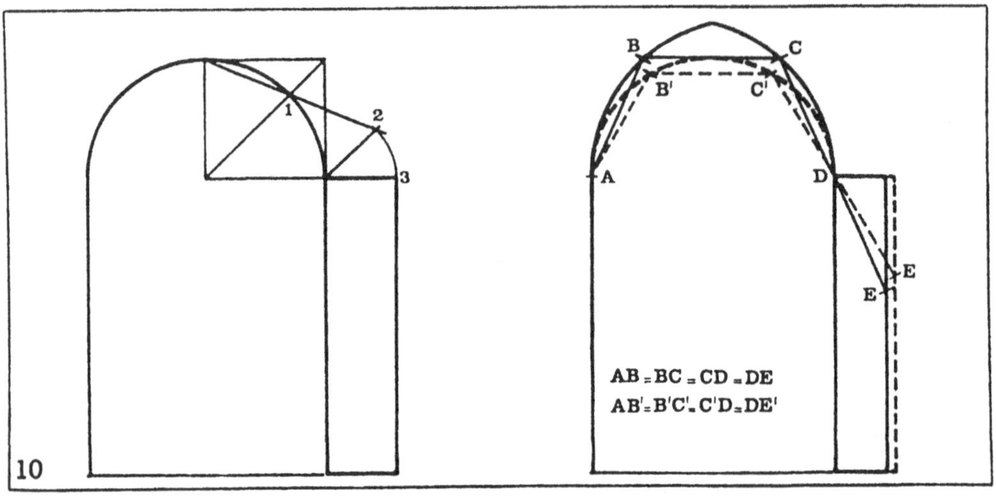

10, *two geometric constructions to determine the size of buttress needed by a given vault; that on the left according to Rodrigo Gil de Hontanon, and that on the right Blondel's rule.*

the available interpretive framework.

Was that framework merely the *scientia geometriae* and its underlying geometric symbolism? Despite the evident empiricism and the failures that marked its course, this seems unlikely. Most of the failures can, in fact, be attributed largely to errors in detailed proportioning or to faults in construction, particularly to undue haste in the latter and insufficient care over materials and workmanship,[37] rather than to basic errors in the structural concept. Structures like St. Sophia, 7, the great Gothic cathedrals, Brunelleschi's dome in Florence, 8, the dome of St. Paul's and even the more modest churches of Guarini, 9, have suggested to many that their designers possessed a 'superb statical intuition.' Is there any truth in this and, if so, what was this intuition and whence did it come? Did the early statical theory from which modern theory has sprung, or something akin to this, bridge the gap between the *scientia* and *ars* already discussed where these alone were inadequate?

THEORY (2)

Most in evidence in surviving records alongside the *scientia geometriae* proper, and not always sharply distinguished from its operational rules, are numerous purely empirical rules for determining the correct dimensions of foundations, walls, piers, buttresses, arches and vaults. Some are expressed geometrically; some arithmetically. They are found already in Vitruvius and in all subsequent architectural treatises,[38] and they are not unknown even in modern codes of practice. Sometimes, as in the case of the

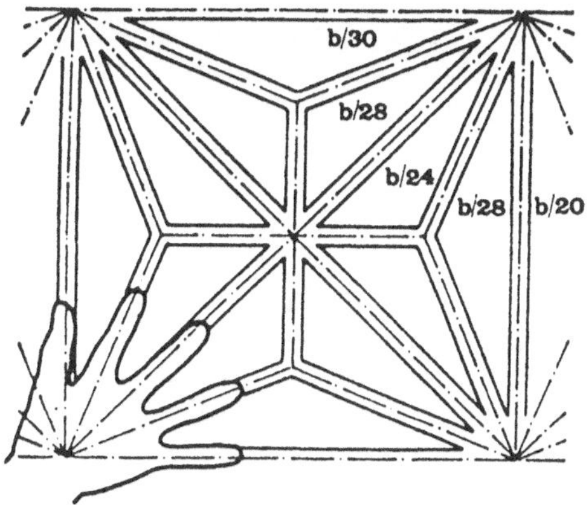

11, *Rodrigo Gil de Hontanon's rule for the section depths of the ribs of a late Gothic vault in terms of the mean bay width 'b', these depths being notionally scaled in proportion to the outstretched thumb and fingers of the hand.*

later rules for determining sizes of buttresses from the profiles of the vaults they support,[39] 10, the rules have considerably more generality than the *scientia geometriae.* More commonly though, as in the case of a late Spanish rule for the sizes of vault ribs,[40] 11, they do no more than codify established practice within a closely confined tradition. They might then be useful as simple definitions of the starting points for innovation, but they would be of no further use.

The much more general theory of statics can be traced back at least to the Greeks. Two Greek roots can be distinguished; Aristotelian and Archimedean.[41] Of the two, the *Mechanics* of Pseudo-Aristotle[42] was essentially dynamic in approach, while Archimedes' *Equilibrium of Planes*[43] was essentially static, and they differed also in other important respects. The first was largely descriptive and non-quantitative and therefore of little direct practical value. But, in its treatment of the movements of weights, it introduced the germ of the

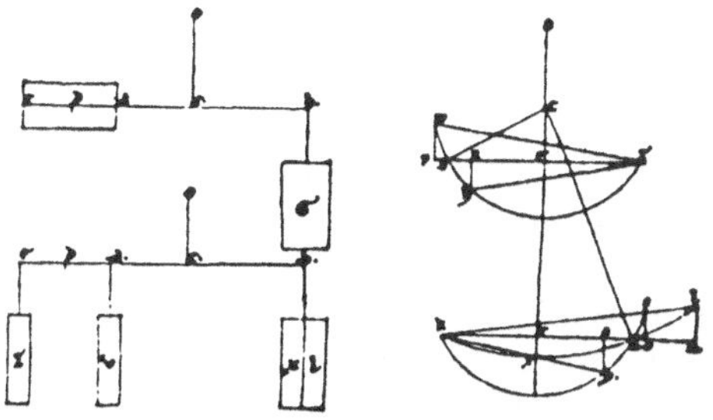

12, *static moments and the law of the balance according to Jordanus.* 13, *static moments and the law of the balance according to Leonardo.*

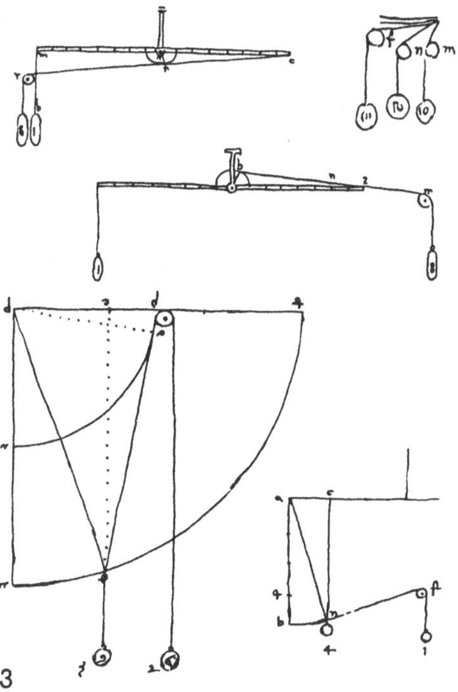

13

modern principle of virtual work and opened the way to a quantitative study of simple machines and their mechanical advantages. The second, with its Euclidean basis and tight Euclidean logical structure, was rigorously quantitative throughout and first introduced the concept of the centre of gravity and furnished the means for calculating its position.

Later Hellenistic writers, notably Hero of Alexandria, developed the work of Archimedes and the Aristotelian school which then passed, with the Islamic conquests, into Arabic scientific thought and and, in turn, became known in the West through Sicily and Spain in the twelfth and thirteenth centuries.[44] In the medieval statics of the enigmatic Jordanus, 12, and his followers, the two roots further interacted, and precise concepts of static moment, of the resolution of a vertical force along an inclined plane and of virtual work were arrived at.[45] Yet, despite the contributions of people like Simon Stevin[46] and Galileo[47] and a keen interest on the part of Leonardo,[48] 13, it was not until the late seventeenth century that a fully generalized concept of a force acting in any direction was reached and the general conditions of equilibrium stated and applied to problems such as the stability of the voussoirs of an arch. Similarly it was not until then that the earlier speculations of Leonardo and Galileo on problems of strength and deformation bore fruit in a manner that permitted quantitative calculations on the basis of tests on materials.[49]

Vitruvius was well aware of early Greek statics and mechanics. He implies, though, that in the first century B.C. a work like Archimedes' *Equilibrium of Planes* was of interest chiefly to mathematicians and used little, if at all, in design.[50] By the early fourth century A.D., however, as Pappus of Alexandria indicates in the preface to the eighth book of his *Collection*, the mechanician—one highly skilled in geometry, statics and mechanics—had assumed a role superior to that of the architect in the design of important structures.[51] In sixth-century Constantinople, for example, the leading designers of St. Sophia, Anthemius and Isidorus, are both described by Procopius as mechanicians;[52] and it is probable that both were professors of geometry or mechanics[53] though less clear to what extent their knowledge of statics, as distinct from that of geometry, was used in structural design. For almost a millennium after this—until the fifteenth century—there is, on the other hand, no clear evidence that theoretical statics played any significant part in design in the West. In the crucial early stages of the development of the Gothic structural system in the twelfth century the theories were, indeed, virtually unknown there and, for a variety of reasons, it seems much more likely that the practical achievements of the mature style in the thirteenth and fourteenth centuries spurred on the theoretic enquiry than that they were materially assisted by it.[54] No quantitative application of statical theory is recorded before the time of Wren.[55]

Since, as a quantitative guide to design, the Archimedean theory was, like the *scientiae geometriae*, framed in terms of the structural geometry alone, its validity was equally dependent on the close identity of structure and form in actual construction. Any practical application called, moreover, for

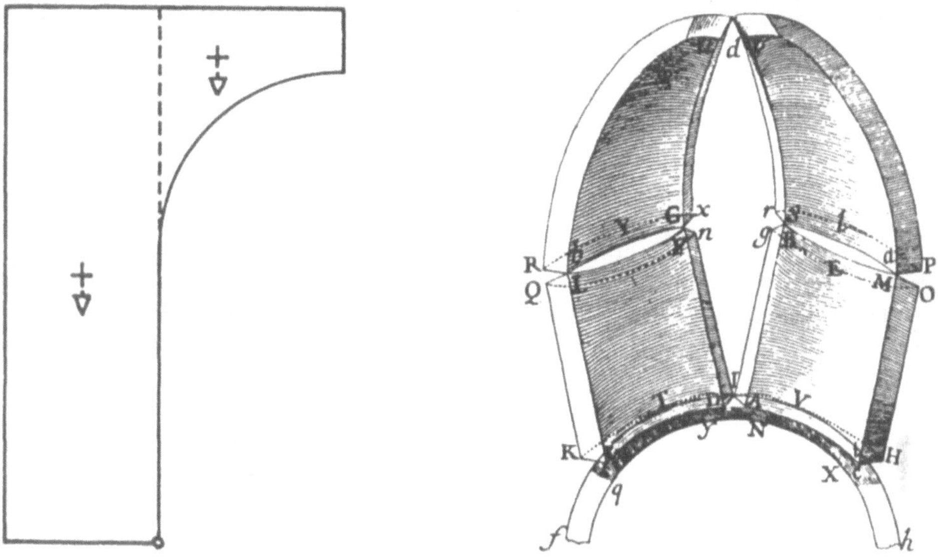

14, above left: the basis of Wren's static analysis of the size of buttress needed by a given vault. Above right: the mode of collapse of the dome of St. Peter's visualized by the three mathematicians. Below: an identical mode of collapse (smaller sketch) and a converse one in which each rotation takes place in the opposite sense (larger sketch) as visualized by Leonardo.

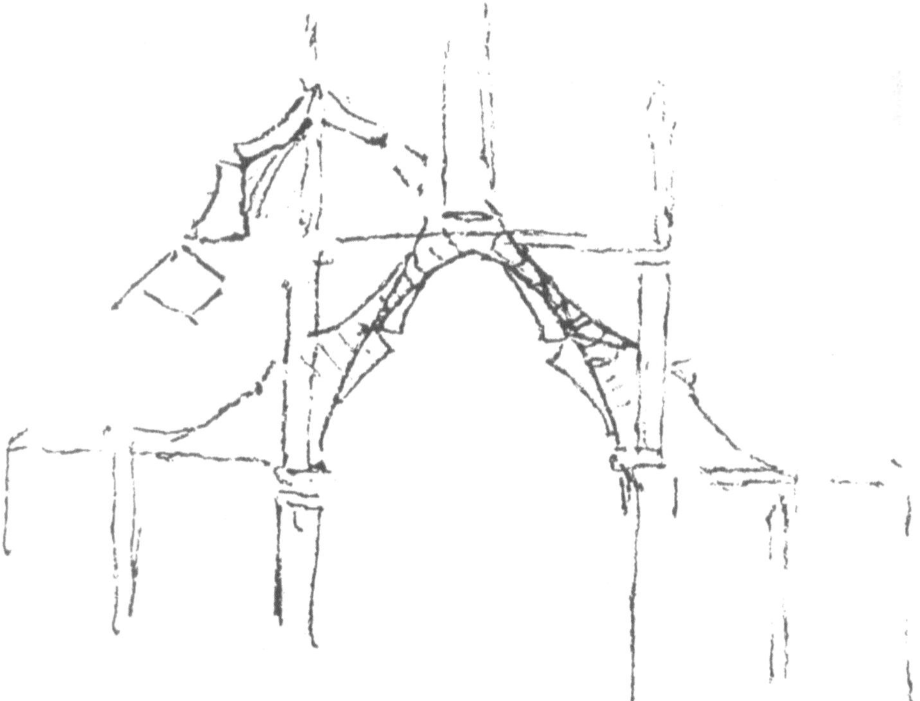

14

a high level of abstraction from the full complexities of behaviour, for which the theory itself provided no guidance. It could be used only to analyze the equilibrium of notionally rigid isolated elements of a structure under the action of their own component weights and other purely vertical applied loads or reactive forces. Whenever it was necessary to isolate such an element, for the purpose of the analysis, by an arbitrary slicing of a more complex continuous structure, the validity of the result would depend also on whether the slicing was done in a manner which truly respected these limitations, and the assurance that this was so could come ultimately only from a wider understanding of the conditions of equilibrium than the theory itself offered. Without that wider understanding the theory could as easily mislead as help. Even Wren was apparently misled, in the analysis of the buttressing required by different types of vault,[56] to such an extent that it would be difficult to believe that he took the analysis seriously were his choice of vault form in St. Paul's not based on his conclusion from it that this form required the least buttressing.[57] He considered a slice of vault up to the crown to constitute, with its supporting pier, an isolated rigid body tending to tip forward about the inner edge of the pier. No account was taken of any horizontal thrust from adjacent slices of the vault, so the abutment was assumed to provide a contrary moment to the inward tipping moment of the weight of the cantilevered vault slice, 14. It was argued that, if this slice was so proportioned as to be in equilibrium, 'the Whole will stand as well as the Halves.' Among the criticisms that could be made of this analysis perhaps the most significant, for the present purpose, is that it implies as the converse of equilibrium (though as purely statical analysis it does not explicitly assume) a mode of collapse that is impossible.

In contrast, the analysis of the dome of St. Peter's in 1742–43, which has been chosen to terminate the present study, was able to draw on the much wider understanding of the conditions of statical equilibrium reached near the end of the seventeenth century and it brilliantly anticipated a modern collapse analysis. Assisted also by direct observation of the cracks already visible around the base of the dome, which were then giving cause for alarm,[58] 15, the three mathematicians called in to advise on remedial measures visualised a plausible collapse mode,[59] 14. Applying to this the principle of virtual work they then calculated the chain tension required at the level OP, QR to prevent collapse. But they still succeeded only in obtaining a somewhat unrealistic upper bound.

DESIGN (2)

It cannot be concluded that either the 'rules of thumb' that existed alongside the *scientia geometriae* or the early theory of statics played much direct part in guiding structural innovation; but the visualization of a plausible mode of collapse of the dome of St. Peter's hints at another possiblity. That visualization, though guided by observa-

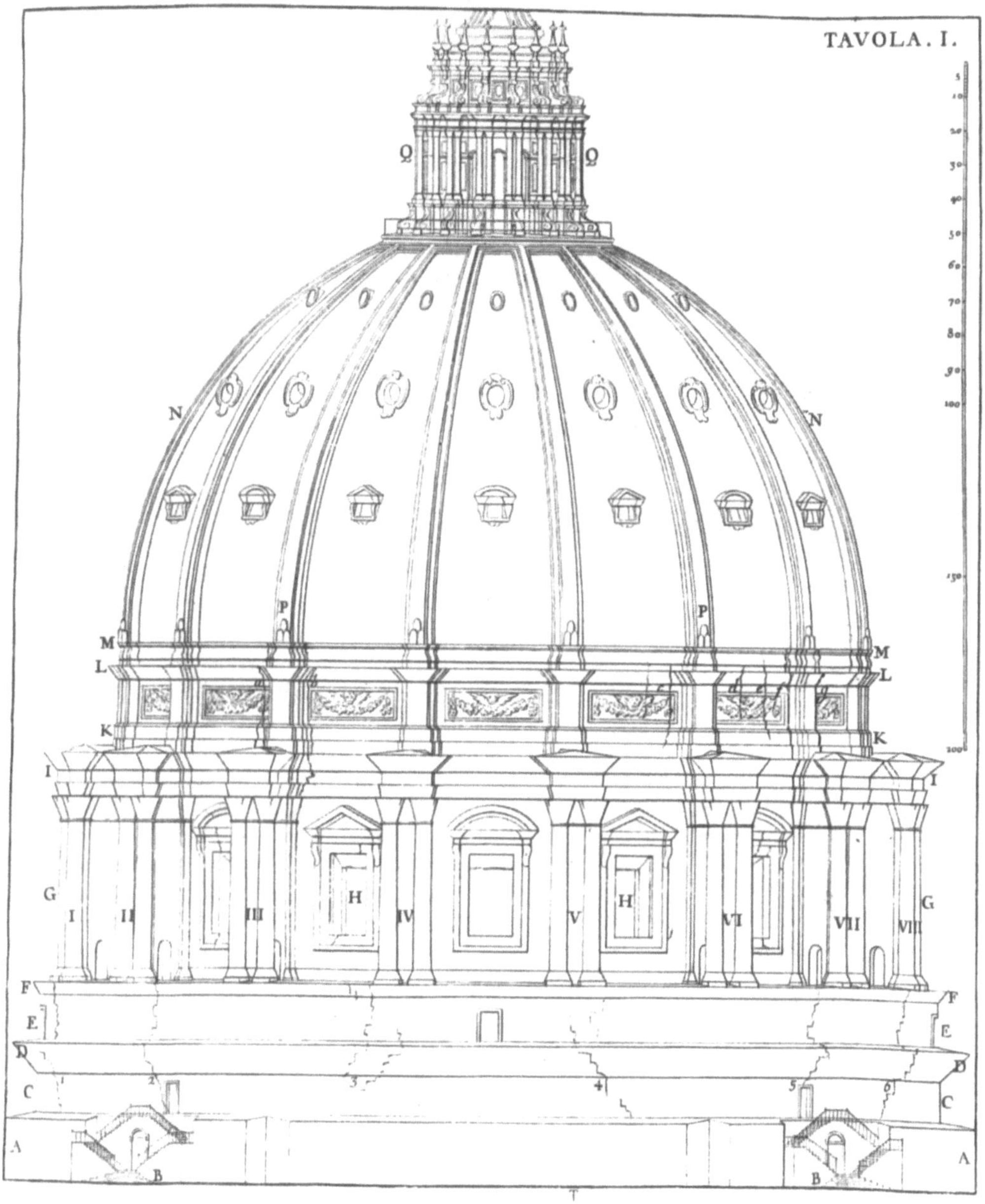

15, *cracking around the base of the dome of St. Peter's in 1742.*

tion of the cracks that had already resulted from the structural actions, as Leonardo had been guided earlier by observing the splitting of the skin of a squashed pomegranate or orange,[60] was a purely geometrical one. Concepts of force and resistance entered only into the calculation of the necessary size of chain.

The complementary and equally geometrical or stereometrical vi-

16, *shrine on Mount Cynthium, Delos.*

sualization of the conditions of structural integrity is possibly the oldest and most fundamental of all insights into structural behaviour;[61] and there is little doubt, on the basis of much circumstantial evidence, that it was, in fact, the chief guide to designers. It clearly lies behind the initial conception of the most primitive of all structures like the simple slab roof of an ancient shrine in a natural rock cleft on the island of Delos, 16, and, at a more sophisticated level, behind the idea of slightly more complex structural forms like the voussoir arch or dome. The great inherent strength and stiffness of the dome and other similar forms with doubly curved surfaces can still be most readily intuited in purely geometric terms by trying, in vain, to imagine modes of deformation that do not involve considerable extensions or contractions of the surface.

Even the dual roles, as both vault and buttress, of the greater and lesser semidomes in St. Sophia, 7, could most readily have been visualized in these terms.

In relation to this more intuitive structural insight, the role of the *scientia geometriae* or the theory of statics was to give quantitative precision where it was otherwise lacking and where it was not a natural outcome of an overriding concern with the practical problems of construction, particularly with the stability of the incomplete structure during construction as in the case of Brunelleschi's dome in Florence,[62] 8. Partial anticipations of the fuller statical understanding of today are seen playing this role in determining arch and vault profiles in some of Leonardo's sketch projects for the tiburio over the crossing of Milan Cathedral,[63] 17, and in two of Wren's sketches for the dome of St. Paul's,[64] 18. More frequently, though, it is the basic geometric thinking behind the *scientia geometriae* which is seen in action here, most notably in the widespread preference for the semicircular profile for an arch or dome, which even appears—in the form of semicircular relieving arches over the existing Gothic pointed ones—in the accepted design for the Milan tiburio by Francesco di Giorgio,[65] 19. The geometric reasoning behind this preference is well illustrated by Alberti. Of the full semicircular arch he wrote: 'That the entire arch is strongest of all appears not only from experience but reason; for I do not see how it can possibly

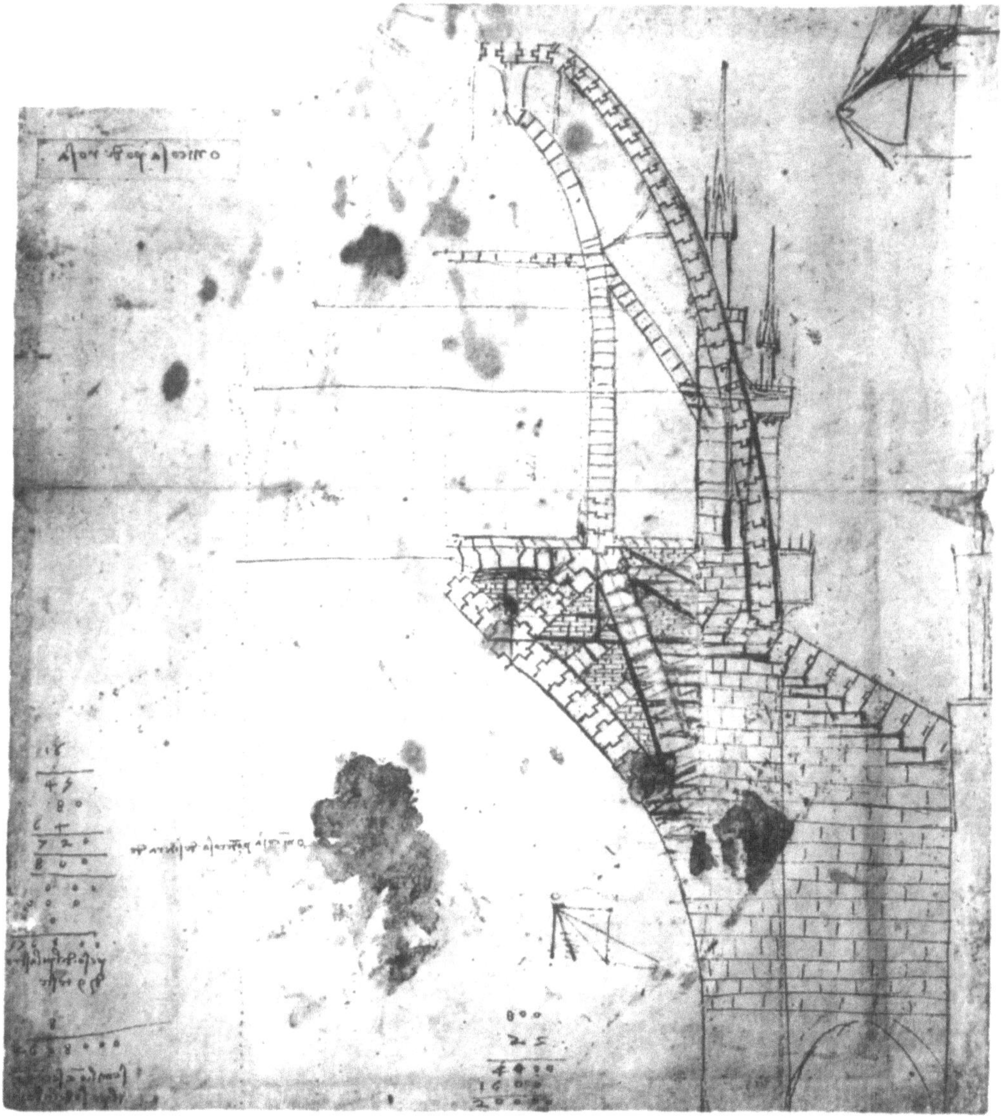

17, *unexecuted project by Leonardo for the tiburio over the crossing of Milan cathedral.*

disunite of itself unless one wedge shoves out another which they are so far from doing that they assist and support one another.'[66] And, of the 'perfect cupola': '. . . it is composed not only of arches, but also, in a manner, of cornices. . . . And when these cornices or arches are thus built one upon the other, if the work were inclined to ruinate, where should it begin, when the joints of every stone are directed to one centre with equal force and pressure?'[67]

18, *two sketches by Wren of the dome of St. Paul's showing approximate catenary curves as guide lines.*

The preoccupation with geometry is thus seen to go much further than the *scientia geometriae,* and there is much to suggest that that theory acquired its structural authority not only in the manner suggested above but also, more directly, through a deep-seated geometric concept of structure akin to that just discussed. The classic statement of that concept, which both antedates Archimedes' *Equilibrium of Planes* and goes far beyond it in offering a rational explanation of the physical world, is to be found in Plato's *Timaeus.*[68] Here both microcosm and macrocosm—the structure of the elements and the form of the universe and the inter-relations that guarantee cohesion and maintain life—are alike visualized in terms of the geometric perfection of the five regular polyhedra and the sphere and the cognate symmetries of the simple 'musical' numerical proportions. Here also, therefore, the equilateral triangle, the square, the regular pentagon and the circle are given a structural significance. The parallel can even be extended. For, just as the

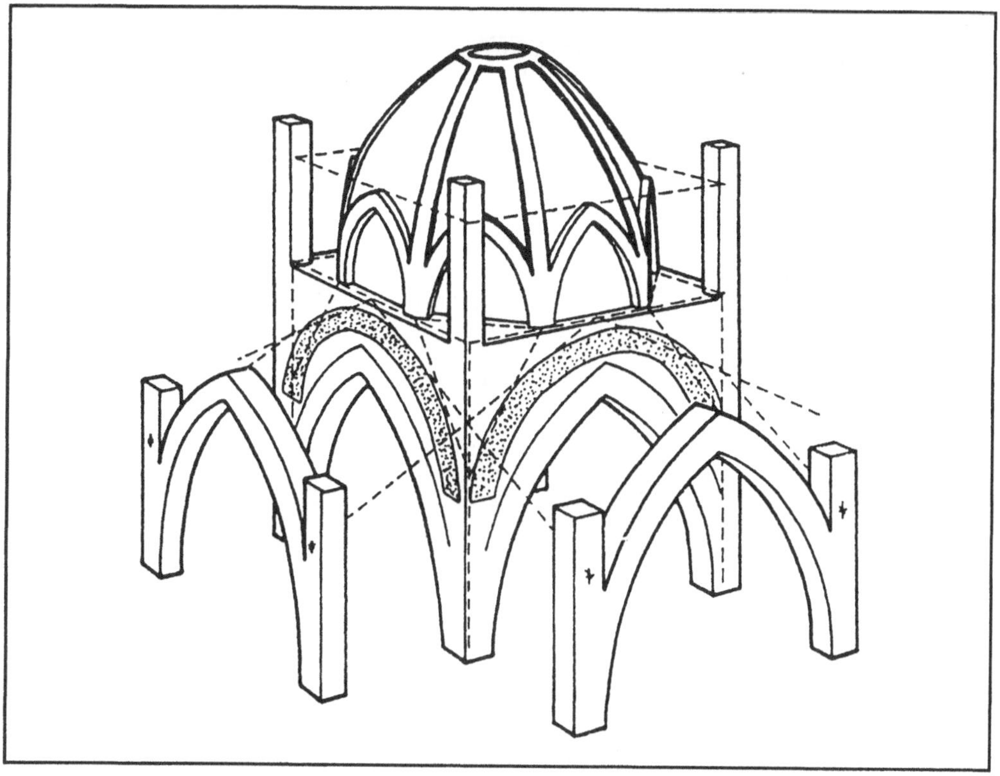

19, *executed project by Francesco di Giorgio for the tiburio of Milan cathedral. Above the existing pointed arches of the crossing the semicircular relieving arches proposed in 1490 by Francesco are shown stippled.*

perfect geometry of the regular polyhedra is immanent rather than openly manifest in the finished built structure, it defines, for Plato, the unseen reality of 'being' rather than the visible shadow of everyday experience. In his universe only the circle and the sphere are visibly present in the dome of the stars and the circular orbits of the planets, as they are in the constructed dome and the circular arch.

It is an important corollary of this that what would now be called the structural aspect of design was hardly ever, in the past, clearly distinguished from what would now be called the formal or aesthetic. The *scientia* was equally relevant to both, to such an extent that, when neither the structural nor the aesthetic aspect was (as at Milan) of overriding concern, it is difficult to know to which it was considered to be primarily applicable. Vitruvius' recommendation of certain proportions 'for convenience, beauty and strength (*ad usum et ad speciem et ad firmitatem*)'[79] is echoed time and again in later treatises.

There were inevitably shifts of emphasis over a period of some two thousand years, and some are fairly obvious in the surviving

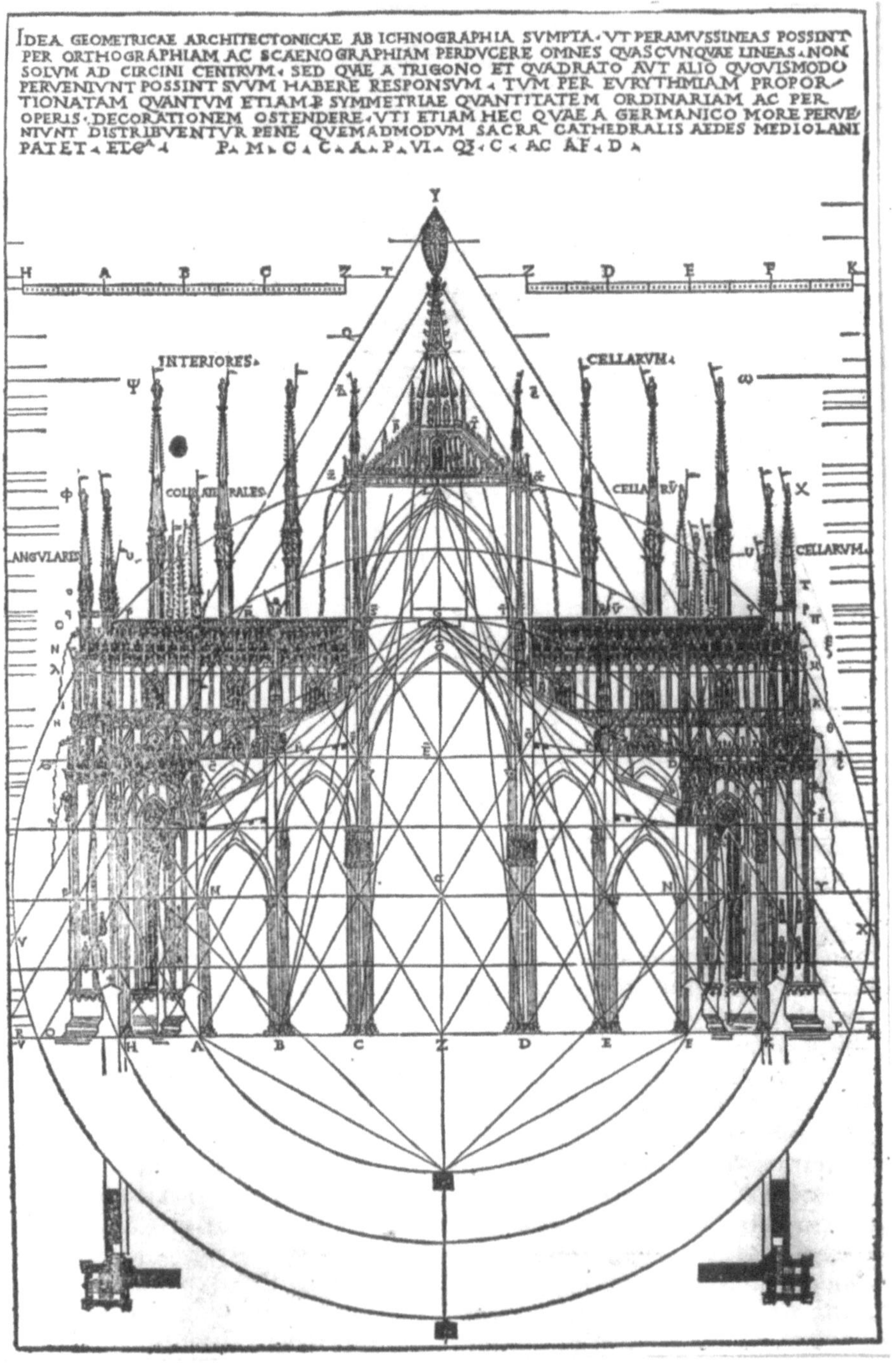

20, *the cross section of Milan cathedral as represented in Cesariano's edition of Vitruvius in* 1521.

21, *Christ Pantocrator in the dome of the monastery church at Daphni.*

structures themselves. A concern for stability must have been dominant (in the minds of designers if not of their patrons) in the early development of the brick domical structures of the later Roman and early Byzantine periods and of the stone vaulted structures of the Gothic West. But it clearly yielded prior place to a greater interest in pure form at some other times. Almost as significantly, though less obviously to modern eyes, the latter interest in pure form alternated between one not very different from the present concern for appearance and quite different concerns for symbolic meaning which have no counterpart today.

Symbolic meaning was even more directly related to Plato's *Timaeus* and similar cosmologies than was the structural authority of the *scientia geometriae*. Through its absorption into Christian theology as a rational extension and interpretation of the biblical narrative of the creation,[70] the *Timaeus*, for instance, profoundly influenced the ideas of those ecclesiastical patrons, closely associated with the school of Chartres, who provided much of the impetus for the initial development of Gothic construction.[71] It had no less influence, though a rather different one, placing more emphasis on the Pythagorean musical element, on Renaissance architectural theorists and patrons.[72] At both times these men were seeking, in their different ways, an ideal form for the church which would adequately mirror the divine creation, and they found that form in one or another of the perfect symmetries of the Platonic myth, 20. Simi-

22, *the vertical curve of the stylobate of the Parthenon. Other refinements include similar vertical curves in the superstructure, entasis and inclinations of the columns, slight thickening of the corner columns in relation to the others, and other slight inclinations.*

larly, throughout the Byzantine period, the dome represented the spherical heaven, and this symbolism was frequently reinforced by gold mosaic and a representation, in the crown, of the ascending Christ or of Christ Pantocrator,[73] 21. The alternative more directly visual interest may be seen in the 'optical refinements' of the Doric temple, 22, and again in the High Renaissance and Baroque and Rococo, 23, when the main architectural commissions (including St. Peter's) passed largely to men trained initially as painters or sculptors.[74]

Philibert de l'Orme, in his treatise published in 1567, marks very well the transition from the Gothic and early Renaissance concern for symbolic meaning to the later concern for simple appearance with its associated desire for a more complex geometry. In seeking an ideal form for the church he aligns himself squarely with his predecessors and appears, in his design for a church facade, 24, to have copied an earlier design in Francesco di Giorgio's treatise of about 1482.[75] But in his major

23, *interpenetrating vaults with warped binding arches in the Abbey Church at Neresheim.*

emphasis on complex stereometry he returns, knowingly or unknowingly, to an earlier precedent—that of Hero's *Stereometrica*[76] and his lost treatise on vaulting—and heralds the Baroque and Rococo interest in the mathematics of complex vault surfaces and their warped intersections illustrated in 23 and best exemplified in some of Guarini's lost churches and his *Architettura Civile.*[77] The excitements of the fifteenth-century Italian interest in perspective, again marking a return to a much earlier past, were portents of the change but had initially too many Pythagorean-Platonic overtones to accomplish it.[78]

In analysing and attempting to understand the structural achievements of the past, either structurally or aesthetically, all these shifts of emphasis must be borne in mind. At the same time it is necessary to remember the underlying constancy of the close interrelation between structure and form, both in initial conception and in the realized structure. This interrelation, coupled with an inescapable but, in retrospect, wholly beneficial restriction of choice of materials and other technical means, and the sort of care for detail seen in the Doric temple, imparted much of the unity and 'style' which are now so highly prized. But it should also be remembered that the major works were not achieved in the isolation which may now seem to characterize many of them, and that they represented, in terms of the total resources of the times, almost unbelievable concentrations of sustained effort. They were, in this sense at least, *tours de force* which we, with our different senses of values, can never hope to equal. Nor should we, without the religious and other convictions of their builders, necessarily try to emulate qualities which stemmed from those convictions. The structures themselves remain to awaken interest on many levels, to challenge and even inspire, and to help us, through an attempt to learn something of the way in which they were conceived; to come a little closer to the elusive but essentially unchanging marriage of art and science in the process of creative design.

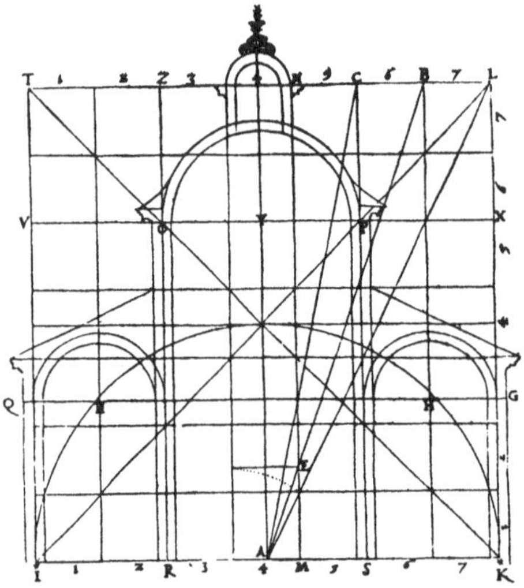

24, *project for a church by Philibert de l'Orme.*

NOTES

[1] *Annali della fabbrica del Duomo di Milano dall'origine fino al presente*, I, Milan 1877, p. 209.

[2] Vitruvius, I, i, 15: 'ex duabus rebus singulas artes [sunt] compositas, ex opere et eius ratiocinatione,' etc. For later writers see, for instance, Pappus of Alexandria, Collection, VIII, preface, quoted by G. Downey, 'Byzantine architects, their training and methods,' *Byzantion*, XVIII, 1948, pp. 106–7, and Ibn Khaldun, 'Muqadimmah,' translated F. Rosenthal, Bollingen Series No. 43, New York, 1958, II, p. 363.

[3] T. Le Seur, F. Jacquier and R. G. Boscovich, 'Parere di tre mattematici sopra i danni, che si sono trovati nella cupola di S. Pietro' Rome 1743. See below.

[4] *Annali* (op. cit.) p. 209. The best commentary on these discussions is that of J. S. Ackerman, 'Gothic theory of architecture at the cathedral of Milan,' *Art Bulletin*, XXXI, 1949, pp. 84–111, but see also P. Frankl, 'The secret of the mediaeval masons,' *Art Bulletin*, XXVII, 1945, pp. 46–60 and Idem, 'The Gothic, literary sources and interpretations through eight centuries,' Princeton, 1960, pp. 57–86 (referring also to similar discussions at Chartres and Gerona).

[5] P. Frankl, 'The Secret' (op. cit.) pp. 52–55, E. Panofsky, 'An explanation of Stornaloco's formula,' *Art Bulletin*, XXVII, 1945, pp. 61–64 and J. S. Ackerman (op. cit.) pp. 89–90.

[6] Vitruvius, VI, iii, 3 and IX, preface, 4, 5. There is no comprehensive detailed account of the development of the theory from a structural point of view, but R. J. Mainstone, 'Structural theory and design,' *Architecture and Building*, XXXIV, 1959, pp. 106–13, 186–95, 214–21 gives an earlier general account with the emphasis on the contrast with modern theory.

[7] There is no published corpus of early drawings, but some of the most interesting, those of Villard de Honnecourt and from the Reims palimpsest, are reproduced and discussed in H. Hahnloser, *Villard de Honnecourt*, Vienna, 1935, T. Bowie, *The sketchbook of Villard de Honnecourt*, New York, 1959, and R. Branner, 'Drawings from a thirteenth-century architect's shop: the Reims palimpsest,' *Journal of the Society of Architectural Historians*, XVII, 4, 1958, pp. 9–21. Among publications of later Gothic drawings may be noted particularly O. Kletzl, *Planfragmente aus der Dombauhütte von Prag*, Stuttgart, 1939. The Villard drawings are the most valuable for the present purpose and include three sheets of masons' geometry (Hahnloser XXXIX–XLI, Bowie 55–7) and several (though mostly non-architectural) of controlling geometric grids. The danger of inferring such grids from finished drawings (where they are not discernable as scribed lines on the drawing itself) is illustrated, however, by the conflicting interpretations of Villard's plan of the tower of Laon cathedral (Hahnloser XVIII, Bowie 39) by W. Ueberwasser, 'Nach rechtem Mass,' *Jahrbuch der preussischen Kunstsammlungen*, LVI, 1935, pp. 250–72 and M. Velte, 'Die Anwendung der Quadratur und Triangulatur bei der Grund- und Aufrissgestaltung der gotischen Kirchen,' Basle, 1951.

[8] Measurements of buildings are even less reliable guides to the theory behind their design, unless the measurements are very

precise, and there is evidence of a similar precision in construction and of unusual consistency in the practical application of the theory. (There is no *a priori* reason why designers in the past should have followed the indications of their theory any more slavishly than designers do today, but this in no way necessarily belittles the importance of the theory.) The most convincing recent measurements are possibly those reported in K. J. Conant, 'New results in the study of Cluny monastery,' *Journal of the Society of Architectural Historians*, XVI, 3, pp. 3–11 and in M. Detoni and T. Kurent, 'The modular reconstruction of Emona, Situla No. 1,' Ljubljana, 1963. See also O. Von Simson, 'The Gothic cathedral,' New York, 1956, pp. 207–10 and E. Levy, 'On the proportions of the south tower of Chartres cathedral,' ibid, pp. 235–65, but the comments on these latter analyses by S. McK. Crosby in *Art Bulletin*, XLII, 1960, pp. 149–60 are very pertinent.

[9] G. Milanesi, *Documenti per la storia dell'arte Senese*, I, Siena, 1854, pp. 144–253 passim.

[10] G. Gaye, *Carteggio inedito d'artisti dei secoli* XIV, XV, XVI, III, Florence, 1840, pp. 487–510. Discussed by P. Booz, *Der Baumeister der Gotik*, Munich, 1956, pp. 54–61.

[11] 'Trattato di architettura,' written in Milan 1461–4, MS reproduced with English translation in J. R. Spencer, *Filarete's treatise on architecture*, New Haven, 1965. Discussed in H. Saalman, 'Early Renaissance architectural theory and practice in Antonio Filarete's Trattato di architettura,' *Art Bulletin*, XLI, 1959, pp. 89–106; and P. Tigler, *Die Architekturtheorie des Filarete*, Berlin, 1963, pp. 46–68 and 141–77.

[12] *Trattato di architettura civile e militare*, written about 1482–92, printed by C. Promis and C. Saluzzo, Turin, 1841. Discussed in R. Papini, *Francesco di Giorgio architetto*, Florence, 1946, pp. 49–67 and 193–6; and H. Millon, 'The architectural theory of Francesco di Giorgio,' *Art Bulletin*, XL, 1958, 257–61.

[13] *Regole generali di architettura*, Venice, 1537; *Il primo libro d'architettura* and *Il secondo libro di perspettiva*, Paris, 1545, etc. Serlio's work is less explicitly concerned with theory than the others referred to but he demonstrates the Vitruvian and mediaeval quadrature; for instance, on p. 3b of Book I, shows it as the *proportione diagonea* on p. 15a, and uses it frequently both in the perspective demonstrations of Book II and in deriving the proportions of the orders in Book IV (Regole generali).

[14] *Le premier tome de l'architecture*, Paris, 1567. Discussed in A. Blunt, *Philibert de l'Orme*, London, 1958, pp. 124–33 but with the emphasis very largely on de l'Orme's later theory of divine proportion.

[15] Reproduced in R. K. Heideloff, *Die Bauhütte des Mittelalters in Deutschland*, Nürnberg, 1844, pp. 101–16. Discussed in W. Ueberwasser (op. cit.); P. Frankl, 'The secret' (op. cit.); M. Velte (op. cit.), pp. 17–28; and (mentioning also several similar late Gothic treatises) P. Booz (op. cit.), pp. 41–7.

[16] The use of the equilateral triangle at Milan has already been noted. That of the regular pentagon is hinted at by Villard de Honnecourt (Hahnloser XXXVI, XXXVII, Bowie 36, 37) and again, at the end of the period, by the 'Geometria deutsch' of 1472 (also reproduced in R. K. Heideloff (op. cit.), pp. 95–9), but it is nowhere so conclusively demonstrated. See also, in this connection, W. Ueberwasser, 'Spätgotische Baugeometrie,' *Jahresbericht der öffentlichen Kunstsammlung Basel*, 1928–30, pp. 79–122; O. von Simson and E. Levy (op. cit.); B. G. Morgan, *Canonic design in English mediaeval architecture*, Liverpool, 1961; P. Frankl, *Gothic architecture*, Harmondsworth, 1962, pp. 37, 156; a forthcoming paper by K. J. Conant, 'The afterlife, Vitruvius in the middle ages'; and relating to much earlier periods, J. Hambidge, 'The Parthenon and other Greek temples: their dynamic symmetry,' New Haven, 1924; and N. Detoni and T. Kurent (op. cit.).

[17] Philibert de l'Orme's theory of divine proportion referred to in note 14 is one example of the change, but primarily numerical systems of commensurate proportions can also be traced back much earlier as shown, for example, by K. J. Conant (op. cit.). See especially R. Wittkower, *Architectural Principles in the Age of Humanism*, 3rd ed., London, 1962; Idem, 'Systems of proportion,' *Architects' Yearbook* V, 1953, 9–18; and P. H. Scholfield, *The theory of Proportion in Architecture*, Cambridge, 1958.

[18] H. Hahnloser (op. cit.); R. Branner, 'A note on Gothic architects and scholars,' *Burlington Magazine*, XCIX, 1957, 372–3; and B. L. Ullman, 'Geometry in the mediaeval quadrivium,' *Studi di Bibliografia e di storia*, IV, 1964, pp. 263–85.

[19] O. Kletzl (op. cit.), pp. 16–18; P. du Colombier, *Les chantiers des cathedrales*, Paris, 1953, pp. 64–5; and P. Booz (op. cit.), pp. 74–79.

[20] D. Knoop and G. P. Jones, 'The decline of the mason-architect in England,' *Journal of the Royal Institute of British Architects*, XLIV, 1937, pp. 1004–7; P. Frankl, 'The secret' (op. cit.), pp. 48–50; L. F. Salzman, *Building in England down to* 1540, Oxford, 1952, pp. 16–17; P. du Colombier (op. cit.), pp. 65–6; and P. Tigler (op. cit.), pp. 141–171.

[21] *Annali* (op. cit.), pp. 209–10.

[22] e.g. the discussions at Bologna referred to above.

[23] e.g. D. Barbaro, *I dieci libra dell'architettura di M. Vitruvio*, Venice, 1556, p. 24 (commentary on Vitruvius I, ii, 4) 'ne si puo lodare abastanza l'effetto della proportione, nella quale e posta la gloria dell'Architetto, la fermezza dell'opera e la maraviglia dell artificio'; and Philibert de l'Orme, "Le premier tome" (op. cit.), p. 234a, referring to the purely geometrical basis of the design for a church illustrated in 24, 'Quant à la pente et sa couverture ayant une poussée et boutée suffisante contre la grande voute . . . vous les prenez sur la ligne horizontale qui fait la quatrieme partie de la hauteur . . .'

[24] See below and note [illegible].

[25] The literature on methods of construction is very extensive, the most useful sources being: E. E. Viollet-le-Duc, *Dictionnaire*

raisonné de l'architecture française, Paris 1858–68 especially the article on construction IV, pp. 1–279; A. Choisy, *L'art de bâtir chez les Romains*, Paris, 1872; Idem, *L'art de bâtir chez les Byzantins*, Paris, 1883; M. E. Blake, *Ancient Roman construction in Italy from the prehistoric period to Augustus*, Washington, 1947; Idem, *Roman construction in Italy from Tiberius through the Flavians*, Washington, 1959; G. Lugli, *La tecnica edilizia Romana*, Rome, 1957; J. B. Ward-Perkins, 'Notes on the structure and building methods of early Byzantine architecture,' *Second report on excavations in the Great Palace of the Byzantine Emperors at Constantinople*, Edinburgh, 1958, pp. 52–104; W. MacDonald, 'Some implications of later Roman construction,' *Journal of the Society of Architectural Historians*, XVII, 4, 1958, pp. 2-8; J. Fitchen, *The construction of Gothic cathedrals*,' Oxford, 1961; and N. Davey, *A history of building materials*,' London, 1961.

[26] This point is well made in J. Heyman, 'The stone skeleton, *International Journal of Solids and Structures*, II, 1966, pp. 249–79, and in a forthcoming paper, 'On the rubble vaults of the middle ages and other matters,' *Gazette des Beaux-Arts*. Heyman proceeds to analyze the stability of masonry arches and vaults in very largely geometric terms, and the analysis can be criticized only on the grounds that it does not go far enough. In taking the stability of the vertical supports for granted he ignores the main problem of the designers of the structures considered here—that of ensuring that the horizontal thrusts generated by the superstructure would not push aside the supports sufficiently to permit collapse—and in assuming that voussoirs never slide on one another he also ignores an important contributory factor to most of the near-collapses that have come to the writer's notice.

[27] Wren's many reconsiderations of the design, especially of that of the dome, while construction proceeded, are well illustrated by the surviving drawings. See *Wren Society*, I–III and XIII–XVI, 1924–6 and 1936–9; E. Sekler, *Wren and his place in European architecture*, London, 1956, pp. 110–45; V. Furst, *Sir Christopher Wren*, London, 1956.

[28] The cathedrals of Milan and Bologna have already been mentioned. Florence Cathedral and St. Peter's in Rome are other outstanding examples. For an excellent discussion of the earlier phases in Florence see H. Saalman, Santa Maria del Fiore: 1294–1418,' *Art Bulletin*, XLVI, 1964, pp. 471–500. The design and construction of the dome are discussed by P. Sanpaolesi, *La cupola di Santa Maria del Fiore*, Rome, 1941; F. D. Prager, Brunelleschi's inventions and the 'Renewal of Roman masonry work.' *Osiris*, IX, 1950, pp. 457–554; and H. Saalman, 'Giovanni di Gherardo da Prato's designs concerning the cupola of Santa Maria del Fiore in Florence,' *Journal of the Society of Architectural Historians*, XVIII, 1, 1959, pp. 11–20. The literature on St. Peter's is very extensive. See, for the present purpose, J. S. Ackerman, *The architecture of Michelangelo*, London, 1961, I, pp. 89–102 and II, pp. 83–112.

[29] M. Aubert, *Notre-Dame de Paris*, Paris, 1920, pp. 86–107; Idem, 'Les plus anciennes croisées d'ogives, leur rôle dans la construction,' *Bulletin monumental*, XCIII, 1934, 5–67 and 137–237; M. E. Lefevre-Pontalis, 'L'origine des arcs-boutants,' *Congrès archéologique de France*, LXXXII (Paris 1919), 1920, pp. 367–96. The cathedral of Orvieto provides a well documented, though slightly different later example outside France. See R. Bonelli, *Il duomo di Orvieto e l'architettura italiana del duocento trecento*, Città di Castello, 1952, pp. 80–2. The more usual Italian expedient, seen for instance in Florence Cathedral, was the addition of internal tie rods. The scissor arches in the crossing at Wells illustrate yet another analogous expedient.

[30] K. J. Conant, 'Observations on the vaulting problems of the period 1088–1211,' *Gazette des Beaux-Arts*, 6th ser. XXVI, 1944, pp. 127–34.

[31] G. Vasari, *Le vite de piu eccelenti architetti, pittori et scultori . . .*' Part III, 1, Florence, 1550, p. 880 and Part III, 2, Florence, 1568, p. 751.

[32] *The Wren Society*, XIV, 1937 (Building Accounts), pp. 85, 89 and 98.

[33] Added, as suggested by the three mathematicians and Poleni, by Vanvitelli in 1743–4. G. Poleni, *Memorie istoriche della gran cupola del Tempio Vaticano*, Padua, 1748, pp. 410–423. See also note [3] and below.

[34] R. J. Mainstone, 'The structure of the church of St.Sophia, Istanbul,' to be published in *Transactions of the Newcomen Society*, XXXVIII, 1965–6. The study of the structural history of the church by Mr. R. L. Van Nice and the writer is still proceeding and will be reported more fully in the text volume accompanying the Dumbarton Oaks publication of Mr. Van Nice's survey drawings.

[35] In addition to models of the dome itself, one of which appears to have been a substantial structure of brick and stone (C. Guasti, *La cupola di Santa Maria del Fiore illustrata con i documenti dell'Archivio dell'Opera secolare*, Florence, 1857, docs. 17, 18, 19, pp. 17–18 and 68, 69, pp. 34–35), Brunelleschi built two smaller but similar cupolas before starting work on the dome (Manetti, *Vita di Ser Brunellesco* (c. 1485), ed. E. Berti-Toesca, Florence, 1927, p. 37 and G. Vasari (op. cit.), Part II, p. 310).

[36] F. D. Prager (op. cit.), pp. 495–6.

[37] Much of the trouble at St. Peter's and St. Paul's necessitating the remedial works referred to above while construction was still in progress may probably be attributed to undue haste or insufficient care in the choice of materials. G. Vasari (op. cit.), Part II, p. 597, for instance, suggests haste on the part of Bramante—'et era tanta la furia di lui che faceva, et del Papa, che aveva voglia, che tali fabriche non si murassero, ma nascessero . . .'—and C. S. Peach and W. G. Allen, 'The preservation of St. Paul's cathedral,' *Journal of the Royal Institute of British Architects*, XXXVII, 1930, pp. 656–76, comment on the excessive use of uncoursed rubble fillings to the walls and piers of St. Paul's. Procopius, *Buildings*, I, i, 77, hints similarly at haste in construction at St. Sophia.

[38] Vitruvius, III, iv, 1; V, i, 3, 5; VI, viii, 5–7; L. B. Alberti, *de re aedificatoria*, Florence,

1485, III, 3, IV, 6, VII, 5, 10, 14, 15; A. Palladio, *Quattro libri dell'architettura*, Venice, 1570, I, 11, 13; and Philibert de l'Orme (op. cit.), II, 7.

[39] E.g. both the popular 'Blondel's rule' illustrated here (F. Derand, *L'architecture des voûtes*, Paris, 1643, pp. 16–17, and F. Blondel, *Cours d'architecture*, Paris, 1698 (2nd ed.), p. 419, and those recorded by Rodrigo Gil de Hontañon (c. 1500–1577) and incorporated in S. Garcia, 'Compendio de arquitectura,' 1681–5 (published in *Arte en Espana*, VII, 1868, and discussed in G. Kubler, 'A late Gothic computation of rib vault thrusts,' *Gazette des Beaux-Arts*, 6th ser. XXVI, 1944, pp. 135–48).

[40] From the same source as the rules for buttresses referred to above. See also Philibert de l'Orme (op. cit.), IV, 8.

[41] The extensive literature on Greek statics and mechanics and their subsequent Islamic and later western developments is best summarized for the present purpose (with a representative selection of source material) in M. Clagett, *The science of mechanics in the middle ages*, Madison, 1959.

[42] Published with English translation in *Aristotle, minor works*, W. S. Hett ed., London, 1936, pp. 330–411. Discussed in M. Clagett (op cit.), pp. 4–9.

[43] English translation in *The works of Archimedes*, T. L. Heath ed., Cambridge, 1897, reprinted New York, 1950, pp. 189–220. Discussed in M. Clagett (op. cit.), pp. 9–13.

[44] A. C. Crombie, *Augustine to Galileo*, London, 1952, and M. Clagett, op. cit., pp. 14–23, 38–51 and 56–72.

[45] E. A. Moody and M. Clagett, *The medieval science of weights*, Madison, 1952, pp. 182–90, and M. Clagett, op. cit., pp. 72–84 and 104–8.

[46] *Hypomnemata mathematika*, IV, *Mathematicorum hypomnematum de statica*, Leiden 1605, pp. 3–51 and 79–108.

[47] Chiefly in *Discorsi e dimostraziomi matematiche intorno a due nuove scienze*, Leiden, 1638. English translation by H. Crew and A. de Salvio, New York, 1914.

[48] *I libri di meccanica*, A. Uccelli ed., Milan, 1940, collects together all the notes on statics and mechanics and includes a bibliography of facsimile editions of the manuscripts. English translations in *The literary works of Leonardo da Vinci*, J. P. Richter and I. A. Richter ed., 2nd ed., Oxford, 1939. There is, as yet, no adequate critical study, paying proper attention to the sequence of the MSS, of the development of Leonardo's ideas and their dependence on and advance beyond those of his predecessors.

[49] It is beyond the scope of the present paper to review these later developments, since they first found practical application only at the end of the period under review. S. B. Hamilton, 'Charles Augustin de Coulomb,' *Transactions of the Newcomen Society*, XVII, 1936–7, pp. 27–49 and Idem 'The historical development of structural theory,' *Proceedings of the Institution of Civil Engineers*, I, Part II, 1952, 374–402 cite the principal sources.

[50] Vitruvius, I, i, 17. The *Equilibrium of planes* is not mentioned explicitly here but Archimedes is named. In X, iii, Vitruvius shows that he was familiar with the less mathematical Aristotelian *Mechanics* (note [42]).

[51] G. Downey (op. cit.); H. A. Meek, The architect and his profession in Byzantium,' *Journal of the Royal Institute of British Architects*, LIX, 1952, pp. 216–20.

[52] *Buildings*, I, i, 24, 50 and 76.

[53] T. L. Heath, *A manual of Greek mathematics*, Oxford, 1931, reprinted New York, 1963, pp. 518–9, G. L. Huxley, *Anthemius of Tralles, a study in later Greek geometry*, Cambridge, Mass., 1959.

[54] Perhaps the only leading figure in medieval science who was closely connected with an important structure and in a position to influence its design was Robert Grosseteste, Bishop of Lincoln; but his scientific interest was primarily in optics and there is no evidence that he brought any knowledge of statics to bear on the reconstruction of the Lincoln choir vaults after the collapse of 1237 or 1239 or on the construction of the chapter house. F. Nordström, 'Peterborough, Lincoln and the science of Robert Grosseteste,' *Art Bulletin*, XXXVII, 1955, pp. 24–72, sees, on the contrary, evidence in the choir vaults of Grosseteste's interest in perspective illusion and in the chapter house of the Pythagorean-Platonic overtones of his theory of light (C. C. Riedl, *Robert Grosseteste on light*, Milwaukee, 1942, p. 17).

[55] C. Wren, 'Second tract on architecture,' in S. Wren, *Parentalia of the Wren family*, London, 1750, pp. 356–8, in which it is argued that 'the Design, where there are Arcades, must be regulated by the Art of Staticks, or the Invention of the Centres of Gravity, and the duly poising all Parts to equiponderate; without which, a fine Design will fail and prove abortive.'

[56] C. Wren (op. cit.), pp. 357–8.

[57] C. Wren (op. cit.), p. 358.

[58] Reproduced here from one of a series of plates in G. Poleni op. cit. Poleni was also called in to advise, at about the same time as the three mathematicians. He presented his report in 1743 and it is reproduced in the volume cited. The three mathematicians (T. Le Seur, . . . op. cit.) reproduce a single composite external, internal and sectional elevation of the dome and drum showing the same crack patterns.

[59] The present illustration is taken again from Poleni (op. cit.) since it represents the assumed collapse mode more fully than the corresponding sketches in the report of the three mathematicians, which show only the rotations of a single radial cross section such as dRLQK about d, L and K. The two small sketches reproduced in fig. 14c suggest that Leonardo may, some 250 years earlier, already have been thinking, qualitatively at least, along very similar lines. In 1732 Danisy had made tests on the collapse of model arches and these were published in A. F. Frezier, *La théorie et la pratique de la coupe des pierres*, III, Strasbourg, 1739, p. 389.

[60] Arundel MS, 141v, with sketches of a pomegranate and a cracked semidome.

[61] R. J. Mainstone, 'The springs of structural invention,' *Journal of the Royal Institute of British Architects*, LXX, 1963, pp. 57–71.

[62] The need to ensure the stability of the

incomplete structure at all stages of construction may be expected always to have had a predominant influence on the form when temporary supports were dispensed with or reduced to a bare minimum. See, for instance, P. Sanpaolesi (op. cit.), A. Godard, 'Voûtes Iraniennes,' *Athār-é-Irān*, IV, 2, 1949, pp. 187–360; and J. Fitchen (op. cit.). Beyond this the search for practical convenience and economy in construction will have had its influence in the past as much as today.

[63] Leonardo's projects are discussed in L. Beltrami, *Leonardo da Vinci negli studi per il tiburio della cattedrale di Milano*, Milan, 1903, in which a selection of the many sketches in the MSS is reproduced. Apart from the intrinsic evidence of sketches such as that reproduced here (Codice Atlantico 310Br) and the many which explore more generally the stability of the arch (including those mentioned in note [59]), there is further evidence of Leonardo's approach in the draft of a letter of submission of his proposals for, presumably, the tiburio. He writes here that the sick cathedral has need of 'uno medico architetto, che 'ntenda bene che cosa è edifizio, e da che regole il retto edificare diriva; e donde dette regole sono tratte e 'n quante parti sieno divise, e quale sieno le cagioni, che tengono lo edifizio insieme, e che lo fanno permanente, e che natura sia quella del peso, e quale sia il desiderio de la forza, e in che modo si debbono contessere e collegare insieme, e, congiunte che effetto partoriscino . . .' (*Codice Atlantico* 270A, printed in J. P. Richter and I. A. Richter (op. cit.), II, pp. 330–1). In this connection some reference must also be made to Guarini who, apart from his architectural work, wrote inter alia on celestial mechanics. Unfortunately, however, his 'Architettura civile' (see below and note [77]) hardly mentions the domes which play so large a part in his major works and gives no clue to the structural bases of his designs which can only be inferred from the confident daring of, for instance, San Lorenzo, in Turin, 9. Though the dome of San Lorenzo and others like it may owe something to Borromini and have precedents in hispanomoresque and other islamic vaults and in a few Romanesque and Gothic vaults inspired by these, the writer strongly suspects a more direct influence of Leonardo's Milan projects and his closely related projects for centralized churches, and, perhaps through these, of German High-Gothic rib-vaults, towers and baldacchinos. The only reference to Guarini in a purely structural context known to the writer is a report on the stability of the dome of Florence Cathedral in 1695, to which he was one of the signatories (C. Guasti (op. cit.), doc. 391, pp. 177–183). It was recommended that four iron chains should be placed round the dome, but the quantitative analysis which supported a similar recommendation at St. Peter's is lacking here.

[64] In these sketches the approximate catenary curves superimposed on cross sections of the dome recall the entry in Robert Hooke's diary for September 26, 1675: 'riddle of arch, ut pendet continuum flexile, sic stabit (inversum) rigidum.' The same (very conservative) approach to the choice of a stable profile for a dome is illustrated more explicitly in G. Poleni op. cit. (pp. 42–50 and fig. XIV, the latter reproduced in R. J. Mainstone, 'Structural theory' op. cit., p. 188) and is discussed mathematically in a forthcoming paper by J. Heyman, 'On shell solutions for masonry domes.'

[65] G. Milanesi (op. cit.), II, pp. 431–5; L. Beltrami (op. cit.), pp. 57–61; R. Papini (op. cit.), pp. 180–5 and 261–4.

[66] op cit., III, 14 (Leoni translation).

[67] op. cit., III, 13 (Leoni translation).

[68] English translations and commentaries in F. M. Cornford, *Plato's cosmology*, London, 1937; H. D. P. Lee, *Plato Timaeus*, Harmondsworth, 1965.

[69] Vitruvius, III, iii, 6.

[70] R. Klibansky, *The continuity of the Platonic tradition during the middle ages*, London, 1939.

[71] E. Panofsky, *Abbot Suger on the Abbey Church of St. Denis and its art treasures*, Princeton, 1946; O. Von Simson (op. cit.).

[72] R. Wittkower, *Architectural principles* (op. cit.).

[73] K. Lehmann, 'The dome of heaven,' *Art Bulletin*, XXVII, 1945, pp. 1–27; O. Demus, 'Byzantine mosaic decoration,' London, 1948; E. B. Smith, 'The dome: a study in the history of ideas,' Princeton, 1950; C. Mango, 'A twelfth-century description of St. Sophia,' *Dumbarton Oaks Papers*, XIV, 1960, pp. 233–45. For further valuable discussions of this iconographic aspect of design see R. Krautheimer, 'Introduction to an "iconography of medieval architecture",' *Journal of the Warburg and Courtauld Institutes*, V, 1942, pp. 1–33, and P. Frankl, 'Gothic architecture' (op. cit.), pp. 229–40.

[74] Doric refinements are discussed, with bibliographies, in W. B. Dinsmoor, 'The architecture of ancient Greece, 3rd ed., London, 1950, pp. 164–9, and A. W. Lawrence, 'Greek architecture,' Harmondsworth, 1957, pp. 169–75. Later Renaissance design is discussed in W. Lotz, 'Das Raumbild in der italienischen Architekturzeichnung der Renaissance,' *Mitteilungen des kunsthistorischen Institudes in Florenz*, VII, 1956, pp. 193–226, and J. S. Ackerman, 'Architectural practice in the Italian Renaissance,' *Journal of the Society of Architectural Historians*, XIII, 3, 1954, pp. 3–11. Though a direct visual interest seems to have been unusual in the Gothic period see also F. Nordström (op. cit.).

[75] Le premier tome (op. cit.), pp. 233b–4b. Though introduced as an illustration of the new theory of divine proportion to be set out more fully in a promised subsequent treatise which never appeared, the design undoubtedly reflects the practices seen at Milan and Bologna (see note [33]). Francesco di Giorgio's design is illustrated and discussed in H. Saalman, *Early Renaissance architectural theory* (op. cit.), pp. 92–3 and R. Papini (op. cit.), p. 194.

[76] Published with German translation in *Heronis Alexandrini: opera quae supersunt omnia*, V, 'Heronis quae feruntur stereometrica et de mensuris,' J. L. Heiberg ed., Leipzig, 1914, pp. 1–162.

[77] Published posthumously, Turin, 1737. At

the outset (p. 2) Guarini states that l'Architettura come facultà dipende dalla Geometria' and the more theoretical part of the treatise is entirely devoted to it. In this acceptance of the primacy of geometry per se (rather than in its wider but less precisely defined role in the Gothic *scientia geometriae*) Guarini may be considered as the successor to Borromini and the model of Neumann, 23.

[78] R. Wittkower, 'Brunelleschi and "Proportion in perspective",' *Journal of the Warburg and Courtauld Institutes*, XVI, 1953, pp. 275–91. For the later interest in perspective see, in addition to the last two works cited in note [74], A. Pozzo, *Perspectiva pictorum et architectorum*, Rome, 1693, and C. Wren, '*First tract on architecture*, S. Wren (op. cit.), p. 352—'The Architect ought, above all Things, to be skilled in Perpective; for everthing that appears well in Orthography may not be good in the Model, . . . ; and everything that is good in the Model may not be so when built.'

V

Developments in structural design since the seventeenth century especially in Great Britain†

The more closely we look at the developing art of structural design, the more aware we become of innumerable interwoven threads stretching back even beyond recorded history to man's earliest attempts to build, perhaps some 10,000 years ago. Development was not always continuous. Hard won lessons and skills might be lost, only to be learnt and mastered again, later, in some other place. But there was no significant break in Western Europe after the surge in building activity in about the 11th century. The design of the great Gothic cathedrals developed out of that of the Romanesque (or Norman) churches that preceded them. And a detailed study of these cathedrals still guided and inspired leading designers like Wren in 17th-century England and Soufflot in 18th-century France.

By the latter part of the 17th century however, changes were afoot in the understanding of structural behaviour, and they gained momentum during the 18th century. It is clear the Gothic master masons, responsible for the design of the cathedrals, must have had a keen awareness of the way in which, for instance, the weights of the high vaults of these structures pressed downward and outward on their supports and of the action of the flying buttresses in countering the outward pressures. Yet all the evidence we have shows that it was primarily an intuitive understanding, akin to that which we have of the pushes and pulls acting on, and exerted by, our own bodies, and of the ways in which these are balanced.

To build, it was necessary to quantify – to choose particular sizes and spacings of supports, particular shapes of vault and arch, and the particular materials of which they were to be built. This was done directly in terms of past experience with similar structures, rather than indirectly in terms of forces and strengths and abstract concepts of the conditions of balance between forces (fig. 1).

The changes were the bringing to bear, in various ways, of abstract concepts of just this sort. In themselves, they had had a long history, seemingly almost wholly independent of practical construction, and stretching back at least to Archimedes who, in the 3rd century BC, first codified the laws of balance for vertically acting weights. By the latter part of the 17th century, this sort of understanding was no longer limited to vertically acting weights. Men like Robert Hooke turned their attention to such questions as the curve assumed by a hanging chain to maintain equilibrium between the tensions in the links and the weights of the links – and, by analogy, the ideal profile of an arch formed of smooth balls simply resting on one another (fig. 2). At the same time, Hooke and others began to measure the strengths of different materials and the extent to which some of them were (like a spring) stretched by an applied force.

Wren knew Hooke well, so it is probable that Hooke's ideas influenced Wren in, for instance, his final design for the dome

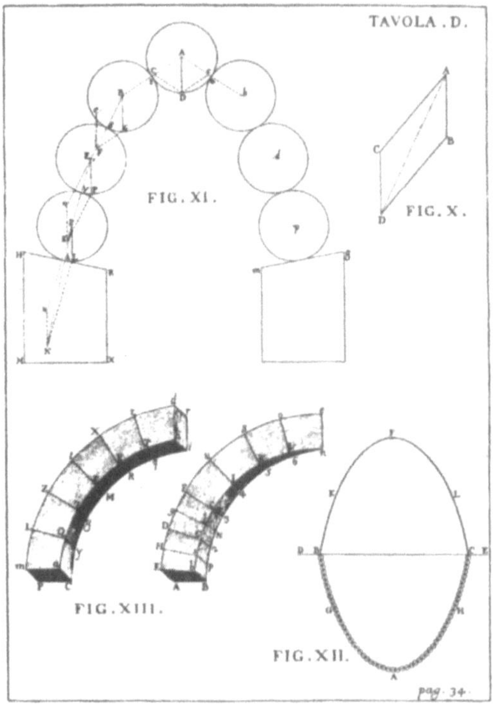

1 (left). The 'critical mechanism' of Salisbury Cathedral according to Price, *A series of particular and useful observations ... upon that admirable structure the Cathedral Church of Salisbury*, 1753.
2 (right). The stability of an arch and a hanging chain according to Poleni, *Memorie istoriche dell gran cupola del Tempio Vaticano*, 1748, pl. D.

These two plates symbolise the old approach to structure and the new one that was taking its place in the mid-18th century. Price, like the Gothic designers, still saw the 'critical mechanism' essentially in terms of the visible form and its geometry. Poleni shows it, in his figures X and XI, in terms of the weights of the balls and the pressures they exert on each other.

of St Paul's. What is beyond doubt is that, around 1742, several analyses were made of the stability of the dome of St Peter's in Rome (then alarmingly cracked near the foot) as a guide to deciding on measures to ensure its future safety. Broadly similar analyses were made a little later to support and guide Soufflot in his designs for what is now the Pantheon in Paris.

In parallel with these changes, the 18th century saw also the emergence of civil engineering as a distinct profession. From the Middle Ages onwards the term 'engineer' had been used intermittently (and sometimes almost interchangeably with 'architect') to denote someone engaged in the design of military engines and defence works. In late 17th century France the military engineer was given formal status by the creation of a *Corps des ingénieurs du Génie militaire*. This corps of military engineers was followed, in 1720, by a corps of bridge and road engineers *the Corps des ingénieurs des ponts et chaussées*. Equally significantly, a systematic training based on the latest scientific knowledge and understanding was introduced (fig. 3). Among the various schools that were founded was the highly influential *École des ponts et chaussées*. In this lively milieu, backed by the State, the greatest French engineer of the century, Perronet, carried the design of masonry arched bridges to a new and technically unsurpassed peak (figs 4 and 5). And it was Perronet's successor as chief engineer, Gauthey, who acted as Soufflot's, and later Rondelet's, advisor on the Paris Pantheon.

In England, the profession may be said

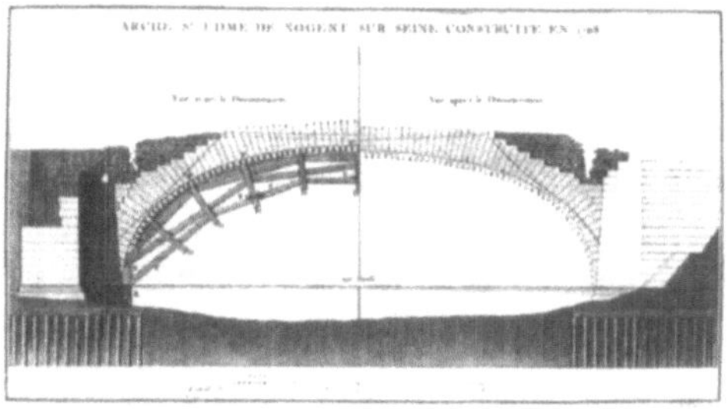

3 (above). The frontispiece to Belidor's *La science des ingénieurs*, 1729.

The first major text that attempted to place structural design on a rational basis, probably written for use primarily by Belidor's students at the school of military engineering at La Fère and concerned chiefly with fortifications.

4 (right). Measured settlements on decentring an arch, 1768. 5 (below). Decentring the arches of Perronet's bridge at Neuilly, 1772 from Perronet, *Description des projets et de la construction des Ponts de Neuilly ..., 1782-89.*

Perronet, chief engineer for French bridges and roads from 1764 to 1794, made a careful study of all aspects of the construction of masonry arch bridges and applied the lessons learnt in the design of bridges like that at Neuilly. Very flat arches with splayed haunches were adopted there, carried on piers wide enough only to support their weights acting vertically downwards so as to offer the minimum resistance to the flow of water past them. All arches had therefore to be constructed together and their centres struck simultaneously so that side thrusts were always balanced on each side of a pier.

to have emerged in the person of one man, John Smeaton, who combined scientific interests with practical skill in design in much the same way as Perronet. The only counterpart of the French *Corps* and *Écoles* was, however, a very different body, the Royal Society, founded in 1661. In spite of its royal patronage, this was essentially a group of individuals active in many fields but brought together by a shared scientific curiosity and (to a greater extent than today) a desire to turn new knowledge to useful ends. Both Wren and Hooke had been members. Smeaton was elected a fellow at the age of 28 in recognition of early work on, and with, scientific instruments. He also undertook notable research on the power of wind and water to drive mills and was recommended by the then President of the Society to prepare designs for the rebuilding of the Eddystone Lighthouse, the commission that really launched him on his civil engineering career (fig. 6). Fifteen years later, in 1771, a Society of Engineers was founded, with Smeaton as its leading light. Their interests ranged much more widely than those of the structural engineer today. But they included these as far as they were relevant.

Other important factors around this time were the introduction of the first major new construction material since Roman concrete and a rapid growth of new industries.

The new material was cast iron. Iron had long been used in small quantities in another form for such important structural members as tie-rods, just as materials with some of the characteristics of Roman concrete had been used well before the 2nd century BC. Now the cast form became available in much greater quantities. Although it was used initially as little more than a straight substitute for timber or masonry (fig. 7), it soon began to be exploited more fully in the hands of men like Telford (fig. 8).

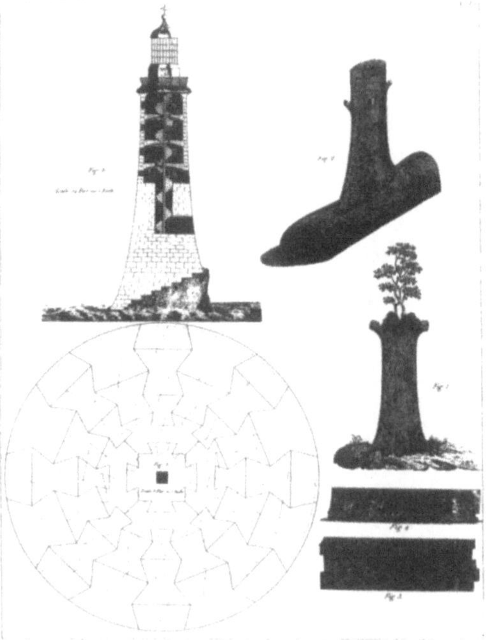

6. Ideas for the Eddystone Lighthouse, from Smeaton, *A narrative of the building and a description of the construction of the Eddystone Lighthouse with stone*, 1791.

The third Eddystone Lighthouse, built between 1756 and 1759, was an achievement without precedent. Charged with its construction, Smeaton made careful inquiries into everything that could be relevant, including making extensive experiments on suitable mortars for use in a very exposed situation at sea. The profile was suggested by that of the trunk of an oak, well adapted to resist the force of the sea. The interlocking of the blocks of stone, as seen in the plan view, was suggested by other types of interlocking illustrated by Belidor in his Architecture Hydraulique *as seen at the bottom right.*

The parallel rapid growth of new industries, especially in England, both generated new requirements and provided the resources with which to meet them. Important among these requirements were those for bridges and for fireproof mill buildings. Both (and particularly those for bridges

7. Detail of the iron bridge at Coalbrookdale, 1779.

This bridge was the first major structure to employ cast iron. But it was essentially a stone bridge in its adoption of a arch of semicircular profile and it large followed traditional timber practices in its detailing.

8. Thomas Telford, frontispiece to the Atlas, *Life of Telford*, 1838.

This great British civil engineer, first President of the Institution of Civil Engineers in 1820, is seen against the background of his Pont-y-Cysylte Aqueduct – an iron structure carried on tall masonry piers.

after the coming of the railways) called for much more consideration of the loads additional to its own weight that were imposed on the structure in use. Previously, with more massive forms of construction and smaller imposed loads, there had been little need for this, and that had been a crucial factor in permitting the development of structures like the Gothic cathedral without any explicit consideration of the forces acting.

In the following sections we look at the ensuing developments chiefly in terms of the developing relationship between science and engineering, and related developments in professional practice. Since the emphasis is on Great Britain, we conclude by comparing achievements there and elsewhere.

Science and engineering

Structural engineering is concerned primarily with the activity of getting things built; science is concerned primarily with knowledge. Science is useful to the engineer both when it clarifies and extends his essential intuitive understanding of structural behaviour (without which he would be reduced to blindly following precedent in choosing what to do) and when it enables him to make quantitative analyses that draw more effectively on past experience for the purpose of selecting precise dimensions, cross-sections, materials, etc., to meet the given requirement. The chief branch of science with which we are concerned here is structural mechanics. At its most fundamental, this comprises theories about the action and balance of forces and theories about the ways in which materials deform under load and their limits of strength. At another level, these fundamental theories are applied to particular situations such as the bending of a beam or the behaviour of a complex frame. Usually, the theories are expressed in the language of mathematics, without which most of this application of the fundamental theories to increasingly complex situations would be virtually impossible. Such, indeed, is the importance of mathematics that much of the application has been largely a mathematical exercise. To relate the results to the real world, there has, however, been also a constant need for testing.

Some early developments of the fundamental theories have been referred to already, as have some of the first analyses of particular structures. Over the last two centuries the dependence of the engineer on the insights and possibilities of analysis furnished by the scientist has grown continually because of the increasingly wide and demanding range of needs that the engineer has sought to satisfy and the ever-broadening variety of materials, techniques, and forms, on which he has drawn to meet these needs. Almost anyone can design a structure for a small house, using bricks, mortar, and timber, in a traditional way, without any need for modem science. But to design an acceptable and economical structure for a very tall building, making the best use of the materials and techniques now available, can tax to the utmost both the skill of the engineer and the latest scientific knowledge and resources of theory and calculation.

Nevertheless, as practical men with the end always in view and time at a premium, engineers have never espoused science for its own sake while acting as engineers. When they could get along nicely without a new theory, they have, typically, ignored it. On the other hand, when faced with a new and otherwise intractable problem, they have sometimes called in the scientist or mathematician and stimulated him to further effort. Thus there has been a very variable time lag between advances in science or theory and the use of these advances in design. In reviewing past achievements, we must also remember that knowledge can inhibit as well as help. Where it saves us from error it should, of course, be regarded as a help more than an inhibition. But it is rarely complete. It becomes an unfortunate inhibition when its incompleteness means that it instils an unnecessary caution – a consequence that is commoner now than it used to be because of the extent to which concerns for public safety have led to control of design by regulations.

Early British engineers were not wholly ignorant of scientific developments across the channel. Telford, like Smeaton before him, possessed an extensive library which included most of the major French publications of the time. But they always seem to have placed more trust on direct observation and test than on calculation, and they were less inhibited than any of their successors by an inability to calculate how a new structure would behave, by a fear of failure, or by a need to seek the approval of any academy or regulatory body. Remarkably little theory went into the development of the suspension bridge by Telford and Brown in relation to what we now know of the complexities of the problems it presents, and even in relation to the analyses made by the their French contemporary, Navier, after his inspections of their bridges. On the other hand, Stephenson's venture into the unknown with his Conway and Britannia tubular bridges called for a major scientific investigation.

The one type of structure that was widely used in the 19th century, and that could not be designed efficiently without a clear theoretical understanding of the way in which the loads were carried, was the trussed railway bridge. In the earlier part of the century, large numbers were built, especially in North America, without this understanding. Some were built wholly timber, some of iron, and some partly of timber and partly of iron. All were highly inefficient, with an unnecessary multiplicity of members (fig. 9). There were also many collapses. Once the behaviour was better understood, much simpler forms were adopted with just enough members to create a rigid assembly in each span. The forces could then be calculated in each member and the necessary strengths and stiffnesses provided. The calculation had to assume that, wherever several members met, they were connected by means of a frictionless pin. Care was therefore taken to introduce such pins until it was realised that they were usually unnecessary and that a better structure might result from the use of simpler, more rigid, joints.

9. Savage Bridge, Maryland, 1852.

A Bollman truss, one of many types patented in the USA during the first half of the 19th century and built in large numbers on the new railways. A better understanding of the structural action permitted great simplification in later designs.

The calculations here were purely 'statical' ones, i.e. they were based only on the simple conditions of balance of forces (those in the truss members plus any imposed loads) acting together at a joint. The calculations could be carried out entirely in terms of the imposed loads and the geometry of the proposed truss. Most subsequent structural design has called for more complex calculations because there have been alternative possibilities of balance in terms only of geometry and imposed loads, just as there were in the truss illustrated in fig. 9. The actual behaviour in such cases depends also on the relative stiffnesses of the alternative paths for each load – i.e. on the relative lengthening or shortening under load – and hence on the characteristics of the materials employed. It is said to be 'statically indeterminate'.

Statically indeterminate structures have many practical advantages, including the obvious one that the existence of alternative paths for loads makes sudden collapse less likely. Strictly speaking, moreover, they include all structures in which different materials are, constrained to act in unison, as they are in reinforced concrete. Thus there are few 20th-century structures that are not statically indeterminate in some important respect, and the analysis of such structures has been a major concern of engineering science since the latter decades of the 19th century. Mathematically, it is inevitably more complex than purely statical analysis, so it is hardly surprising that designers have not infrequently fought shy of its full rigour and sought simplifications where they seemed to be justified by experiment or experience. All assumptions about the properties of the materials in an as-yet unbuilt structure, the loads that will, in future, act on it, etc., are, in any case, open to some doubt, and the engineer's objective is a safe and economical structure rather than an elegant and rigorously correct analysis.

From the late 19th century onwards, major areas of interest (after the solution of the problem of the bending of a beam that is continuous over several supports) were the behaviour of continuous frameworks with rigid joints and the behaviour of members and structures of reinforced (and later prestressed) concrete. In the past few decades, more attention has been given to fully three-dimensional structural behaviour and to the general question of how to ensure, at the time of design, that a structure will be both safe and serviceable.

During this same period, powerful electronic computers have come to the engineer's aid. They have greatly reduced the need for the simplifications and the ingenious computational techniques that facilitated analysis in the middle part of the century. It is no exaggeration to say that they now make feasible the design and safe execution of structures that could scarcely be dreamt of before. Yet it remains as true as ever that it is the engineer who must conceive the form and envisage clearly enough how it will carry its loads to decide what analyses need to be made and by what criteria the predicted performance should be judged acceptable. This is an onerous responsibility whenever new ground is being broken, as periodic failures during construction or in service continue to testify. The individual engineer is only partly relieved of it by the Codes of Practice devised by his peers to guide him and the regulations devised by legislators to safeguard the public.

Engineer and architect

It was noted above that the terms 'engineer' and 'architect' were sometimes used almost interchangeably in the past. We are today struck by the structural invention to be seen in the Gothic cathedral. But most designers – whether of buildings or fortifications, bridges or harbour works – either stayed well within the limits of proven practice in construction or moved beyond it very circumspectly. The chief skill they were called on to display was the ability to conceive a form and convey their concept to others through drawings – a skill referred to in Renaissance Italy as *disegno*.

Engineering began to appear as a specialism within the wider ranks of designers as certain types of construction began to call for a more detailed knowledge and understanding of structural and mechanical matters. But Belidor could still follow his *Science des Ingénieurs* (fig. 3) with a longer work entitled *Architecture Hydraulique* which set out to place on a similarly scientific footing the design of canals, harbour works, underwater foundations, etc. And Telford, whose early work was certainly architectural in the modern sense, moved easily to the design of bridges and roads and canals when opportunity arose and he recognised 'a stronger disposition for executing works of importance and magnitude than for the details of house architecture'.

By the mid-19th century the civil engineer had a virtual monopoly of these 'works of magnitude and importance', which now included all major works in iron and all that called for structural analysis or test. He was relatively unfettered by aesthetic precedent because there was little that was relevant, but his designs often showed considerable aesthetic sensibility and are often today valued as much for this as for their technical merit. This left to the other side of the construction-design profession – to the architect, as he continued to be known – most buildings and anything of a more decorative or monumental, than utilitarian, character. He became, on the contrary, increasingly interested in the past and its aesthetic canons, as he saw them. Thus arose a sharp divergence of the two sides of the profession, nowhere better exemplified than by the juxtaposition of Scott's Gothic-style Midland Hotel at St. Pancras and Barlow's iron-roofed train shed – equally fine works on their own, very different, terms.

By this time the civil engineer had already largely dropped the earlier interest of men like Smeaton in machines, leaving these to the mechanical engineer. Further specialisation followed with the introduction of new materials and new structural forms and the need to master more and more theory. Some civil engineers continued to practice over the whole range of construction. Others limited themselves to reinforced concrete or steel structures, for instance, and became known, more precisely, as structural engineers. These soon found themselves entering into a new relationship with the architectural profession. The introduction of steel and reinforced concrete into buildings for the main load-bearing elements made their structural expertise necessary in design, while the architect remained the overall designer responsible for both planning and appearance. Hardly surprisingly, the results were not always happy. It is difficult not to regard as the finest works of the early part of this century those – like the Swiss bridges of Robert Maillart – in which an engineer with a good eye remained in sole control.

From around 1910, some architects in continental Europe – first in Germany, later in France and the USSR and elsewhere – began, however, to see the architectural possibilities of some of the new structural forms, as indeed several Chicago architects had already done a few decades previously in the case of the multistorey steel frame. In the 1930s it was no longer necessary to speak chiefly of continental Europe. Thanks partly to a number of architects and engineers who came over from the continent including Arup and Samuely, but also to men like Sir Owen Williams, a similar recognition appeared in Britain. Since 1945 that recognition, and the mutual understanding by architects and engineers of each others' roles and abilities, has become much more widespread, with wholly beneficial results. Many of the best works of these postwar decades are the product of true professional collaboration. Yet one must also admit that there has been a tendency, on the part of some architects, to be seduced too easily by the novelty and excitement of new structural possibilities like the cantilever and some shell roofs, and too great a readiness, on the part of their engineer consultants, to give them what they ask for. In a large bridge, a dam, or an offshore oil rig, structural considerations may be dominant; in buildings, they should rarely be allowed to become so.

Britain and elsewhere

No art has developed for long in complete geographic isolation. It seemed right to celebrate a British anniversary by concentrating on British achievements. But it will already have become clear that these have owed a good deal to achievements in other countries. British engineers have, in return, contributed to, and influenced, developments abroad, from the examples they set by early iron bridges onwards.

It would be out of place here to review all borrowings, leads, influences, and other interactions. Broadly speaking, the interactions become more numerous, and certainly more immediate in their effect, the nearer we come to the present. In recent years, there has, for instance, been far more contact, and more continuous contact, between both engineers and scientists on a world wide basis than ever before.

We may, nevertheless, ask how, or why, there comes to be a British – or any other national – achievement that may be distinguished by more than mere location. Why was the development of the suspension bridge carried so far in Britain in the 1820s, only for the lead to pass, for over a century, to the USA and then to be taken up again in Britain around 1950? Why did the pioneering development of the iron-framed building in Britain lead nowhere after 1860, leaving it to a few architects and engineers in Chicago to demonstrate the real potential? Why do we then find the frame apparently exploited so very differently in the post war decades in the USA, the USSR, and Britain (figs 10 to 12)? And why, at this time, was a novel shell form taken up so enthusiastically in Central America when, across the border in the

USA, the chief focus of similar enthusiasm was the very different framed geodesic dome (figs 13 and 14)?

There is never a single simple answer. Obviously, the skill and knowledge of the designer, and sometimes the testing and analytical support available to him, are important factors. But at least as important are the opportunities presented to him, the resources of men, money, and materials, and the whole situation in which he works. Unlike the proverbial artist in a garret, he cannot do as he pleases but must do as some client, and ultimately the society in which he lives, asks him and allows him. Major advances have been made in response to new needs and challenges and their direction and pace determined partly by when and where these have arisen.

Thus each major advance in suspension

10 (above). Lever Building, New York, 1952.
11 (above right). Moscow University, 1952.
12 (below right). York University, 1963.

All these buildings have a steel frame as the basic structural support. The very large differences between them probably stem more from differences in what the architects were asked to do than from anything else. The Lever Building might, as one of the most successful buildings of its time in the spirit of the pioneering modern architecture of the 1920s and 1930s, be regarded as the product of enlightened private patronage. It was also a commercial slogan skilfully exploiting a recent change in planning legislation as well as new construction techniques. Moscow University was more a slogan of the state and put into effect a very different concept of state planners for that city. York University was also, in effect, state sponsored, and employed a structural system developed during the 1950s for use in state schools. Its unassertive character stems from a primary concern, on the part of both its architect and the state, for its educational role.

13. Church of the Miraculous Virgin, Mexico City, 1954.

In the 1950s Candela made extensive use of the hyperbolic paraboloid form of thin reinforced concrete shell, sometimes to great architectural effect, as in this church. Nowhere else was it used to the same extent. It uses little material to cover a given area, but much labour is required to construct the formwork on which the reinforcement and concrete are placed. In Mexico, this labour was relatively cheap. In Europe, where the form was first used, it became too expensive, and the daring of some of Candela's structures might have been unacceptable to the watchdogs of public safety.

bridge design was a response to a challenge at a particular place and time and against a different background of resources, prior experience, knowledge, and restraints on venturing into the unknown. Similar challenges were at least partly responsible for the fact that it was in Chicago rather than, say, London that the tall steel-framed building first appeared in essentially its present form, and for rapid developments in steel-bridge design in Germany in the 1950s. The differences seen in figs 10 to 12 stem chiefly from different social priorities and values, while the different approaches to roofing a large single space seen in figs 13 and 14 reflect not only different visions on the part of the designers but also very different national economies, construction industries, and building costs.

In late 18th and early 19th century Britain, the conditions under which engineers worked must have been close to ideal for fruitful innovation. There was every encouragement to try something new and few restraints, a new material in plentiful supply and ample labour, and the designer was able to specify and control the whole process of construction if he wished. By the end of the 19th century, and in the early 20th century, there were fewer challenges and considerably more restraint in the form of regulations, so that there was more scope for initiative in the USA and on the continent. In recent years, there have been fresh challenges both at home and abroad and stimuli to innovate, both to meet new

14. Climatron, St Louis, Missouri, 1960.

Fuller's development of the geodesic dome, used on a large scale in the Climatron, stemmed from a desire to enclose space with the minimum resources in the very different economy of a highly developed country. The dome was therefore designed for easy assembly on site from prefabricated components which were madeas near identical with one another as possible.

requirements and to achieve more with limited resources. The engineer's task has not, however, been an easy one. Concern for public safety has grown rather than diminished and still tends to be expressed through regulations that are too restrictive and costly to comply with. The easy route to economy through large-scale repetitive production that was long followed in the post-war period in the USSR and some other Eastern European countries for mass housing, was denied by a justifiable desire for greater choice and variety. And the organisation of the building industry, with its separation of responsibilities for design and construction, has deprived the designer of the advantage enjoyed by many designer/ contractors abroad of being able, like his 19th-century predecessors, to specify the construction process and take full advantage of it.

What of the future? The possibilities are now so wide in terms of the engineer's skills and the materials and techniques available to him that the answer seems to turn less on these than on what he is asked to do and with what means. Professor Happold foresees a greater use of organic materials, modelled, perhaps, on forms in nature with their characteristic initially compliant response to load.[1] Will this sort of response be acceptable? Or shall we continue to require most of our structures to give way as little as possible when loaded? Natural structures have innate powers of regeneration to cope with ageing or damage. In the absence of similar powers in their manmade counterparts, what limitations on life span shall we accept?

It is on the answers to questions such as these, and more generally on the nature of the demands made of him and the conditions under which he will work, that the nature of the engineer's future achievements will now surely chiefly depend.

† This is a slightly edited and abridged version of a paper originally published as the Introduction (pp. 5-26) to *Structural Engineering: two centuries of British Achievement*, ed. A. R. Collins, published by Tarot, 1983, on behalf of the Institution of Structural Engineers to mark its 75th anniversary.

1. Ibid, pp. 171-179, chapter on 'The Future'.

VI

Structural Analysis, Structural Insights, and Historical Interpretation

In the eighteenth century, structural analysis as we now know it began to be used to assess the safety of standing buildings and then to assist in the design of new ones. The first recorded assessments of a standing building by such means were those made of the stability of the dome of St Peter's in 1752–1753 [*Figure 1*].[1] The objective was to dispel fears for its safety or, if this could not be done, to show what additional tying might be desirable to safeguard it. Uses in the design of new buildings followed. Developments in construction then prompted fresh developments in analysis and the underlying theory. The emphasis was initially on masonry arches and vaults, then on new uses of iron, and more recently on an ever-growing number of new forms in steel, reinforced and prestressed concrete, and other materials.

By the mid-nineteenth century the new analytic tools began to be used also to satisfy by simple calculation a more disinterested curiosity about how earlier structures like Gothic cathedrals stood up to gravity, wind, and other loads. The deeper personal understanding to which they led also gave engineers like Robert Willis and Auguste Choisy new qualitative insights into aspects of behavior that were less amenable to calculation.[2]

In recent decades the quantitative analysis of historic structures has received fresh stimulus from rapid advances in the numerical processing capacity of computers and especially from the increased versatility of the tool known as the finite element method. A recent paper in these pages, after listing some of its advantages, even claimed that, with minimal training, it enabled non-engineer historians to participate directly in the analysis.[3] Further claims have been made about the wide-ranging conclusions that can be drawn.

At the joint meeting of the American and British societies of architectural historians in 1973 and again at the annual symposium of the British society in 1979, I took a more critical look at the relevance of structural expertise and structural analysis to some of the questions raised by historians.[4] This article develops further the ideas then put forward, taking note of the subsequent advances in computation. It looks first at the nature of the underlying structural theory, the related computational possibilities and the qualitative understanding of the individual engineer, and at the ways in which these are used in designing new structures and in assessing the safety of recently built ones. In this light it then critically considers some possible ways of analyzing historical timber and masonry structures before turning to five main topics of historical inquiry and illustrating possible approaches to them.

PRESENT STRUCTURAL THEORY, ANALYSIS, AND UNDERSTANDING

Both the theory and the related understanding of structural behavior can be traced back to very general ideas about the nature of the physical world. On the one hand these are the intuitions of pushes and pulls and other physical actions which, in the late seventeenth century, developed into precise concepts of forces and the conditions of balance between them. On the other hand they are our recognitions of the different ways in which materials respond to load and of the parts played by their different strengths and stiffnesses in determining how loads are shared between elements of a structure when there is more than one way in which they could be carried.

The conditions of balance are now stated simply and precisely in the laws of static and dynamic equilibrium. Laws of strength and deformation cannot be stated with the same universally valid precision. Only idealized approximations are possible here. Hooke's law of the direct proportionality of stress and strain—otherwise known as linear elasticity—was the first and is the best known. The other usual idealizations today are those of perfect plasticity—continuing strain at constant stress—and of perfect elasticity followed, at some limit of stress, by perfect plasticity. Various assumptions are made about the conditions of stress or strain that lead to local failures of the materials and thereby determine the strengths of structural elements made from them.

Over more than two centuries these very general ideas have been applied to an ever expanding range of structural forms and types of behavior.[5] Well into the nineteenth century, the emphasis was on behavior that could be explored using only

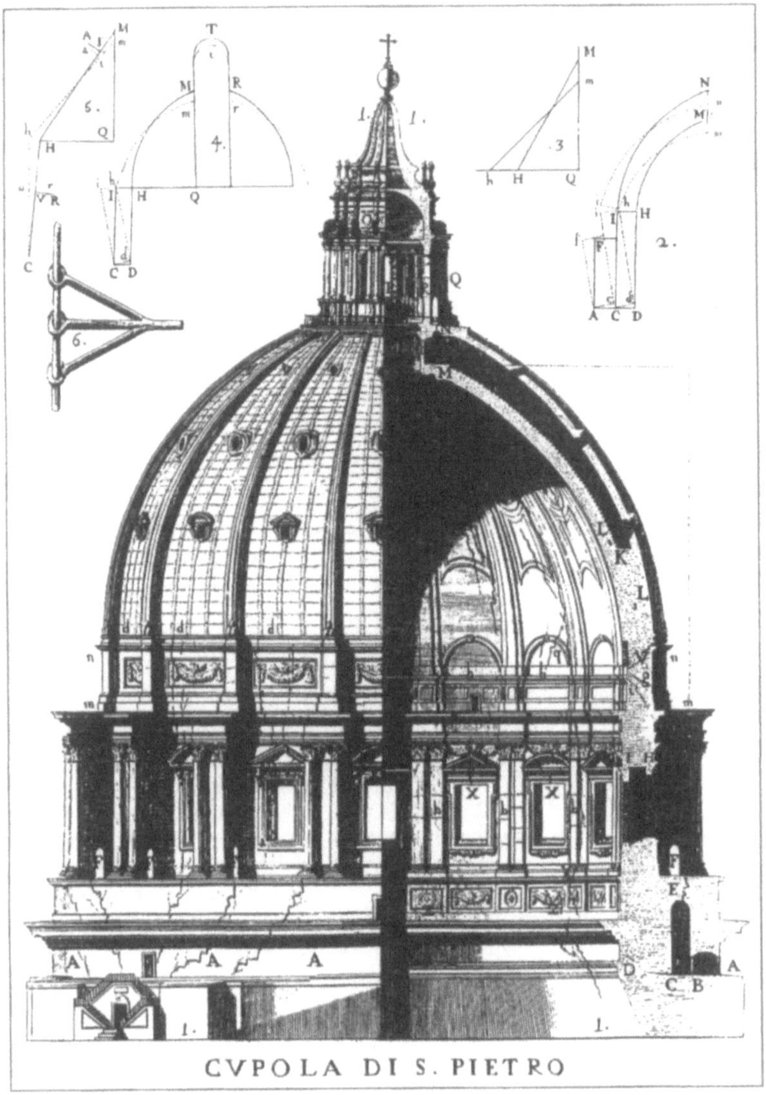

FIGURE 1: Observed cracking of the dome of St. Peter's in 1742 and, in the small sketch at the top right, the analytical model based on it by Le Seur, Jacquier, and Boscovich

the laws of static equilibrium —behavior that was said to be statically determinate. Much effort was devoted to studying the equilibrium of masonry arches and then that of various types of truss. Analysis was facilitated in the latter part of the century by the introduction of graphical methods in which lines were drawn to represent the forces in both magnitude and direction. Where the behavior of interest was not truly determinate in this way—i.e., statically indeterminate or hyperstatic—it was notionally made determinate. Elements not thought to be of primary importance and restraints to relative rotations at joints were ignored. Or assumptions were made about the ways in which forces would split where there were alternative paths for them to follow.

To avoid the need for such simplifications it is necessary to distribute the loads according to the relative stiffnesses of the alternative paths, with more load following the stiffer paths. Initially this was done solely on the basis of assumed linear elasticities of the materials; later on the basis of assumed plastic yield stresses and the corresponding calculated load-bearing capacities for yielding over the whole cross section;

still more recently on the basis of combined elastic and plastic behavior during the approach to maximum load. Calculation based on linear elasticity gave the deflections and internal stresses to be expected under normal working conditions if certain initial stresses were ignored. Calculation based on full plasticity at critical sections gave the ultimate strength. But it did so only if the associated deflections (which were ignored in these calculations) did not lead to instability before the full load was reached. To overcome this limitation, calculation had to be based, directly or indirectly, on elastoplastic behavior in the later stages of loading.

Calculation of statically indeterminate behavior in any of these ways was always more difficult and time consuming than the calculation of statically determinate behavior. Direct "exact" solutions for elastic or elastoplastic behavior were feasible only for relatively simple situations. For more complex ones it was necessary to converge on the solution by successive approximation. Sometimes this process had a physical counterpart, as when the joints in a column-and-beam frame were notionally restrained from rotating at the outset to make the behavior of the individual beams and columns determinate and the restraints were then progressively relaxed. Sometimes, as in the finite difference approach, it was a purely mathematical procedure to facilitate the solution of large numbers of equations.

Recently the finite element method has come to play a major role.[6] Here, if the structure does not already consist of members like struts, ties, and beams, which can be readily idealized as discrete linear elements meeting at dimensionless points, its actual members are notionally broken down into smaller elements connected only at a limited number of nodes. For this purpose it is possible to choose between a wide range of elements, including linear, two-dimensional, and three-dimensional ones with different characteristics and numbers of nodes. The approximation now stems from the fact that these elements are of finite size, as distinct from the infinitesimal elements assumed in arriving at the governing differential equations for an exact solution. The program first calculates the displacements of the nodes for the specified loading and conditions of support. As the calculation proceeds, the sizes of the elements may be reduced where there are large local variations in strain in order to arrive at results of an acceptable accuracy for the chosen model. A further program then derives and displays the aspects of behavior that are of prime interest, usually overall deformations (greatly magnified for clarity) and internal stresses.

This is a great improvement on the earlier approximations and an increasingly versatile and powerful one. It allows more realistic idealizations of the ways in which materials behave and more realistic modeling in other respects. But it is not wholly automatic. The user must not only allocate the relevant properties and conditions of support but must also choose appropriate types of elements and element meshes. To do this properly calls for considerable prior understanding of the behavior being studied. The program itself cannot provide this understanding. Nor can it do anything to draw attention to unwise choices or to errors in the output.

To guide the idealizations and simplifying approximations made in the different calculation procedures referred to, there has for many years been much testing of specimens in the laboratory and some linked field testing of full-scale structures.[7] Despite continuing improvements in the procedures, these tests have usually disclosed marked differences between the actual and calculated behavior of all but the simplest specimens. Measured deflections have usually been less and measured strengths greater than those calculated because some load-bearing possibilities were ignored in the calculation, but not invariably so. Measured stresses have diverged even more where they have included stresses due to fabrication and to the temporary loads that arise during construction. Moreover, nominally identical laboratory specimens rarely behave in precisely the same way.

THE DESIGN OF NEW STRUCTURES

When designing a structure and assessing its likely future safety such divergences have to be taken into account, both those inherent in the selected calculation procedure and those arising from the way in which the structure has to be idealized to allow this procedure to be followed. Geometry will usually have to be simplified. Members that have width and depth may, for instance, have to be treated as having none and the joints between them treated either as frictionless pins or as allowing no relative rotation. Simplifying assumptions will have to be made about the conditions of support. Some types of behavior such as shear, bending, or "second-order effects" may be ignored. However accurate the calculation is in itself, it will relate only to this idealized model and be subject also to the errors that it introduces.

Fortunately it is not essential for purposes of design that the model be faithful in all respects and the calculations free of all error, since the prime objective is usually to produce a safe and economical structure rather than to know precisely how it will behave. Errors can be reduced by careful modeling. Suitable criteria of acceptability can then be adopted to allow for the errors that remain and for inescapable uncertainties about such factors as future loadings and the conformity of what actually gets built to what was specified.

Choice of model, of the types of behavior to be analyzed, of calculation procedures, and of criteria of acceptability all call for skillful judgment, whether wholly that of the individual designer or partly the collective judgment of a building code committee. Such judgment requires a knowledge of the theory and calculation procedures but cannot come directly from them. Like the imaginative leaps in the conceptual phases of design it has to be based on wider experience and personal understanding.

Personal understanding of how structures behave is thus an essential counterpart of formal theory and calculation methods, and its importance cannot be overemphasised. It must be rooted in personal observation and a search for significance in this. Basic theoretic ideas do provide the necessary interpretive schemata. Tests and calculations add depth to it. But it is not necessarily limited to aspects of behavior that have been theoretically formulated. It can go beyond these limits, and only by doing so does it stimulate and guide fresh developments in the theory.[8]

HISTORIC STRUCTURES AND THEIR ANALYSIS

When a structure is already standing, the role of analysis changes somewhat. The evidence that it already provides of its strengths and weaknesses and modes of action becomes a useful starting point that is not available to the designer. This is so whether the structure is a new one or an old, historically important one. There are different requirements, though, for assessing present safety and for satisfying a disinterested curiosity about the structure's present or past behavior or seeking to understand better some other aspect of its history. Cautiously conservative criteria of acceptability and assumptions about unknowns are out of place when attempting either of the latter. The model must then adequately represent not only the geometry but also all relevant details of its actual construction if it is to serve its purpose fully.

When the structure is an old one this can be difficult because it will have been built of materials like timber and masonry which are less homogeneous than those used today and have unknown and more variable strengths and elasticities. Different methods of assembly and jointing will have been employed. Different modes of deformation are likely to predominate. And behavior may have changed somewhat over a long life.

Timber Structures

Apart from its lower strength, its susceptibility to moisture movements, and its shrinkage on initial drying out, timber may behave in a manner broadly comparable with steel when subjected to direct tension or bending. It can resist both, as well as some axial compression. However, in built-up members and complete timber structures, the manner of jointing is crucial. Only modern glued joints come near to matching the joints now usual in steel. All early timber joints were weaker than the members they joined because of the cutting involved. Usually they were so much weaker that they, rather than the full cross sections of the members, determined the types of action that could be resisted and the loads that could be carried [*Figures 2, 3*].

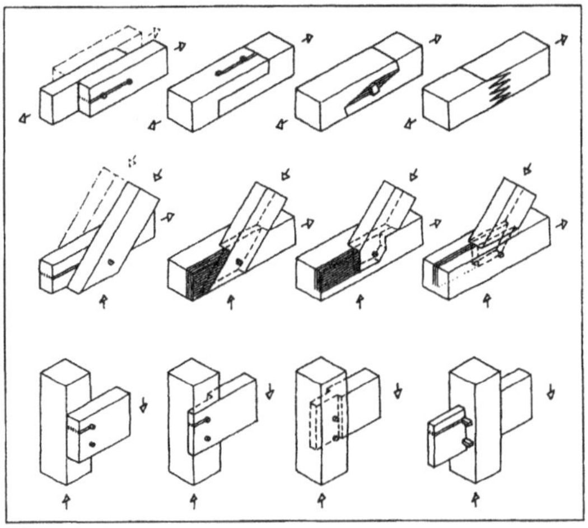

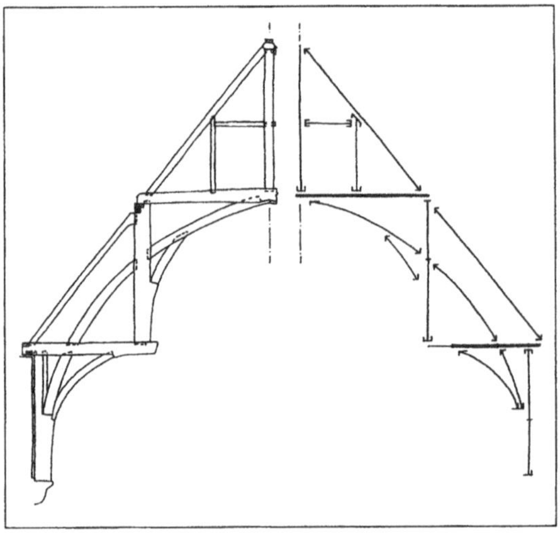

FIGURE 2: Typical early joints compared with a modern machine-cut and glued one (last in the top row). The early joints were much weaker, being liable to fail by shearing along the grain as indicated.

FIGURE 3: Transverse framing of the roof of Westminster Hall. The principal members and their joints are shown actual size on the left. On the right they are shown diagramatically so as to indicate the limitations on the forces that can be transmitted. Only the continuous members shown in heavier line could transmit significant tension.

Modeling and Analysis of Timber Structures for Static Loading

In those timber structures that are of greatest interest—wide-span roofs and bridges—and in most early iron counterparts, struts, ties, and beams are discrete elements. There is therefore no need for a notional breakdown into smaller elements to facilitate analysis. The principal need is for a correct modeling, not only of the actual members, but also of all joints. It is necessary to recognize both the restraints that they offer to relative movements of the members and the limitations that they impose on the forces and moments that the members can carry. Usually some relative rotation will be possible except where a member is continuous past a joint.

To calculate a unique distribution of forces throughout the structure where there is significant statical indeterminacy it would be necessary also to know how closely the members would fit together if free of all load. Any excess length of a compression member would necessitate forcing it into place and would attract more load. Conversely, any initial slackness would reduce the load carried. For a tie the opposite would be true. In practice this is likely to mean that the manner in which the loads were carried would also be partly dependent on the erection sequence and the extent to which joints were initially wedged tight. Close inspection of the joints may allow a likely sequence to be deduced. But there is bound to be a large measure of uncertainty about the initial fit. Rot and other decay apart, the main subsequent influences on behavior will be movements of the supports and loosening of joints due to shrinkage of the timber.

Even a relatively simple structure will have had many possible states, and it is unlikely that there will ever to be enough data to pinpoint the precise state at any particular time in the structure's past. This has not prevented numerous attempted analyses of one of the most interesting early timber structures, the roof of Westminster Hall. All have concentrated on the arched hammerbeam transverse frames that are its principal elements [*Figure 3*]. A. J. Sutton Pippard, Jacques Heyman, Mainstone, and Robert Mark and their coauthors, being interested chiefly in the primary behavior, each considered only a simplified loading with all the weight of the roof reaching these frames via the ridge beam and main purlins.[9] They all discounted some members as being of secondary importance but made different assumptions about support and jointing and therefore arrived at different results. Pippard and Heyman treated all joints as pinned, ignoring in particular the very different actual situation at the complex and mostly indirect jointing of upper rafter, collar beam, main purlin, hammer post, and lower rafter. Pippard's preferred solution showed the crown post and arched ribs as the main load-bearing members but most other members (including upper and lower rafters) implausibly in tension. Heyman, making a different assumption about support, found the rafters and collar beam to be the main members, all acting in compression with the largest compression in the lower rafters. The arches then played only a secondary role. My analysis, recognizing the flexural continuity of the hammerbeams but the limited capacities of most joints and the consequent unlikelihood of the lower principal rafters playing a significant role in the primary action, reduced the problem to a statically determinate one. The arches now appeared to be the main members and the corbels at their feet the main supports. Calculated deflections confirmed the reasonableness of the omission of the lower principal rafters from the model. Mark, using finite elements, obtained broadly similar results except that a different modeling of the complex joint referred to above led to these lower principal rafters being assigned a greater role. More recently E. Toby Morris, R. Gary Black, and Stephen O. Tobriner have repeated the analysis for a different distribution of load from the roof, again with broadly similar results if the tracery is ignored.[10]

The similarity of the results of all the later analyses proves nothing, but suggests that the overall picture obtained from them was as good as can be hoped for. Though numerous more precise analyses could be made, their end product would not be a unique pattern of loads in the members. It would be a series of patterns corresponding to some of the innumerable small variations that could have occurred in the original seasoning, cutting, erection, and jointing of the timber.

Morris, Black, and Tobriner did indeed go further in exploring also the possible role of the tracery. Their analysis ignored, however, the way in which it was housed and the closeness of the initial fit. Their results could therefore do no more than illustrate the limits between which the contribution of the tracery might lie. If it was a fairly loose fit (as seems likely unless the tracery was intended to be load bearing), it would initially carry very little load.

Though Heyman insisted that his analysis of the roof most closely reflected its actual behavior, despite having unrealistically idealized it as a pin-jointed truss, he and Elizabeth Wade were rightly more cautious in their analysis of the three-dimensional timber roof of the Ely octagon.[11] Detailed inspection of this roof is now difficult. Since also it was twice subjected to substantial strengthening, there are some doubts about both the original form and that after the first modification. What (for good reason) were judged to be the principal members of the presumed original structure were first isolated. Simple statical analyses assuming pinned joints then showed how they could have carried the lantern but drew attention to an unexpected consequence of the geometry of the roof, namely, the slightly precarious situation that would have existed if they had been left to do so without the actual semi-stiff joints of the inner ring and support from secondary bracing elements. Heyman and Wade also noted the difference between the stability of the curved ribs of a masonry vault subjected to transverse load throughout their length and the weakness of similarly curved ribs of a timber vault subjected instead to a few isolated transverse loads.

The full analysis of the majority of timber-framed buildings would present greater problems than even the roof of the Ely octagon on account of the large numbers of bracing members or multiple interactions between frame and infillings. Thus, the firm of Harris and Sutherland found it impossible to identify the precise behavior of the wide-span timber roofs erected over English dockyard slipways in the early nineteenth century.[12]

Masonry Structures

Unlike timber, masonry (whether of brick or stone) is much weaker in tension than in compression, and even a single block of stone has only limited bending capacity. Masonry structures therefore stand largely by virtue of internal compressions which carry their weights and all other loads to the ground. The cracking that almost invariably occurs influences the paths taken by these compressions but does not usually threaten stability. Collapse may be precipitated only by excessive deformations (to which the cracking contributes but of which it is not the only cause) or by local compressive failures (strictly speaking, failures under the secondary orthogonal tension that always accompanies a primary compression).

Masonry structures differ from timber structures in other important ways. The tightness of initial fit that is important for timber is relevant only for any ties that are incorporated and for any additions or insertions of new masonry. The slow gain in strength and stiffness of any mortar does nevertheless lead to an equally if not more important counterpart, namely, a dependence of stresses and deformations on the rate at which and the manner in which superincumbent load increases during construction. They will be affected by both sequence of construction and the extent and duration of any temporary support that is provided. So will the extent of cracking. These effects are all the more important because self weight is usually the predominant load. Further changes will occur after completion of construction through shrinkage and creep of the mortar and the alternate opening and closing of cracks under the influence of changes in temperature and humidity. While some cracks may heal themselves through continued chemical action between the mortar and the atmosphere, most will slowly grow through the progressive jamming action of falling debris.

In addition, the cross sections of piers, walls, vaults, etc. are far from homogeneous, and relative settlements of the ground on which the structure stands are likely to have a more significant effect than for timber roofs on the ways in which loads are distributed between different supports. Lack of homogeneity considerably reduces strengths and stiffnesses. Even the best ashlar has much less strength than a monolith of the same stone because weaknesses are introduced by the vertical joints. Usually further reductions in strength and stiffness result from nonuniform bearing on the bed joints and the greater compressibility of any mortar used in them. A closely jointed face may hide much thicker joints behind it and/or a comparatively weak rubble fill. What appears to be a single mass may even consist of an original core with later unbonded additions [*Figure 4*]. Highly nonuniform internal stress distributions will then be accompanied by large reductions in overall strength and stiffness.[13] Much depends on bonding and on the nature of the mortar. With a good pozzolanic or similar mortar, considerable effective homogeneity may be achieved, giving to the core properties not far short of those of some reinforced concrete today, as is shown by the cohesion of some large fallen masses. But in most structures of post-Roman date this is the exception rather than the rule.

To make matters worse from the point of view of analysis, surface appearances may give little clue as to what lies behind. Detailed analysis of the main vaulting system of S. Lorenzo in Turin is, for this reason, virtually impossible. Elsewhere, probes into the mass or other local tests may give some clues as to what to expect but will often leave a considerable measure of doubt. Moreover, any structure of much interest will have had to stand in numerous incomplete forms while being built and on completion will have been capable of standing in many more ways than one until it is on the point of collapse. In the long run a progressive narrowing of the range of possibilities will be brought about by the growth of cracks, the opening of joints, deterioration of materials, and similar changes. As the narrowing takes place the distribution of stress and strain will continually change.

For all these reasons the behavior of historical masonry structures can never be fully knowable, whatever improvements there may be in analytical methods and computing power. It is indeterminate per se, irrespective of shortcomings in analytical procedures and computing power. On the other hand, present structural behavior is usually much more readily legible than that of many modern structures on account of the relative absence of hidden reinforcement and the prevalence of cracking.

Modeling and analysis of Masonry Structures for Static Loading

Two types of statically indeterminate model have nevertheless been tried in recent years: photoelastic models and numerical finite element models. Robert Mark applied the first extensively in the 1960s and 1970s.[14] I also must confess to having contemplated its use for similar studies of Gothic structures when a photoelastic bench was installed to explore the behavior of brick walls at the UK Building Research Station ten years earlier. However this type of model has insuperable drawbacks, which would remain even if we knew everything we wished to know about the physical characteristics of the structure being modeled. One is a serious lack of precision, even discounting

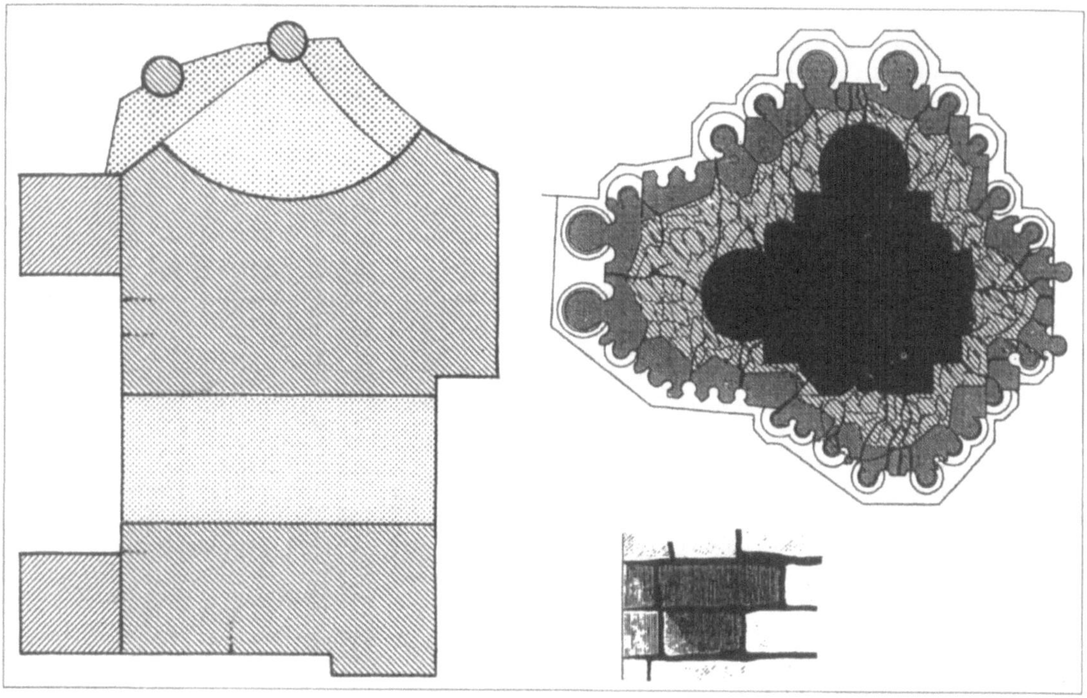

FIGURE 4: Details of masonry piers: Left (at half the scale of the other two), horizontal section of the southwest pier of Hagia Sophia at gallery level. Hatching shows two stages of the original construction and stippling successive additions. Major cracks on the only two surfaces where the masonry has been partly exposed are also shown. Upper right, horizontal section of the Romanesque core and fourteenth-century enlargement of the northwest crossing pier of Bayeux cathedral. Lower right, a partial sectional elevation of two courses of the original southwest crossing pier of the Panthéon, Paris, as observed in a test hole.

the errors introduced by the need to simulate widely distributed loads such as wind and self weight by pulls exerted at a few isolated points on a virtually weightless model. More fundamental are the neglect of all influences of the construction process on the development of stresses and deformations and the representation of the actual structure by an elastic continuum that is immune from cracking and from the large inelastic deformations that arise in practice. In dismissing the failure to allow for cracking on the basis of success in similarly modeling a reinforced concrete shell, Mark ignored the difference made by the reinforcement of the shell and a different construction procedure.

Today finite element models have superseded photoelastic models. They are most readily applicable when the notional finite elements are discrete elements of the real structure, such as the individual blocks or bricks and beds of mortar of which it is composed. But it is feasible to proceed in this way only for purposes of research—to investigate the deformation and cracking of the masonry in a wall, for instance.[15] As soon as the elements in the model represent arbitrary assemblages of the individual blocks or bricks, it becomes possible only to predict approximate deformations and averaged levels of stress and to locate regions in which cracking can be expected. Refinement of the mesh will improve computational accuracy for a given set of assumed properties but cannot compensate for a poor choice of elements. Incorporation from the outset of known or likely cracks and other discontinuities such as lacks of bond will help further to arrive at a realistic answer. But appreciable margins of possible error will remain because of the difficulty of assigning the mechanical properties correctly. A particular difficulty here is that of recognizing the weakening effect of the cracking that can occur under purely compressive applied load as a result of the secondary tensions, already referred to, that arise at right angles to a primary compression.

Domes and other vaults have hitherto received the fullest attention. Analyses by Rainer Barthel of groined vaults, for instance, have shown the influences of vault profiles and separation of the springing points on the flow of stress and have simulated the cracking that typically occurs.[16] Similar analyses could be made of ribbed vaults. But it is doubtful whether they would add much to what can be concluded from observation of actual vaults, from the simple fact that loads will always be attracted to the stiffer elements; and from the fairly obvious further dependence of the structural role of the ribs on the detailing of their junction with the severies, on the manner and sequence of construction, and on the rate at which it proceeds in relation to the stiffening of the mortar.

To analyze similarly a complete bay of a Gothic cathedral, or even an apparently simple complete domed structure like the Roman Pantheon, would call for far more elements and computational time and far more assumptions about unknown characteristics and details of construction to achieve anything like the same accuracy. Analyses based solely on a simplified structural geometry and assumed constitutive relations for the masonry can do no more than indicate one theoretically possible behavior strongly influenced by the simplifications and assumptions made. More often than not, in analyses undertaken hitherto, the behavior so indicated would, in practice, have been impossible on account of early cracking. Nor is there much value in attempting to determine equivalent constitutive relations by matching calculated deformations directly to estimates of the present total deformations. These will have accrued over long periods under changing conditions of load and stress, and they will be the combined manifestation of present elastic deformations, early plastic deformations, long-term yield, and relative displacements at cracks and other discontinuities. In Hagia Sophia, for instance, the principal present deformations are an order of magnitude greater than those to be expected for purely elastic behavior without cracking.

Statically determinate models avoid the need for a correct recognition of all the nonhomogeneities of the cross sections and changing characteristics of the materials. Three-dimensional as well as two-dimensional models are possible, though only for certain assemblages of linear elements and certain doubly curved forms that are free of bending and suitably supported [*Figure 5*].

Such statically determinate models do not give the unique answers given by finite element models. They accept the actual indeterminacies of the true behavior and its continual variations in that they indicate instead only the limits between which this behavior must always lie. However, they do so directly in terms of the possible paths of the principal internal actions (whether compressive or tensile) and allow direct calculations of the limiting magnitudes of these forces. This direct indication of load paths or thrust lines has the merit of being more closely akin than plots of stress magnitudes to what we intuitively sense if we place ourselves in the position of the structure. It can therefore help directly in developing intuitive understanding.

Heyman has published numerous such analyses, both two-dimensional and three-dimensional, for arches of different

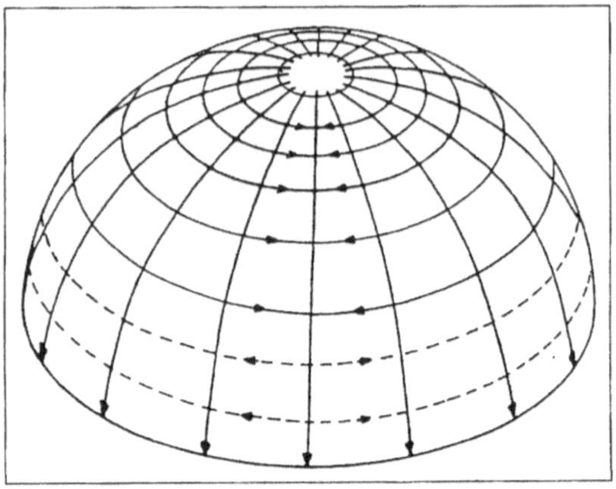

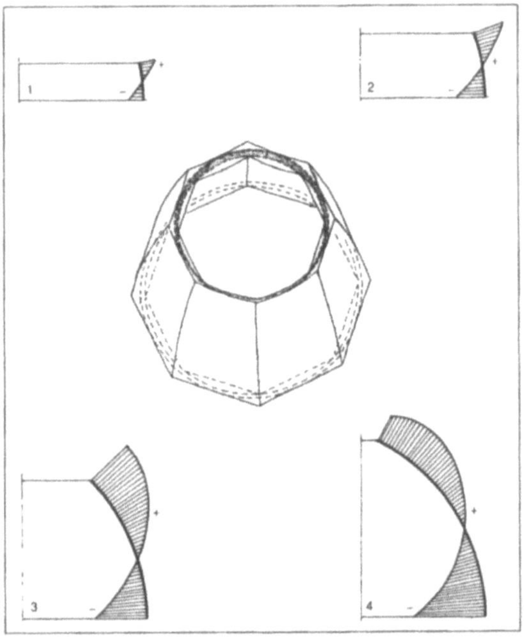

FIGURE 5: A statically determinate model for the structural behavior of a suitably supported hemispherical dome able to resist tension. The radial compressions are held in equilibrium by circumferential compressions in the upper part and circumferential tensions in the lower part. The behavior changes if there are tensile failures and the support allows the base to spread.

FIGURE 6: The development of circumferential stress in the inner dome of S. Maria del Fiore calculated for the circular dome contained within it as shown in the central sketch. The successive stages of construction to which the plotted stresses relate are (1) the change from stone to brick, (2) the start of the *spinapesce* brickwork, (3) the stage at which the stress in the top ring was greatest, and (4) completion to the base of the lantern. The stress always varies from tension (−) at the bottom to compression (+) at the top.

FIGURE 7: Thrust-line analyses of the nave of Strasbourg cathedral for self weight only (dashed line) and self weight plus wind (chain dotted line)

FIGURE 8: Transverse sections of a Romanesque church shown undeformed on the left and with typical deformations and cracks on the right

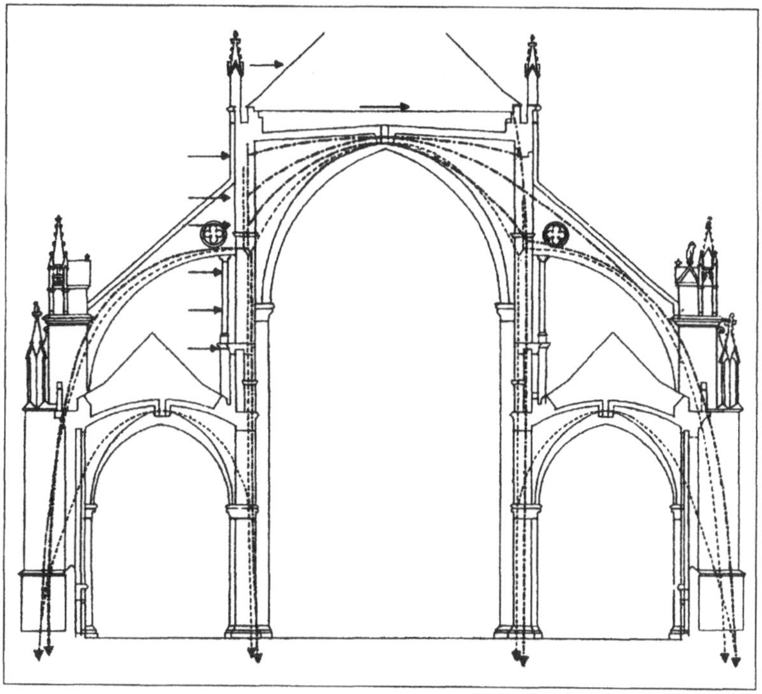

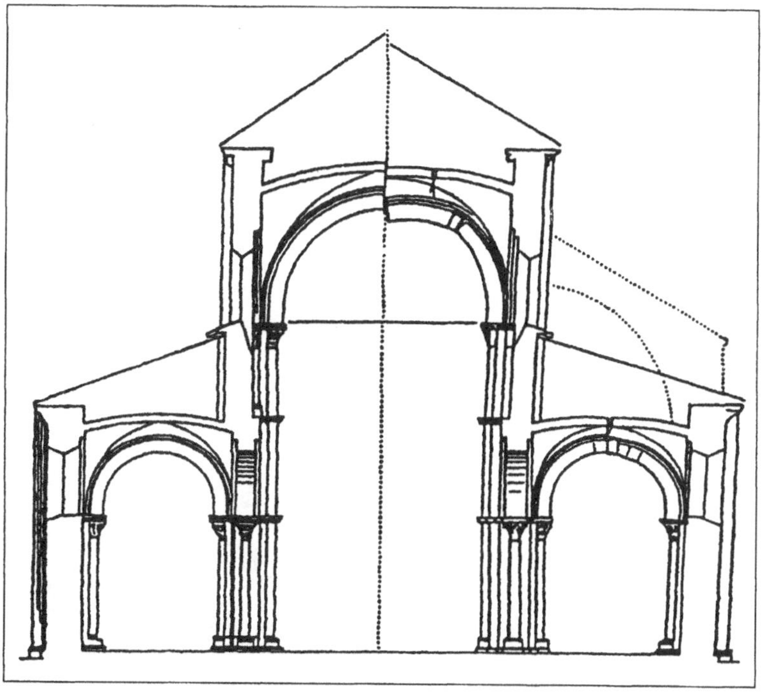

profiles and for flying buttresses, groined vaults, fan vaults, domes, semidomes, and spires.[17] From some of them he has deduced the minimum depths or thicknesses required for stability, assuming that the masonry is not overstressed and that no slips occur. All assumed modes of behavior should, however, be checked against what would actually be possible taking into account discontinuities in the masonry and the necessary compatibility of vault and support movements. His ingenious model for the semidome (seeing it as a set of independent adjacent arches all spanning parallel to the free edge) can pass this test rarely if at all.

Another example, an analysis of the dome of S. Maria del Fiore [*Figure 6*] took as the model the pointed circular dome contained within the thickness of the inner octagonal shell.[18] Its objective was to explore qualitatively the consequences of construction without centering. For this purpose the circumferential stresses were calculated at each level as construction proceeded, allowing tension at the lower levels to see what tensions would arise. Here the chosen model was justified by a number of unusual features of the actual construction that had been previously ascertained by detailed inspection.[19]

Analyses of complete structures have hitherto usually been limited to the plotting of possible thrust lines on undeformed representative two-dimensional sections of the completed structure [*Figure 7*], though they could equally well be profitably performed on deformed cross sections where these can be established and on cross sections of the incomplete structure during construction.

A further advantage of statically determinate models of existing structures is that the indications of possible behavior that are deduced from them can readily be supplemented and their ranges narrowed by direct observation of deformations and cracking [*Figure 8*]. Though the cracks will all be associated with tensile strains beyond the capacity of the masonry, they will be of two kinds. Separation cracks extend back into the mass with a fairly constant width and are usually the most conspicuous unless they have been filled. Since they occur roughly at right angles to the tensions responsible for them and since the tensions are at right angles to the principal compressions, they directly indicate the paths that these principal load-bearing stresses follow [*Figure 9*]. Cracks of the other kind are splayed to a maximum on one face but disappear on the other face, where there may be local splitting, crushing, or slippage. They are usually found in arches or vaults roughly at right angles to the curvature but can occur less noticeably in eccentrically loaded columns and slender piers. In both places they result from the tensions associated with bending and constitute effective hinges. Their value is to locate within

FIGURE 9: S. Spirito, Florence, with cracking slightly strengthened for greater legibility

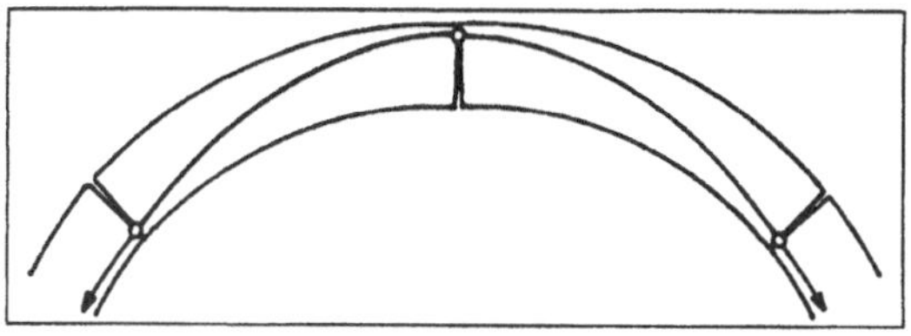

FIGURE 10: Typical hinging deformations of an arch and the corresponding thrust line

narrower limits the lines of action of the thrusts by showing roughly at what depth the compression is concentrated [*Figure 10*].

Of course, filling of cracks can hide and often has hidden or distorted some of the evidence. The cracks in S. Spirito were, for instance, hidden by fresh plaster and redecoration a few years after the photograph in Figure 9 was taken in 1972. When this has happened it may nevertheless be possible to deduce their presence from present deformations. Hinging in arches and piers is betrayed directly by marked discontinuities in curvature. Where accurate survey data or good face-on photographs are available it should therefore be possible to locate the approximate lines of action of the principal thrusts in arches and vaults almost as readily as if the splayed joints could be directly observed. The presence of separation cracks can be inferred where overall separations are greater than the masonry would have been able to accommodate without cracking. In studying Hagia Sophia, for instance, hinging regions in the main arches were identified from their profiles in order to calculate the thrusts exerted on the piers.[20] The overall pattern of cracking in the high vaults [*Figure 11*] was initially predicted from separations and deformations deduced from survey data. The presence of cracks similar to those predicted was then confirmed by limited probes and has been further confirmed by more recent exposures. A very similar pattern was also laid bare in Hagios Panteleimon in Thessaloniki.[21]

Behavior in Earthquakes

In the analyses so far described it has been tacitly assumed that all loads were static. The usual predominant load is self weight, which is truly static, and the assumption is not unreasonable for wind load, which is usually relatively unimportant for historic masonry structures. For the loads due to earthquakes it is less reasonable.

These loads are highly dynamic, and both they and the structural response to them are dependent in complex ways on characteristics of the structure that are not relevant to static loading. Because the response is highly nonlinear except at very small amplitudes, a realistic analysis would also have to follow the precise ground movements step by step. In masonry structures it will, moreover, be partly or even largely a rocking response rather than the flexural one that is characteristic of most modern structures. Since rocking frequencies fall as amplitudes increase, this has the beneficial result that damaging resonance with a predominant frequency of the ground movement is unlikely. But it further complicates the analysis by making it even more necessary to include in the model all major cracks and hinge-like articulations. For arched and vaulted structures there will also be a greater need than for most modern structures to take into account vertical accelerations because they will alternately increase and reduce horizontal thrusts. Yet there is at present no proven calculation procedure for anything more complex than the freestanding column.

It has been suggested that a variant of finite element analysis known as distinct element analysis might be used.[22] In this, all elements are actual blocks of the structure bounded by planes of weakness. They are taken as being completely rigid and incapable of cracking but able to move relative to one another under prescribed conditions. It is therefore possible to follow the rigid-body dynamic responses as the shock proceeds. But it is not possible to take into account cracks that develop during it unless they have been foreseen and built into the model in advance as planes of potential relative displacement. No guidance is given for this difficult task.

In the face of this and other difficulties it seems necessary to fall back on the qualitative insights available from basic dynamic principles and from studies of types of behavior characteristic of particular types of structure during and after earthquakes [*Figures 12, 13*].[23]

HISTORICAL INQUIRIES

When we turn to specifically historical inquiry, five main topics call for attention: (1) the indentification of sequences of construction and (where they have occurred) partial reconstructions; (2) the identification of construction procedures; (3) the identification of reasons for partial collapses; (4) the reconstruction of structures now lost and known only through records and/or partial remains; and (5) the reconstruction of the ideas and methods of past designers. Though they will be discussed here in turn, some of them must in practice be considered together.

Structural analysis is never adequate by itself to answer the questions that arise. It is necessary to draw on the historian's usual literary and graphic source material and the evidence of architectural character and detailing. Indeed these must often be the principal sources. But structural insights and calculations can help, just as sequence and manner of construction are relevant to the proper elucidation of structural behavior of most masonry structures. Even the literary and graphic evidence usually call for interpretation to which structural and similar knowledge and insights can contribute by distinguishing between what is reasonable and what is unreasonable.

Establishing Sequences of Construction and Identifying Additions and Partial Reconstructions

Structures like the typical Gothic cathedral whose main structural fabric is largely exposed to view reveal their sequence of construction most readily. No structural expertise is called for to see the significance of changes in moldings or the way in which the masonry is dressed or to spot and interpret clues in the ways in which elements impinge upon one another. This calls more for a keen eye and powers of three-dimensional visualization of the construction process.[24]

It is chiefly where such evidence is hidden that it is necessary to turn again to the evidence provided by the present deformations, looking now for inconsistencies that betray additions made only when some deformation had already occurred. Sometimes these are obvious and their significance is immediately apparent, as in the case of the flying buttress shown dotted in Viollet-le-Duc's section of a typical Romanesque nave (Figure 8). It exhibits none of the distortion suffered by the adjacent pier carrying the high vault. Structural expertise is called for when the telltale inconsistencies and their significance are less obvious. Significant deformations must first be deduced from the present geometry on the basis of an assumed original setting out. They must then be correctly interpreted on the basis of the changing forces that would have been acting and the types of deformation to which they would have led. A good qualitative appreciation of these is called for rather than any calculation. Finally, a time scale must be established from documentary or epigraphic evidence or the *termini ante* or *post quem* given by evidence of other kinds.

The first part of this procedure calls for strict discipline, checking both for three-dimensional consistency and consistency with what is known of the practicalities of setting out at the time. Where deformations are small it also calls for survey data of a precision that is only now becoming more readily attainable through the use of photogrammetry. Even for Hagia Sophia it was necessary to supplement the extensive data that were already available by additional measurements made specifically for this purpose. Incompatible deformations then both clarified the initial sequence of construction and were the initial basis for a more conclusive identification of a complete reconstruction of the tympana and supporting gallery arcades [*Figure 14*].[25] Though I have not investigated it in the same detail, it seems virtually certain on similar grounds that the present dome of SS. Sergius and Bacchus is not the original one. There has been considerable spreading at its springing level and there is apparently none of the cracking that would be expected to result from this spreading if it had been constructed as soon as the gallery was completed. Elucidation of the history of construction of S. Marco has, however, been more difficult because the recent survey is less comprehensive than that of Hagia Sophia and indicates less precision in the initial setting out.[26]

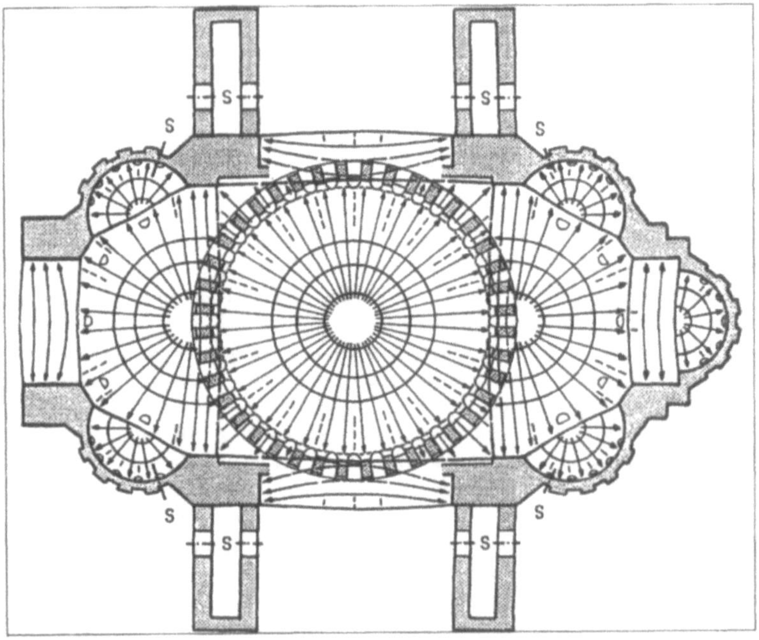

FIGURE 11: Thrust lines (arrowed) in the main vaulting system of Hagia Sophia and the general pattern of associated cracking (broken line). There are major shear failures in the positions marked S.

FIGURE 12: Earthquake damage to the Orphanage, Angra do Heroismo, Azores, in 1980

Establishing Construction Procedures

The aspect of the construction procedure with major structural implications is the need to maintain stability at all times in the as yet incomplete structure. This must always have been one of the main preoccupations, if not the main one, of those in control. Among the questions we may ask about it are the following: Was centering used as temporary support for arches and vaults and if so how was it supported? If not, how was it possible to dispense with it? Was other temporary support provided in the form of struts or ties that were later removed? And were parts of the structure sized primarily to withstand the forces exerted at some critical intermediate stage during construction rather than those to be expected on completion—as commonly happened when bridge piers were sized to resist the unbalanced thrust of an arch on one side only? In answering, structural expertise is valuable chiefly in indicating what would or would not be possible, taking as a starting point whatever more direct evidence there is.

A vertical wall or pier clearly needs no temporary support unless it is unusually tall and slender or is subjected at some stage to an unbalanced force from one side such as that exerted on a nave column by an aisle vault. On the other hand simple statics indicate that an incomplete freestanding arch or arched rib would become unstable without supporting centering once construction proceeded somewhat above a level corresponding to that of the Gothic *tas-de-charge*. For a dome or domical vault the situation is less clear cut. If it is circular in plan and has no openings that are not effectively bridged, the closing circular ring at any intermediate stage of construction could, if able to withstand the compression, serve in lieu of a keystone to the ring of incomplete radial arches and thereby maintain stability. Problems might nevertheless arise in completing successive rings as their inward inclination becomes more nearly horizontal. And there is no similar guarantee of stability without extraneous support for a noncircular dome.

Where statical considerations show that centering or other temporary support would have been necessary, evidence of it sometimes survives in the form, for instance, of seatings for the centering or the sawn-off stubs of temporary tie rods. To build upon this evidence or to proceed in its absence one must place oneself in the position of the designer/builder and explore the possibilities as John Fitchen did in his classic study of Gothic cathedrals, always bearing the structural implications in mind and testing their feasibility where necessary by calculation.[27] Aspects commonly neglected are the facts that timber centering is never completely rigid and that it will never relieve the piers destined to support an arch or vault of all thrust during construction. Indeed there will often be virtually no relief from flying centering, itself carried by the piers.

Where statical considerations show it might have been possible to proceed without centering and there is no reliable documentary or graphic evidence to confirm this or show how it was done, it is necessary to seek clues in the overall form, in the setting of the masonry, or in such evidence as vestiges of supports for working platforms. In a circular dome the adoption of a pointed profile increases the likelihood that centering was not used because it reduces the maximum inward inclination of the topmost ring. So does the leaving of an open central eye. Certain techniques of setting the bricks that are often seen in Byzantine and Islamic domes also increase the likelihood because they also would have eased the construction of successive rings. Most of these clues—plus the further one of the use of quick-setting gypsum mortar—are to be found in the dome of the Mausoleum of Oljeitu at Sultaniya.[28] The procedures adopted for the domes of the Pantheon and Hagia Sophia are less clear. Some centering must have been used for both. But it is statically feasible that, in place of complete centers spanning the whole voids and erected before construction began, only lighter circumferential centers were used, erected in stages as the work advanced. They could have been strutted back to the feet of the coffers in the Pantheon and tied back through the circular openings left at regular intervals in Hagia Sophia.[29]

Even in the rare instances in which documentary evidence is available, it can call for interpretation and corroboration by physical evidence or calculation. The unusually fully documented construction without centering of the dome of Florence cathedral will serve as an example here.[30] Though this dome does have a pointed profile and a central eye (capped later by a lantern), it is not circular but octagonal in plan, and its unprecedented maximum diameter meant that there would have been much higher stresses than in most earlier domes and correspondingly greater potential problems in completing successive rings.

The elucidation of Brunelleschi's achievement began with the observation that the inner shell was thick enough at all levels to contain within its thickness a circular ring, as Alberti seems to have known. It remained necessary to confirm that there was a truly circular ring at all levels rather than one that rose toward the corners of the octagon as it would have done if it had been part of an inclined flat bed behind each inner face. It was also necessary to establish that it would have been feasible to construct the ring unsupported at all levels and that there would be enough strength at each level to resist the circumferential compression that would arise as estimated from the analysis referred to earlier (Figure 6). Close examination of the brickwork where it is visible in the internal walkways showed that it was indeed set in inverted conical rather than pyramidal beds, thereby meeting the first condition. To meet the other requirements, these beds were bonded to one another by bricks inserted vertically at intervals to spiral upward [*Figure 15*] and construction was spread over fourteen years. The slow progress allowed appreciable time for the mortar to gain strength, and the bonding meant that the necessary strength could be developed by several courses acting together. The slow progress and substantial support from the drum below also meant that the tension that developed later near the base (Figure 6 again) could have been resisted for long enough to avoid marked increases in the stresses at the top.

Diagnosing Reasons for Collapses

Collapses no doubt occurred during critical phases of construction at least as frequently in the past as they do today. But they have, with rare exceptions, gone unrecorded. This means that

FIGURE 13: Earthquake damage to the cathedral, Venzone, Italy, in 1976

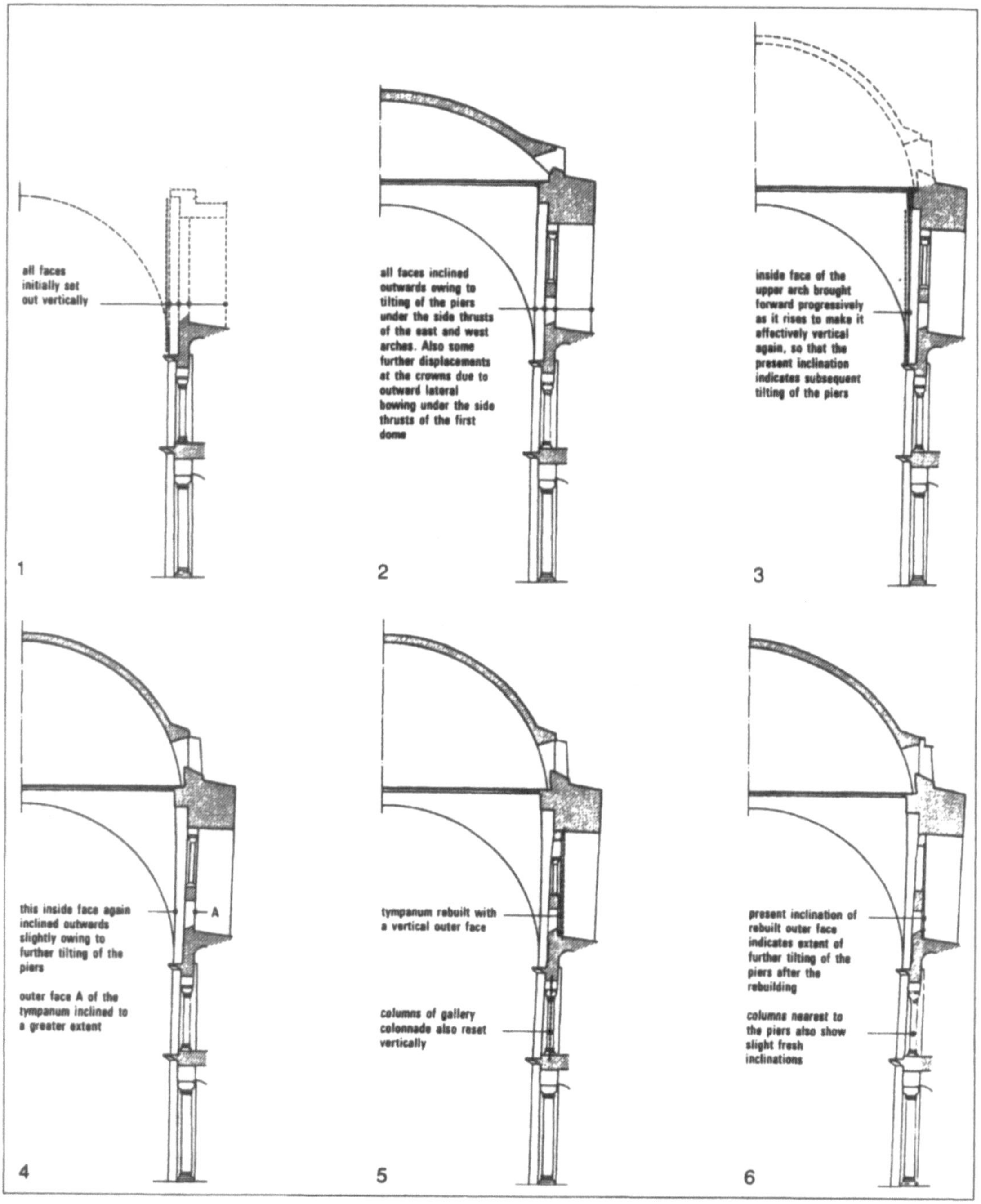

FIGURE 14: Partial sections of Hagia Sophia looking east on its transverse centerline: (1) at the start of construction of the main arches; (2) in 557 shortly before the first partial collapse; (3) in about 560 after preparatory work for rebuilding the dome; (4) a few centuries later; (5) after reconstruction of the tympana and the gallery colonnades below them; and (6) today.

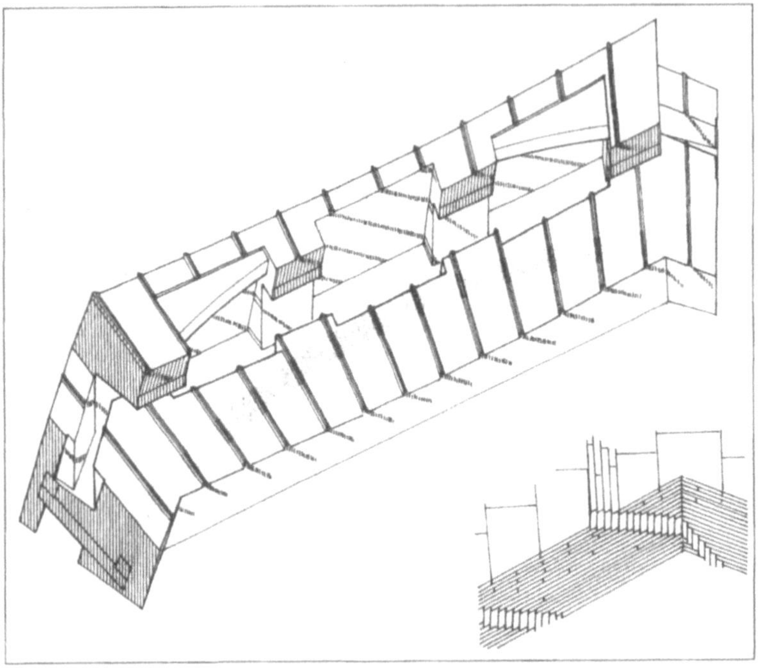

FIGURE 15: Axonometric view of the construction of the dome of S. Maria del Fiore at a level slightly above the second internal walkway with an enlarged detail of the *spinapesce* bond of the brickwork. The circular dome within the thickness of the inner dome is accentuated by a shaded band. One side of the thinner outer dome is also shown, but to a greater height (as it would have appeared at a later stage of construction) in order to show the radial and circumferential ribs that assisted in supporting it during construction.

it is chiefly partial collapses of structures that had already stood for some years that merit detailed inquiry. Moreover, most of the records relate to vaulted masonry structures, so I concentrate on structures of this kind.

The first difficulties in making a diagnosis are likely to be the loss of all direct evidence of what happened and uncertainties about the precise character of the structure that suffered the collapse. For the latter reason any calculation will be even more prone to error than if the structure were still standing. The accuracy needed to establish, simply by calculation and with reasonable confidence, the weaknesses that led to the collapse is therefore unattainable. The fact that the structure was completed and stood for a time thereafter is, of course, incontrovertible proof of its initial stability, at least under self-weight load and normal wind load. It is necessary to ask how it could have deteriorated until some critical limit was reached.

Fortunately other clues are available in the behavior of comparable structures that have gone only part way along the same road. We invariably see several clues if we look closely enough. Principal supports are bowed or inclined outward by the thrusts of the vaults, the crowns of the vaults are correspondingly depressed, and there are hinging deformations and extensive cracking of the kinds seen in Figures 1 and 8. Further progress of the deformations and hinging can lead directly to collapse of the vault with or without part of the supports. Or the concentrated compression near extrados or intrados at one or more "hinges" can precipitate it by leading to local splitting or crushing followed by slips [*Figure 16*].[31]

The best-documented partial collapses of this kind are those of the main eastern and western arches in Hagia Sophia. Each time, the central part of the arch fell and took with it the part of the dome that it supported and part of the adjacent semidome. But enough of the structure survived to permit fairly full reconstructions of the original forms and of the sequences of events leading up to the collapses. By establishing the progress of deformation of the supporting piers it could be

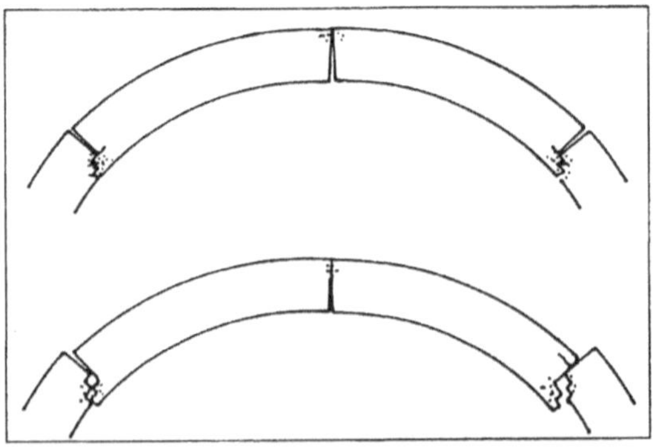

FIGURE 16: Typical later stages in the collapse of a masonry arch

FIGURE 17: A sectional elevation of the collapsed crossing piers of Chichester cathedral looking south, showing also (at the right) the splitting of the masonry skin of the southwestern pier that preceded its collapse

shown that it was their excessive north-south spreading at the springing level that was chiefly to blame (Figure 14). It was then possible to identify the features of the original design and construction procedure that must have been largely responsible for these large deformations and even to reproduce the main mechanisms on a smaller scale in the laboratory. Earthquakes could have played only a relatively minor part. Observed recent responses of similar smaller structures showed that they could nevertheless have widened cracks and facilitated local slips in the crown regions of the arches.

Another possible reason for collapes is effectively a crushing failure of a main supporting pier (effectively because it is in fact a buckling instability precipitated by splitting of the masonry). It seems, on available evidence, to have occurred usually after a longish interval and some deterioration of the masonry. The best-documented instance is the collapse of the central tower of Chichester cathedral.[32] The primary weakness here was the highly nonhomogeneous masonry of the supporting piers. Parts of the ashlar skin would from the start have been subject to compressive stresses far above the average chiefly for this reason, but also on account of side loads imposed by abutting arches. These stresses were increased over the centuries by weakening of the rubble core and some cutting away of the skin. Vertical cracks appeared because of the associated orthogonal tensions [*Figure 17*]. These eventually allowed the skin to crush and buckle in the weakest regions, leading to catastrophic overall collapse. The collapses of the campanile in Venice and more recently of another in

Pavia seem to have happened in broadly similar ways, the local concentrations of stress there being partly due to openings within the thicknesses of some walls.[33]

These are probably the most characteristic modes for masonry structures. A third possible mode is progressive tilting due to differential settlement at foundation level leading to rigid-body instability without prior failure of the structure itself. But this is likely only for freestanding columns or towers.

In diagnosing collapses for which there is sparser evidence it is necessary both to bear these possible modes in mind and to remember that each is merely the limit of a type of behavior exhibited to some degree by all buildings in which such a collapse would be possible. Thrusting arches and vaults have always caused their supports to spread at the springing level. In the taller French Gothic cathedrals typical estimated spreads are 0.5 to 1.0 meter [*Figure 18*].[34] Associated with these spreads are pronounced bends and tilts of the piers carrying the main vaults. The tilts make the compressions in the piers more eccentric than they would otherwise be and increase the maximum local compressive stresses. Whether or not the spreads allow vaults to fall or piers to collapse through local buckling will depend on precise details of construction and particularly of the presence of local weaknesses resulting from poor initial design, poor construction, changes in design during construction, subsequent changes, or some other cause such as a fire. Diagnosis should therefore begin with a careful study of the present fabric to establish as far as possible its history and what it was like before the collapse and to identify possible weaknesses.

Alongside those of Hagia Sophia the collapses at Beauvais have hitherto received most attention. There is little doubt that the second collapse—that of the sixteenth-century tower and spire and the western crossing piers—was caused by excessive additional weight on the piers when these were also bowed westward by unbalanced thrusts from the east. But the situation that led to the first collapse is less clearly apparent. For well over a century commentators have suggested a variety of local weaknesses that could have led to local failures. Viollet-le-Duc assumed that there were pairs of monolithic shafts attached as today to the exterior of the piers at clerestory level to assist in carrying the high vaults. He then suggested that shortening of these piers in the decade after completion of construction threw more load on the shafts than they could bear and thereby precipitated the collapse.[35] Mark reasonably tested this hypothesis by a simple calculation of the possible shortening which, on the assumptions made, showed that it was unlikely to be sufficient.[36] But indications of tension under wind in his photoelastic model were far from adequate to demonstrate an alternative cause. Only by a careful thorough study of the present fabric and the changes in plan during construction did Stephen Murray arrive at a more definitive identification of the critical weakness.[37]

In general a similar approach seems best, drawing on all available evidence and guided by personal observation and understanding of the behavior of similar structures in different states of well-being or distress. Calculation cannot take its place if it is based on highly idealized models of the whole structure. It is better reserved for testing likely weaknesses once these have been identified.

Reconstructing Lost Structures

Where the whole structure has collapsed or where little now remains on account of deliberate demolition, robbing, or lack of maintenance, a paper reconstruction may be attempted, even if it has to remain somewhat speculative. Such a reconstruction has more in common with designing a new structure than what has been considered hitherto. It calls for a similarly wide range of skill and expertise, both architectural and structural. It nevertheless differs in important ways. Materials, techniques, and precedents will all have been different from those now available, and the designer will have visualized

FIGURE 18: Amiens cathedral. W. H. Goodyear estimated the maximum spread in the nave at the springing level of the main vaults as about 1.5 feet (exchange with John Bilson in *Journal of the Royal Institute of British Architects*, 3rd series, 15 [1907]: 45).

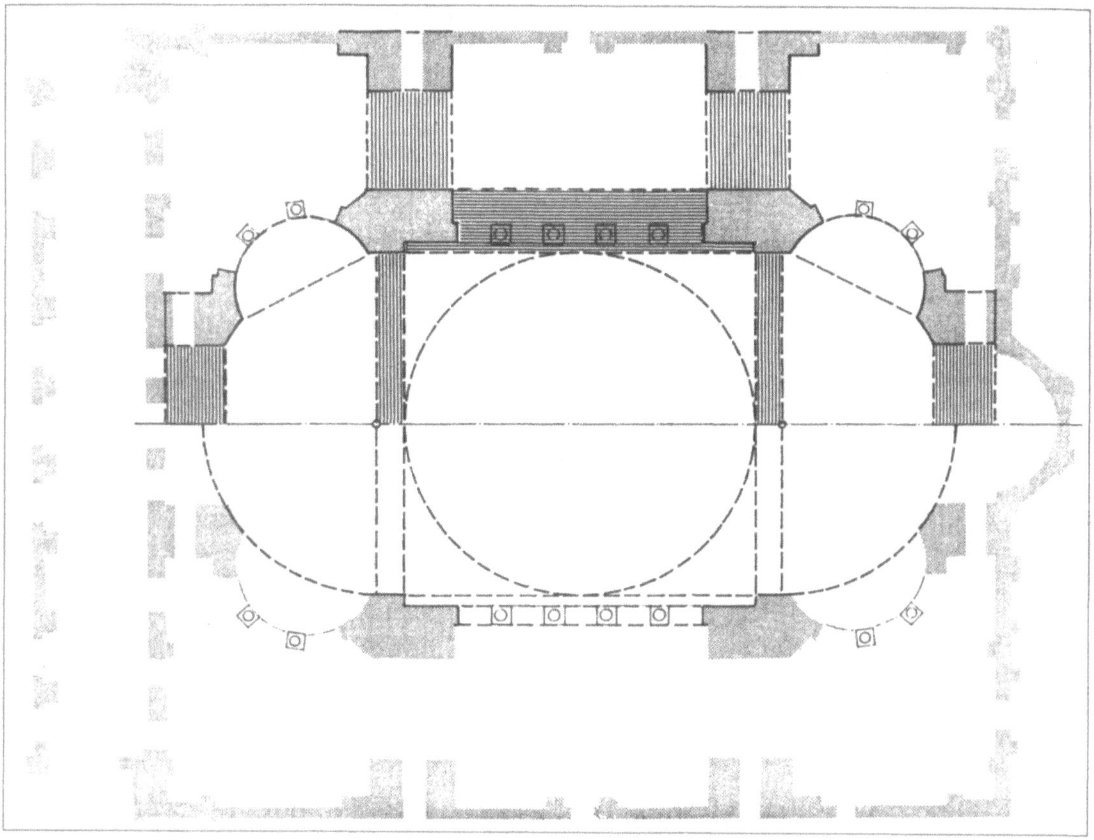

FIGURE 19: The ground plan of Hagia Sophia showing, in dashed line, the only vaulting scheme for which it would have been appropriate. Arches are shown hatched.

feasibility differently. Also there is none of the freedom of choice that exists with a new structure. The only correct design is the one that was chosen previously.

The available evidence varies greatly. Physical remains range from structures still surviving to a considerable height as they were first built to partial foundations disclosed by excavation. Drawings and documents sometimes supplement them and are sometimes the only evidence. Either way they will likewise call for interpretation. Early terminology is often imprecise, especially when it attempts to describe new forms or express new ideas. Drawings present other problems when they were made before our present conventions of orthographic, perspective, and other three-dimensional representation became the norm or when seemingly straightforward representations fail to show important details or to incorporate changes introduced after the drawing was made.

A good understanding of the relevant determinants of structural feasibility and a good knowledge of possibilities and practices of the time are the best starting point. Given this, vaulting or other roofing intentions can often be deduced from no more than a full plan at ground level. Thus the intended superstructure of Hagia Sophia cannot have differed much from that actually built: the unique and distinctive shapes and layout of the piers could have no other structural raison d'être [*Figure 19*]. The ground plan of the fourth-century S. Lorenzo in Milan is less conclusively revealing; but it would be inappropriate for a dome over the central space and so much more appropriate for a groined vault that there can be little doubt that the latter was intended. Equally, the slenderness of their walls and colonnades shows that most early basilican churches could only have had timber roofs, trussed over the larger spans. But some ground plans, like that of the Zwartnots church, disclose less, and what survives of the large podium of Hagios Polyeuctos does not even furnish one. Its broad inner longitudinal walls must have carried the vaulted floor of the nave but could have carried almost any aisled superstructure.

The sine qua non of any reconstruction is an adequate

three-dimensional visualization. Too many proposed reconstructions fall short even here.[38] Given an adequately visualized design that would be feasible in other respects, some structural calculation may then provide useful checks. However, in view of the limitations mentioned above, quantitative assessments of overall safety are of far more questionable value. They are unlikely to be more helpful than intuitive qualitative appraisals and comparisons with surviving structures of similar date. Mostly these were greatly overdesigned, as is shown by, inter alia, their ability to survive centuries of neglect, mutilation, or added burdens. For a masonry structure it may at most be worth estimating certain thrusts to judge whether it would have been feasible at the time to meet them by tying. For a timber truss it may similarly be worth estimating tensions or bending moments in critical members. Simple statically determinate models should usually suffice for this.

I have recently discussed elsewhere the interpretation of the documentary evidence for the lost first dome of Hagia Sophia.[39] As an example of the relevance of structural insights to the interpretation of graphic evidence I therefore turn here to Andrea Palladio's Cismone Bridge—a structure of considerable interest in relation to the development of timber trusses.

This bridge is described and illustrated in both elevation and plan in Book III of the *Quattro Libri* [*Figure 20*]. It is related to a type of trussed beam that was sketched earlier by Taccola and Francesco di Giorgio and depicted as a bridge somewhat later by Verantius [*Figure 21*].[40] Unlike the other timber bridges that are shown immediately after it, it is known to have been built and to have stood for over fifty years. So we know that the final design was a successful one. Yet a structural appraisal seems desirable because neither its detailing nor its mode of action are clear from the woodcuts and accompanying text.

Taking these at their face value, the deck was carried by transverse beams (lettered *I* in the plan and *C* in the elevation) that were suspended at each end from vertical ties. These ties were part of an assembly of substantial timbers of truss-like appearance on each side of the bridge. But not all the joints depicted would be capable of transferring loads as they would be required to do in a true truss. If they are correctly shown, there were really two largely independent systems. The first consisted of the top and bottom chords *GAG* and *DD* and the two longest vertical ties *EE*. The second system, set within it, consisted of the three pairs of shorter inclined struts and the three shorter vertical ties plus the constituent lengths of the

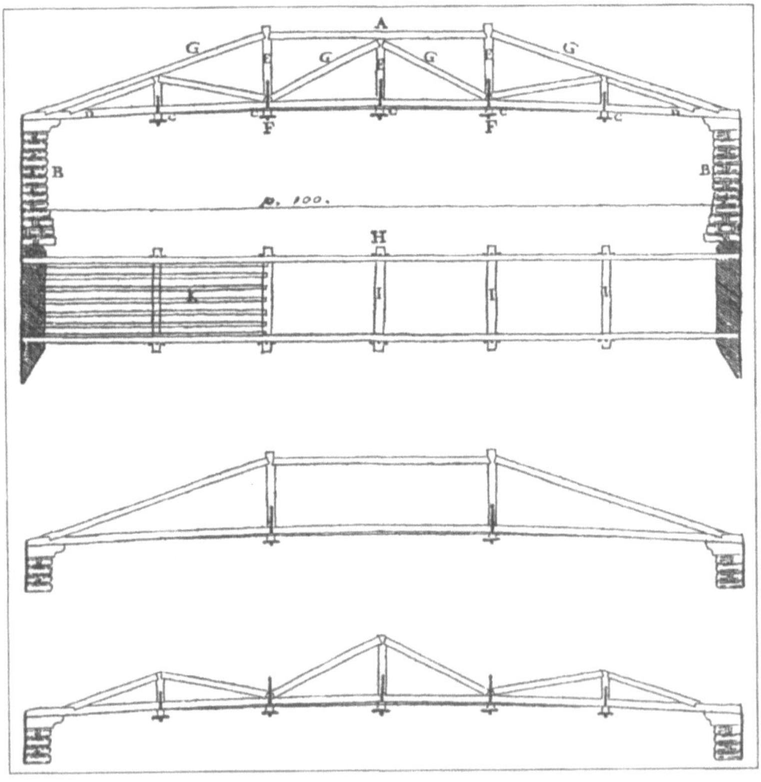

FIGURE 20: Palladio's Cismone Bridge, at the top, as shown in the *Quattro Libri*, and, below this, the two separate systems of which the side "trusses" seem to be composed

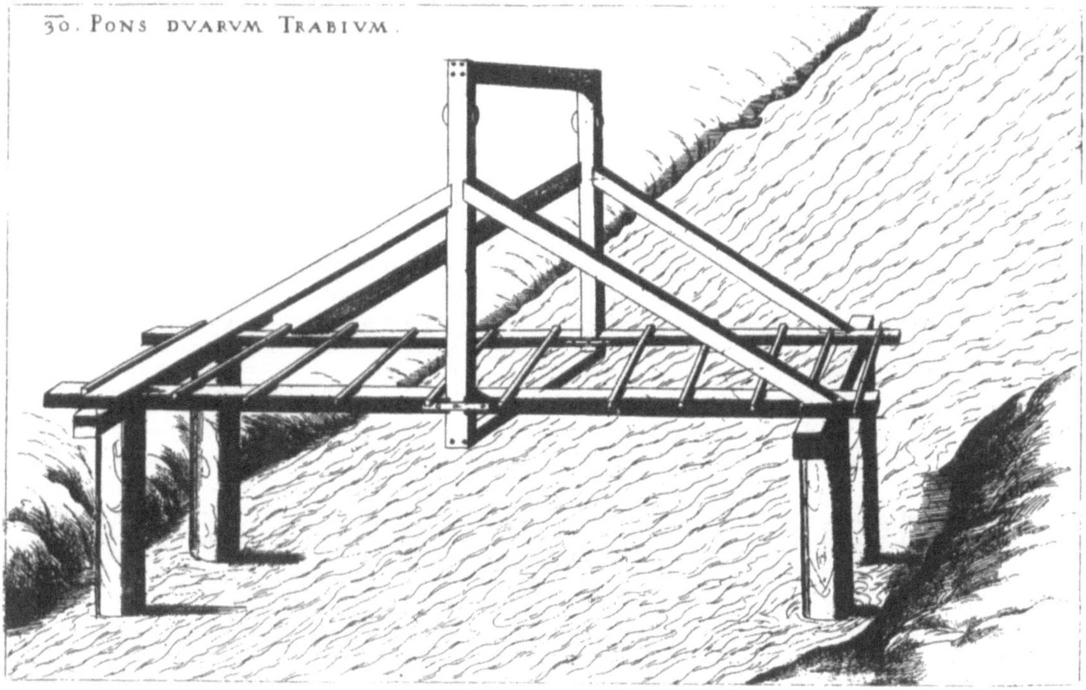

FIGURE 21: A simple trussed-beam bridge depicted by Faustus Verantius

same bottom chord. The second system would have passed half the loads supported by the first and fifth deck beams directly to the abutments and all other loads to the longest vertical ties. The loads passed on in the latter way, together with those supported by the remaining deck beams, would have been carried ultimately by the top chord *GAG* in compression and the whole length of the bottom chord *DD* in tension.

So far, so good. But two questions arise. If this system was to work at all there must have been a continuous transmission of tension from end to end of the bottom chord. Rough calculation shows that the indicated cross section would have been ample if it was a single timber. But could such timbers have been found? The length, if correctly given, was certainly near the limit.[41] Thus it seems more likely that separate shorter lengths with scarfed joints would have had to be used. This would have considerably reduced the safety margin on the strength of the timber at the reduced cross sections at the joints. The second question relates to bracing. True truss behavior would have called for full interconnections between the secondary systems and the principal one at the tops of the three shorter vertical ties. If the two systems merely abutted one another here, there would have been a tendency for at least the central secondary system to move slightly in relation to the principal one whenever the load on the bridge changed. This would have made the bracing of the central bay far from satisfactory. It therefore seems likely that there were some changes either at the time of construction or not much later.[42] Structural understanding and calculation can take us this far. But they cannot take us further on present data.

Reconstructing the Ideas and Methods of Past Designers

The ambiguities in the elevation of the bridge suggest that Palladio may have been unclear in his own mind about how the main framing was intended to work. However, entering the mind of the designer centuries ago is even more hazardous than reconstructing a design. To do it with any assurance of truthfulness calls for more than drawings and more than the expertise of today's structural engineer. Certainly modern calculation can throw no direct light on it.

The general pattern in the past was a slow development of particular forms over long periods, punctuated much more rarely than today by major innovation. Then as now, economy must have been a desirable objective. But because construction procedures were mostly so different this would have called for different choices and would have been seen much less in terms of what we might now loosely call structural efficiency—that is, in terms of minimum use of material, uniform factors of safety, and the like. Antonio Gaudí's funicular designs for the chapel of the Colonia Güell and the Sagrada Familia or Julien Guadet's redesign of the flying buttresses of St. Ouen would have been less economical in the thirteenth century than what was actually built, quite apart from being at odds with the aesthetic attitude of the time.[43]

Surviving early treatises, notebooks, drawings, and records of discussions between those involved do provide both a starting point for the inquiry and a salutary warning to the analytically minded engineer not to attribute to his forerunners twentieth-century objectives, ideas, insights, and procedures. Their chief characteristic is an emphasis on geometry, whether the mystical geometry of the Timaeus; the simple practical geometry of the Roman land surveyors, Villard de Honnecourt [*Figure 22*], and late medieval manuals; or the complex stereometry of the sixteenth, seventeenth, and eighteenth centuries.[44] Even Anthemius and Isidorus were primarily geometers.

This geometry had an obvious use in developing a design and conveying to others what was intended and again in stonecutting and setting out on site. It also provided one means for codifying proportions that experience had shown to be adequate. Such codifications (some of which still survive as "rules of thumb") were more helpful than they would be now because, with self weight as the predominant load and a more restricted choice and use of materials within a particular tradition of building, it was then chiefly geometry that determined whether a structure would be stable and otherwise adequate.

Nevertheless a little thought soon discloses the limitations of a purely geometric discipline. Rules of thumb apart, there was an almost unlimited number of ways in which the basic discipline could be followed, and it provided no basis for choosing between the safe and the unsafe. The arguments about how to proceed with the building of Milan cathedral at the end of the fourteenth century are but one illustration.[45] Particularly when attempting anything new, the builders had to make choices for which the discipline gave no guidance.

To gain a deeper insight into how these choices were made we must ask what other understanding of a more intuitive nature earlier designers could have possessed.[46] It can have come only from personal experience, chiefly that gained by observing how actual structures behaved. This would have been partly observation of the behavior of structures already standing. But more directly relevant would have been the day-by-day observation, as construction proceeded, of structures for whose design they were themselves responsible.

As stone was placed on stone or the members of a large timber frame were assembled, the immediate overriding concern would (as already noted) have been maintenance of the stability of the still incomplete structure. Without our analytical tools to assess in advance any proposed departures from precedent, the test would always be to go ahead and observe what happened. As was stated in the specification of 1420 for building the dome of S. Maria del Fiore, "in building it is practical experience that teaches how to proceed."[47] Any movements that appeared to threaten stability would not only call for measures to halt them to allow work to continue: they would also give some indication of the forces acting and the kind of response called for. As happened more than once in the construction of Hagia Sophia, continued stability was seen to be threatened. Weak vaults were then underpinned, piers were strengthened, and buttressing was improved.[48] In such ways understanding would grow and would help in guiding future design.

I therefore see the chief contribution of present structural expertise to understanding how designers worked in the past as being the help it can give in establishing the situations in which they would have found themselves and then in visualizing what they would then have seen and what they could have learned from it. We are thus brought back to the elucidation of structural behavior, construction procedures, and sequences.

CALCULATION AND INTUITION

It should be clear from the earlier part of this paper that the seeming precision of the structural behavior calculated by the latest analytical tools is somewhat illusory. This is often true even for the structures for which these tools and the underly-

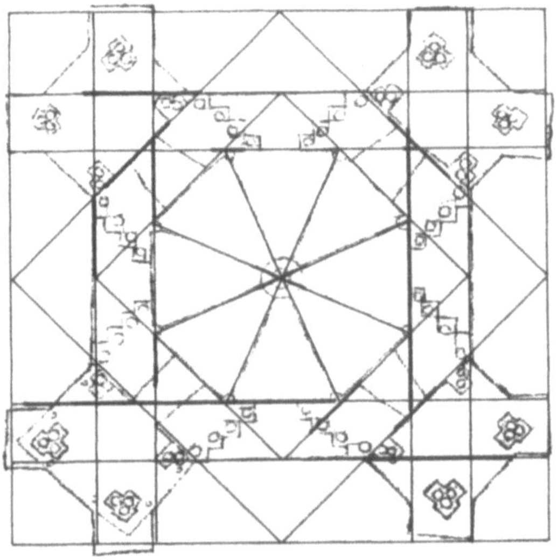

FIGURE 22: Villard de Honnecourt's plan of one of the towers of Laon cathedral with the scribed geometric setting out added. Heavier lines denote scribed depressions in the vellum that can still be observed in a glancing light. Lighter lines complete the pattern. Several different levels of the tower are superimposed.

ing theory were primarily developed. It is much truer for most historic structures. And although, in the design of a new structure, allowances can be made for inaccuracies by appropriate factors of safety, such allowances are irrelevant to the understanding of the behavior of a historic one. The wide ranges of possible error that stem unavoidably from the many unknowns must simply be acknowledged and accepted.

If calculations are undertaken, the choice of the model is paramount if they are not to mislead. The speed and ease with which multiple calculations can now be made does not diminish the importance of that choice. They may do little more than make explicit a particular range of uncertainty and will be worthless if the model is not adequately representative. I cannot, therefore, accept the view that the easy accessibility of finite element analysis now enables architectural historians to use it profitably as an entry point to the understanding of historical structures. Prior personal understanding is called for in many ways: to ask appropriate questions, to see where calculation may assist in answering them, to choose suitable and suitably detailed models on which to base it, and to assess the validity and significance of the results.

Alongside the predictions of analysis we have the evidence that a standing structure provides of its own behavior. To read it fully likewise calls for at least some prior personal understanding. But, properly read, it also offers the best way to develop this understanding.

Thus I do not see quantitative analysis and more intuitive qualitative understanding as alternatives between which we are free to choose. If models for the former are chosen arbitrarily or simply because they have been used previously by someone else, the answers will, more likely than not, be incorrect. If the latter is not sometimes supplemented and tested by calculation it will be weaker and more fallible. Both are required. In historical inquiry as in current design, each has its proper place, as the examples in the later part of this article have attempted to show.

Notes

[1] Tommaso Le Seur, Francesco Jacquier, and Giuseppe Boscovich, *Parere di tre mattematici* (Rome, 1742), and Giovanni Poleni, *Memorie istoriche del gran cupola Vaticana* (Padua, 1748).

[2] For Willis see R. Willis, "On the Construction of the Vaults of the Middle Ages," *Transactions of the Royal Institute of British Architects* 1, part 2 (1842):1–69; *Report on a Survey of the Dilapidated Portions of Hereford Cathedral in the Year 1841* (Hereford, 1842); and *The Architectural History of Chichester Cathedral* (Chichester, 1861). For Choisy see A. Choisy, *L'art de bâtir chez les Romains* (Paris, 1872); *L'art de bâtir chez les byzantins* (Paris 1883); and *Histoire de l'architecture* (Paris, 1899).

[3] E. T. Morris, R. G. Black, and S. O. Tobriner, "Report on the Application of Finite Element Analysis to Historic Structures," *JSAH* 54 (1995): 336–347.

[4] Neither paper was published at the time but a summary of the first appeared in *JSAH* 33 (1974): 166.

[5] There have been far too many relevant publications to refer to here. Development up to 1950 is summarized in T. M. Charlton, *A History of Theory of Structures in the Nineteenth Century* (Cambridge, 1982); S. P. Timoshenko, *History of Strength of Materials* (New York, 1953); and, much more briefly, S. B. Hamilton, "The Historical Development of Structural Theory," *Proceedings of the Institution of Civil Engineers* 20, part 3 (1952): 374–402. There is as yet no comparable summary of the more diverse subsequent development.

[6] For a brief outline of the development of the method up to 1992 by one who has played a major part in it see O. C. Zienkiewicz, "The Finite Element Method: Its Genesis and Future," *Structural Engineer* 70 (1992): 355–360. Its use to analyze the structures typically employed in modern buildings is well covered in I. A. Macleod, *Analytical Modelling of Structural Systems* (London, 1990).

[7] The literature on laboratory testing is also too extensive to be referred to fully here. Major field tests on buildings began early in this century at the University of Illinois and have continued on both sides of the Atlantic since then. The principal ones up to 1955 are referred to in R. J. Mainstone, "Tests of the New Government Offices, Whitehall Gardens," *National Building Studies Research Paper 28* (London, 1960). These tests were particularly relevant in the present context in showing the marked influence of construction procedures, even in a modern steel-framed building, on the final stresses under self-weight load.

[8] This is discussed further in R. J. Mainstone, "Intuition and the Springs of Structural Invention," *VIA* 2 (1973): 42–67. (Earlier version published as "The Springs of Structural Invention," *Journal of the Royal Institute of British Architects* 70 [1963]: 57–71.)

[9] A. J. S. Pippard and W. H. Glanville, "Primary Sresses in Timber Roofs, with Special Reference to Curved Bracing Members," *Building Research Technical Paper 2* (London, 1926); J. Heyman, "Westminster Hall Roof," *Proceedings of the Institution of Civil Engineers* 37 (1967): 137–162; R. J. Mainstone, discussion on the previous paper, ibid. 38 (1967): 788–792; L. T. Courtenay and R. Mark, "The Westminster Hall Roof: A Historiographic and Structural Study," *JSAH* 46 (1987): 374–393.

[10] E. T. Morris, R. G. Black and S. O. Tobriner, "Application of Finite Element Analysis to Historic Structures" (see n. 3).

[11] C. Wade and J. Heyman, "The Timber Octagon of Ely Cathedral," *Proceedings of the Institution of Civil Engineers*, part 1, 78 (1985): 1421–1436.

[12] R. J. M. Sutherland, "Shipbuilding and the Long Span Roof," *Transactions of the Newcomen Society* 60 (1988–89): 107–126.

[13] Field measurements, starting with those on the Victoria Tower, Westminster, reported in N. Davey, "Testing Techniques Used by the Building Research Station to Measure Existing Stresses," *RILEM Symposium on Structural Testing 1955*, C19, 1–16, and continuing to those on the quadripartite piers of S. Marco, reported in P. P. Rossi, "Analisi delle condizioni statiche delle strutture murarie," *San Marco: la cripta il restauro* (Milan, 1993), 81–91, have all shown variations in stress, even in the facing masonry, many times what would be expected for homogeneous cross sections.

[14] R. Mark, *Experiments in Gothic Structure* (Cambridge, Mass., 1982).

[15] S. Chiostrini, P. Foraboschi, and S. Sorace, "Problems Connected with the Arrangement of a Non-linear Finite Element Method to the Analysis of Masonry Sructures," *Structural Repair and Maintenance of Historic Buildings*, ed. C. A. Brebbia (Southampton, 1989): 525–534; C. Ignatakis, E. Stavratakis, and G. Penelis, "The In-plane Behaviour of Masonry: An Analytical Finite-element Model," *Structural Conservation of Stone Masonry* (ICCROM, Rome, 1990): 123–130.

[16] R. Barthel, "Analysis and Structural Behaviour of Masonry Cross Vaults," *Structural Repair and Maintenance of Historic Buildings*, ed. C. A. Brebbia (Southampton, 1989): 119–128; idem, "Computer Simulation of the Structural Behaviour of Masonry Cross Vaults," *Structural Conservation of Stone Masonry* (see n. 15)): 181–191.

[17] J. Heyman, "The Stone Skeleton," *International Journal of Solids and Structures* 2 (1966): 249–279; idem, "On Shell Solutions for Masonry Domes," ibid. 3 (1967): 227–241; idem, "Spires and Fan Vaults," ibid. 3 (1967): 243–257.

[18] R. J. Mainstone, "Brunelleschi's Dome of S. Maria del Fiore and Some Related Structures," *Transactions of the Newcomen Society* 42 (1969–1970): 107–126. Figure 5 of this article is a more comprehensive plot of the calculated stresses.

[19] A later analysis of the uncracked octagonal dome gave essentially the same result. See S. di Pasquale, "Structural Analysis of the Dome in Florence," *IABSE Symposium Venice 1983*, 161–168. More light has now been thrown indirectly on the development in stress at the working level by a finite element analysis of the

dome of S. Vitale. See G. M. Roberti, N. Lombardini, and H. Falter, "Late Roman Domes in Clay Tubes: Historical and Numerical Study of San Vitale in Ravenna," *Spatial Structures: Heritage, Present and Future* (International Association of Shell and Space Structures, Milan, 1995): 1237–1244.

[20] R. J. Mainstone, *Hagia Sophia: Architecture, Structure and Liturgy of Justinian's Great Church* (London, 1988): 88.

[21] Plots of the cracking are reproduced in N. Ioanidou, "Historical Documentation and Intervention Principles of the Conservation Study of the Church of S. Panteleimon at Thessaloniki," *Restoration of Byzantine and Post-Byzantine Monuments*, ed. P. Astrinidou (Thessaloniki, 1986):131–145, and in C. E. Ignatakis and K. A. Stylianidis, "Repair and Strengthening of St. Panteleimon: Preliminary Study," ibid., 147–164.

[22] A. El Shabrawi and T. Verdel, "Modelling of Ancient Masonry Structures by the Distinct Element Method under Dynamic Loads," *Structural Repair and Maintenance of Historic Buildings IV*, ed. C. A. Brebbia (Southampton, 1995), 309–316.

[23] R. J. Mainstone, "Typical Structural Forms and Their Seismic Behaviour," *Building Construction under Seismic Conditions in the Balkan Region: Repair and Strengthening of Historical Monuments and Buildings in Urban Nuclei* (Vienna, 1984): 17–77. The drop in frequency with increase in amplitude was confirmed even at a very low amplitude by recordings in Hagia Sophia of the effect of the 1992 Karacabey earthquake. See A. S. Çakmak, R. Davidson, C. L. Mullen, and M. Erdik, "Dynamic Analysis and Earthquake Response of Hagia Sophia," *Structural Repair and Maintenance of Historic Buildings III*, ed. C. A. Brebbia and R. J. B. Frewer (Southampton, 1993), 67–84.

[24] This approach has been well exemplified in the recent work of Cecil Hewett, John James, Stephen Murray, and Malcolm Thurlby. See, for instance, C. Hewett, *English Cathedral Carpentry* (London, 1974); idem, *English Historic Carpentry* (London, 1980); J. James, "Evidence for Flying Buttresses before 1180," *JSAH* 51 (1992): 261–287; S. Murray, *Beauvais Cathedral* (Princeton, 1989); M. Thurlby, "The Elevations of the Romanesque Abbey Churches of St. Mary at Tewkesbury and Sts. Peter at Gloucester," *Medieval Art and Architecture at Gloucester and Tewkesbury* (British Archaeological Association, 1985): 36–51; idem. "The Romanesque High Vaults of Durham Cathedral," *Engineering a Cathedral*, ed. M. Jackson (London 1993): 43–76.

[25] R. J. Mainstone, "The Reconstruction of the Tympana of St. Sophia at Istanbul," *Dumbarton Oaks Papers* 23/24 (1969/70): 355–368.

[26] R. J. Mainstone, "The First and Second Churches of San Marco Reconsidered," *Antiquaries Journal* 71 (1991): 123–137. Another paper, "The Construction History of the Basilica," presented to the Convegno di scienza e technica del restauro della basilica, Venice, 1995, is in press.

[27] J. Fitchen, *The Construction of Gothic Cathedrals* (Oxford, 1961).

[28] R. J. Mainstone, "Brunelleschi's Dome," *Architectural Review* 162 (1977): 156–166 (chiefly 165 and figures 23–25).

[29] Recent investigation has shown that the rings of lined holes through the thickness of Hagia Sophia's dome extend below the level from which lighting hoops have been and could have been suspended. At the lower level it seems that their only use could have been to accommodate tiebacks for such centering.

[30] Mainstone, "Brunelleschi's Dome of S. Maria del Fiore and Some Related Structures," (see n. 18) and "Brunelleschi's Dome" (see n. 28).

[31] Splitting and local crushing accompanied by slips are commonly seen in arches that have undergone appreciable deformation. Ensuing collapses have been recorded on film both in tests on masonry arch bridges and, at least once (at the Wilson Bridge in Tours in 1978), in collapses that occurred naturally as a result of movements of the supporting piers.

[32] R. Willis, "Introductory Essay on the Fall of the Tower and Spire," *Architectural History of Chichester Cathedral* (see n. 2), i–xxiv and 1–7.

[33] See L. Beltrami, "Fall of the Campanile of St. Mark's, Venice," *Journal of the Royal Institute of British Architects* 9, 3rd ser. (1902): 429–437, and L. Binda, G. Gatti, G. Mangano, C. Poggi, and G. Landriani, "The Collapse of the Civic Tower of Pavia: A Survey of the Materials and Structure," *Masonry International* 6 (1992): 11–20. The latter paper is of further interest in showing the depth of investigation that may be needed to determine the cause of a collapse with reasonable certainty.

[34] Measurements deduced from photographs of suspended plumb lines as recorded in W. H. Goodyear, "Vertical Curves and Other Architectural Refinements in the Gothic Cathedrals and Churches of Northern France and in Early Byzantine Churches at Constantinople," *Memoirs of Art and Archaeology* (Brooklyn Institute of Arts and Sciences), 1, no. 4 (1904).

[35] E.-E. Viollet-le-Duc, *Dictionnaire raisonné de l'architecture française du XIe au XVIe siècle*, vol. 4 (Paris, 1860), 180–181.

[36] R. Mark, *Experiments in Gothic Structure* (see n. 14), 69–71.

[37] S. Murray, *Beauvais Cathedral* (see n. 24), 112–120.

[38] It is useless, for instance, to suggest that the circular structure adjacent to the large Myreleion rotunda in Constantinople was crowned by a pendentive dome on the basis only of a "volumetric" sketch which fails to show the pendentives (S. Ćurčić, "Design and Structural Innovation in Byzantine Architecture before Justinian," in R. Mark and A. S. Çakmak, eds., *Hagia Sophia from the Age of Justinian to the Present* [Cambridge, 1992]: 16–38, figure 27). If the shallow lateral exedrae were vaulted, as shown, by sliced-off parts of a hemispherical dome whose diameter is that of the rotunda, the pendentives would be parts of this same dome and would have no separate existence. Martin Harrison's reconstruction of Hagios Polyeuctos as a more direct domed forerunner of Hagia Sophia (M. Harrison, *A Temple for Byzantium* [London, 1989], figures 167 and 169]) similarly fails to explore the full implications. The crucial relationships between the exedrae and the four main piers and main lateral arches are simply ignored, and the near impossibility of reconciling these forms where they meet in the aisles seems to rule out this reconstruction.

[39] R. J. Mainstone, letter, *JSAH* 55 (1996): 360–362; correction of proofing error, 491.

[40] Mariano Taccola, *De ingeneis* (Codex Latinus Monacensis 197, pt. 2, facsimile ed. G. Scaglia, F. D. Prager, and U. Montag; Wiesbaden, 1984), book 1, folios 37r and 45r; Francesco di Giorgio Martini, *Trattato di architettura ingegneria e arte militare* (Codex Torinese Saluzziano 148, facsimile ed. C. Maltese; Milan, 1967), folio 22v; Faustus Verantius, *Machinae Novae* (Florence and Venice, c. 1615), plate 30.

[41] R. Meiggs, in *Trees and Timber in the Ancient Mediterranean World* (Oxford, 1982), 472–473, refers to a length of larch converted to a beam 130 feet long for Nero's wooden amphitheater and to beams of fir [?] 100 feet long for Agrippa's Dibitorium. But these lengths appear to have been exceptional even then.

[42] It may be significant that, although Palladio's design continued to be reproduced largely unchanged in design manuals at least to the end of the last century, W. E. Bell (*Carpentry Made Easy*, 2nd ed. [New York, 1891], plate 31) did add iron straps to connect the vertical ties of the secondary systems to the top chord.

[43] For St. Ouen see J. Guadet, *Eléments et théorie de l'architecture*, 4th ed. (Paris, n.d. [c.1910]), 3: 342–348 and fig.1096.

[44] R. J. Mainstone, "Structural Theory and Design before 1742," *Architectural Review* 143 (1966): 303–310. I am greatly indebted to the Bibliothèque Nationale, Paris, for the opportunity in 1966 to examine the Laon plan under glancing light, and to Robert Branner, who suggested to me the possible value of doing so.

[45] Discussed in J. S. Ackermann, "Gothic Theory of Architecture at the Cathedral of Milan," *Art Bulletin* 31 (1949): 84–111.

[46] Mainstone, "Intuition and the Springs of Structural Invention" (see n. 8) and idem, "The Springs of Invention Revisited," *Structural Engineer* 73 (1995): 392–399.

[47] "[N]el murare la praticha insegnera quel'che s'ara a seguire"; quoted from the definitive version of the program published by A. Doren in *Repertorium für Kunstwissenschaft* 21 (1898): 249–262.

[48] Mainstone, *Hagia Sophia* (see n. 20).

Illustration Credits

Figure 1. From Tommaso Le Seur, Francesco Jacquier, and Giuseppe Boscovich, *Parere di tre mattematici* (Rome, 1742)

Figure 2. Author, from *Developments in Structural Form* (London and Cambridge, Mass., 1975)

Figure 3. Author, based on the survey by Sir Frank Baines, *Westminster Hall: Report to the First Commissioner of H.M. Works* (London, 1914)

Figure 4. Left, author; upper right, from H. De Dion and L. Lasvignes, *Cathédrale de Bayeux: reprise en sous-oeuvre de la tour centrale par M. E. Flachat* (Paris, 1861); lower right, from J. Rondelet, *Addition au mémoire historique sur le dôme du Panthéon français* (Paris, 1814), plate 16

Figures 5, 6, 9, 10, 11, 12, 13, 15, 16, 18. Author

Figure 7. From G. Ungewitter, *Lehrbuch der Götischen Konstruktionen*, 4th ed. (Leipzig, 1901), plate 42

Figure 8. From the article "Construction" in E.-E. Viollet-le-Duc, *Dictionnaire raisonné de l'architecture française du XIe au XVIe siècle*, vol.4 (Paris, 1860), figure 15

Figures 14, 19. Author, from *Hagia Sophia* (London, 1988)

Figure 17. From R. Willis, *The Architectural History of Chichester Cathedral* (Chichester, 1861)

Figure 20. From Palladio, *Quattro Libri* (Venice, 1570), book III, with author variations

Figure 21. From Faustus Verantius, *Machinae Novae* (Florence and Venice, c. 1615)

Figure 22. Author, based on folio 9v of Bibliothèque Nationale MS français 19093

VII

On the structure of the Roman Pantheon

[Starting in the early 1st century BC, there was a major development in vaulted construction that had no real parallel until new skeletal counterparts began to appear in the late 18th century. On this latter occasion they were made possible by the use of iron on a much greater scale than hitherto. On the earlier occasion the development also coincided with the exploitation of a virtually new material – the concrete or *opus caementicium* that the Romans made with a natural pozzolanic mortar. Was this pure coincidence? Or did this new material actually permit and stimulate the development as has often been assumed?

Partly in the hope of answering these questions, Robert Mark and Paul Hutchinson undertook a series of finite-element analyses of the Roman Pantheon, whose construction in the early 2nd century AD marked the culmination of the development in terms of span.[1]

Its structure is, however, very complex, making valid modelling for such analyses difficult. As seen in the drawing overleaf, neither the cylindrical drum that supports the dome nor the dome itself is a simple solid mass of concrete.[2] The drum consists essentially of eight bays, each with a large opening to the interior and one of them open also to the exterior for access, alternating with eight others with closed internal voids. All these openings and voids are bridged by deep arches. The lower third of the dome, which is coffered internally but appears externally to be a simple continuation of the drum, has corresponding internal voids, similarly arched over. Above these voids, where it has a stepped profile externally, it is constructed in layers with progressively lighter *caementa* as it rises towards an open eye at the top. Even the most realistic models therefore had to simplify it greatly.

When tension was allowed in these models, only low hoop stresses were found in the dome. When no tension was allowed, radial cracking occurred, similar to that which was observed and recorded by Terenzio during restoration work in the early 1930s and which can still be seen to a limited extent today within some of the internal voids.[3]

From this similarity, the authors concluded that the new concrete could have contributed little to the development of the new vaulted forms. They also drew several other conclusions about specific features of the design and the designer's intentions.

After discussion of the modelling, these conclusions are here critically reviewed.]

In their paper on the Roman Pantheon, Robert Mark and Paul Hutchinson have illustrated well the versatility of the finite-element technique of structural modelling and numerical-computer analysis. In principle, this technique allows a far closer simulation of the complex internal actions of real structures than any that has been feasible previously. In practice, however, as in all other kinds of analysis, simplifications are still necessary to reduce effort and computational costs. And, even when analyzing an existing structure, idealizing assumptions may have to be made about characteristics for which the data are incomplete, such as the properties of its constituent materials and the manner and sequence of construction. If valid conclusions are to be drawn, it is essential to bear in mind the limitations introduced by these simplifications and assumptions. While I agree in part with the conclusions drawn in the paper, I

The Pantheon's internal structure seen in elevation and vertical section and in plan at the intermediate level and the springing level of the dome. Two typical quadrants are shown, that at the right being cut obliquely to show a different vertical section. The entrance bay and the bay facing it are more open. The bridging arches are not visible internally. The upper ones are partly recessed behind the dome coffers and the lower ones are hidden behind the internal facings. (Drawing based largely on earlier drawings and photographs, chiefly those of Kjeld de Fine Licht and Alberto Terenzio)

do feel that, in reaching others, insufficient attention has been paid to these limitations. There is, as a result, a risk that previously prevalent misconceptions of the structural behaviour of Roman concrete may merely be replaced by new ones.

Even the most realistic analysis envisages the structure as having, throughout, an equivalent solid cross-section whose width is slightly less than the actual overall width to compensate for ignoring both the voids and the internal coffers of the dome. It is assumed that the structure undergoes only elastic deformations (reversible like those of a spring or elastic band), has uniform elastic properties, and is rigidly attached to a rigid base. It is also assumed that the material has no tensile strength. Since sequence of loading does not affect perfectly elastic behaviour, the further assumption of the instantaneous 'turning-on' of gravity is not really necessary, though implicit assumptions about the development of the elastic properties of the concrete, the absence of shrinkage effects during construction, and the rigidity of the centring are. It is then demonstrated that a structure of this kind would crack in much the same way as the actual Pantheon has done, and concluded from this coincidence of behaviour that 'to all intents and purposes, Roman pozzolana concrete could not be counted upon to exhibit any tensile strength'. But – despite the recognition of the possible significance of transient loadings, such as those due to differential temperature changes, as another source of cracking – there is no discussion of the extent to which the coincidence might stem from assumptions other than that of the absence of tensile strength.

In fact, any structure built with large quantities of wet mortar also undergoes considerable deformations of an irreversible (inelastic) kind as the mortar sets, dries out, and gains strength and stiffness. This means that the manner, speed, and sequence of construction are all also of importance. Stresses in a dome built on a full rigid centring would not, for instance, develop in the same way as in a dome built on a more flexible centring or (as seems to me more likely at the Pantheon) on limited ring-like centres serving just for successive lifts and strutted up from the already completed work after suitable intervals for the mortar to gain strength.

Other likely causes of the type of cracking observed are the major interruptions of the circumferential continuity of the drum by internal recesses and hidden voids, the bridging of these discontinuities in two tiers by brick arches, which must themselves exert some resultant outward thrusts on the 'piers' between them, and rotational movements at the feet of these piers due to differential settlements of their foundations. The model ignores the first two possible causes (the assumption of reduced effective widths not really being an adequate recognition of them) and dismisses the last on the basis of the authors' own interpretation of the visible evidence. I interpret what I have been able to inspect of this evidence and the records of the investigations made some fifty years ago in the opposite sense, and I do not feel that any of these likely causes can be ignored. Indeed the conditions of support are crucial for the structural behaviour of any unreinforced dome, and radial cracking will always ensue if they are less than ideal.

I think, therefore, that it would be wrong to conclude from the analysis presented that Roman concrete had no more tensile strength than other types of mortared rubble or brick or stone masonry. It could also be very misleading. At its best, this concrete has considerably more tensile strength than most other forms, though there is no sharp dividing line (with some later brick masonry in the Eastern Empire having been given very similar characteristics by the incorporation of powdered underfired brick in the mortar). It is, indeed, valid to call the concrete 'monolithic' in the sense that it has much the same internal cohesion as a block of natural stone. The mistake of some of the earlier authors referred to in the article was in over-emphasizing the 'mono' and equating 'monolithic' with an ability, when used in a vault or dome, always to span without cracking and without exerting lateral thrusts. Even the large natural monolithic dome of the Mausoleum of Theodoric in Ravenna is cracked and must, therefore, thrust outwards on its base.

An important but little appreciated consequence of even the limited tensile strength of good Roman concrete is a much improved performance under compressive loading. This comes about because even pure direct compression in one or two directions itself results in smaller tensions at right angles to the compressions. Failure in compression is, strictly speaking, a form of splitting brought about by these orthogonal tensions. Weak masonry undergoes multiple splitting that rapidly leads to complete disintegration. Roman concrete, on the contrary, develops a small number of widely separated cracks and otherwise retains its integrity. Once it was set in place and sufficiently hardened, this was its chief structural merit. Its other merit, hitherto more widely recognized, was that it reduced the labour otherwise called for in cutting and placing.

Neither the Pantheon nor any other large Roman domed or vaulted building could have been built with the type of mortared rubble common in earlier centuries. Even if the domes or vaults had been built with the aid of complete vault-spanning centring, they would have disintegrated as soon as this centring was removed and the full arch-like compressions were developed in the radial directions. If attempts had been made to dispense with the full centring of a dome, taking advantage of the potential ability of each completed ring to key and thereby sustain the incomplete radial arches below, they would have failed on account of the inability of these rings to sustain the circumferential compressions once the inward inclination of the inner surface became pronounced.

There is therefore a real sense in which the new pozzolanic mortar did make possible the Roman architectural revolution in the manner of enclosing space – though this is not, of course, to say that it caused it. I tried to make its role clear in *Developments in Structural Form* (MIT Press, Cambridge, MA, 1975), though perhaps taking too much for granted the reader's appreciation of what has been said above about the significance of small tensile strengths.

Other conclusions were drawn in the paper about the significance of the stepped outer rings of the dome and about the intentions of Roman designers.

Interpretation of intentions long ago is always hazardous. I still hold the view, on the evidence of what the Romans actually built, that they had little appreciation of what we now see as the potential structural value of the doubly curved domical form. I agree with Mark and Hutchinson that they continued to think primarily in terms of unidirectional arch-like spanning, and that there is some analogy between the stepped external weighting of the lower regions of their domes and other vaults, and the spandrel fills over the haunches of their arches. But I see the characteristic stepped external profile as stemming chiefly from their habit of constructing all concrete vaults and domes in successive horizontal lifts. Where the inner surface was rising steeply in the lower half, it was much simpler to finish each lift on the outside with a vertical wall, and it would have been simply a matter of normal caution to step this wall inwards only gradually as they approached a level where a significant proportion of the overall width of the concrete would have to be placed above the internal void. As far as I am aware, there is no evidence that the concrete of the steps in the Pantheon was a later surcharge. My own observations suggest that it is integral with the concrete of what Mark and Hutchinson regard as the dome proper.

1. Mark, R. and Hutchinson, P., 'On the structure of the Roman Pantheon', *Art Bulletin*, vol. LXVIII, 1986, pp. 24-34.
2. Best described and illustrated in Licht, K. de F., *The Rotunda in Rome*, Jutland Archaeological Society Publications 8, Copenhagen, 1968.
3. Terenzio, A., 'La restauration du Pantheon de Rome', in *La conservation des monuments d'art et d'histoire*, l'Institut de Cooperation Intellectuelle, Paris, 1933, pp. 280-285

VIII

The Structure of the Church of St. Sophia, Istanbul

Introduction

The genesis of structural form, particularly in architecture, has for some time fascinated the writer.[1] The processes by which new forms are first conceived and their details elaborated in a manner which permits their practical realisation are, even today, far from being fully understood. Still less do we understand how, without the benefit of modern structural theory, the great innovating designers of the past were able sometimes to break with such apparent confidence and such a large measure of success the bounds of proven practice.

Nowhere else were the bounds so completely broken, not only in terms of structure, but equally in terms of the purely architectural composition of space, surface and light, as in the church of St. Sophia, Istanbul, as rebuilt by Justinian after the disastrous Nike riots of January 532. Indeed, so unprecedented was it that after more than 800 years it could still be written that it was "in size the greatest of all churches under the sun, surpassing in beauty and magnitude all others wheresoever they may be,"[2] and it was almost 1,000 years before it had any comparable progeny in the great Imperial mosques built after the Turkish conquest[3] (Plate X).

Though dedicated and apparently completed less than six years after construction commenced—on 27 December 537[4]—its domed silhouette still rises, little altered, above the Seraglio Point where the Golden Horn meets the Bosphorus. And though its interior is now deprived of much of its former magnificence by the loss of all its original furnishings and much of its mosaic decoration, the space under the swelling vaults mounting up to the dome is still breathtaking, even after repeated visits, as one enters it from the narthex through the Imperial door. As on entering the Pantheon, one steps into an immense open space unobstructed by columns. But, whereas, in the Pantheon, the enfolding boundaries of this space are simply and clearly defined and immediately comprehensible —a simple hemisphere set on a cylindrical drum—the true boundaries in St Sophia are always elusive and appear continuously to change as one moves around. The dome itself is not even seen at first. It hovers high above as the culmination of a finely orchestrated movement rising through the alternately advancing and receding colonnades of aisles and galleries to lesser and greater semidomes and four great pendentives surmounted by a slightly projecting cornice and a low drum of 40 windows (four of which are now blind). (Plate XI). Only with effort does one slowly comprehend the magnitude of this space, and only with much more effort and after exploring the whole building time and

[1] See especially R. J. Mainstone, "The springs of structural invention," *Journal of the Royal Institute of British Architects*, 70 (Feb. 1963), 57–71; (idem), "On construction and form," *Program* No. 3 (Columbia University School of Architecture, Spring 1964), 51–70; and (idem), "Structural theory and design before 1742," *Architectural Review*, CXLIII (April 1968).

[2] Cantacuzenus, *Historiae*, quoted by E. H. Swift, *Hagia Sophia* (New York, 1940) (cited below as *H.S.*), 19.

[3] For descriptions of the principal mosques and comparisons with St. Sophia see A. Choisy, *L'art de batir chez les Byzantins* (Paris, 1883), 139–141; M. A. Charles, "Hagia Sophia and the great Imperial Mosques," *Art Bulletin*, XII (1930), 321–344; S. Corbett, "Sinan," *The Architectural Review*, CXIII (May 1953), 290–297; U. Vogt-Goknil, *Les Mosques Turques* (Zurich, 1953).

[4] E. H. Swift, *H.S.*, 12.

again can one form a composite intellectual picture of its full complexity and visualise in broad outline the structural organism by which the weights and thrusts of the dome and other vaults are carried to the ground. (Plate XII.)

A close study of the structure first became possible for western scholars when, in 1847–49, Gaspar Fossati, assisted by his brother Giuseppe, was appointed by Sultan Abdul Mejid to carry out repairs necessitated largely by prolonged neglect. The Fossati brothers seem to have lacked the training and inclination to take full advantage of an opportunity which must have rivalled that of Viollet le Duc in another field,[1] but the King of Prussia seized the opportunity to introduce the Prussian architect Salzenberg, who, having access to the scaffolds, was able to publish a large folio volume of measured drawings.[2]

Unfortunately, Salzenberg's drawings, which have formed the basis of most subsequent representations and most subsequent attempts to explain the structure and elucidate the tantalising glimpses of its early vicissitudes given in the documentary sources,[3] were grossly misleading in several respects. Not only did they fail, as all such drawings must, to convey the subtle elusive spatial quality of the interior. More seriously, from the point of view of any attempt to interpret the structural actions and the sequence of construction and reconstruction, they failed completely to delineate the present marked irregularities of all forms and, with one exception[4] and even more regrettably, the precise nature of the interconnections between different elements of the structure and the many cracks and dislocations that must now exist.

The secularisation of the mosque—to which use the church had been appropriated on the Turkish conquest in 1453—furnished, in 1935, the opportunity for further investigation at first hand. Much that must have been visible to Salzenberg is now hidden by plaster or marble revetment, but there has been time for much more detailed measurement using, where necessary, the best modern instruments. A precise survey embracing every detail of the structure was initiated by the late Professor William Emerson of the Massachusetts Institute of Technology with the help of Mr. Robert L. Van Nice and has been sponsored since Professor Emerson's death by the Dumbarton Oaks Centre for Byzantine Studies, Harvard University. This survey, now being completed by Mr. Van Nice,[5] must, in its comprehensiveness and accuracy, be unique among architectural measured drawings. (Plates XIII and XIV.) The writer has played little part in it; but he has been privileged in being

---

[1] As a record of his work Gaspare Fossati published only an album of coloured lithographs showing the exterior and interior of the church before and after restoration: *Aya Sofia Constantinople, as recently restored by order of H.M. the Sultan Abdul Medjid* (London, 1852). Other records of the restoration left by the Fossati brothers and now mostly preserved in the Fossati archive at Bellinzona, Switzerland, are reproduced, in part, in C. Mango, "Materials for the study of the mosaics of St. Sophia at Istanbul," *Dumbarton Oaks Studies* No. 8 (Washington, 1962).

[2] W. Salzenberg, *Altchristliche Baudenkmäler von Constantinople* (Berlin, 1854).

[3] The sources are summarised in more detail than is possible here in W. R. Lethaby and H. Swainson, *The Church of Sancta Sophia, Constantinople* (London, 1894) and E. H. Swift, *H.S.* A fuller account by C. Mango will appear in the forthcoming publication mentioned below.

[4] The elevation of the inside of the south tympanum, reproduced in Plate XV (*a*), shows an open crack following approximately the line of the externally exposed soffit at the west side of the lower main arch. Both this figure and the main plates do give some indications of lack of bond, changes in material and irregularities of line, but these cannot be relied upon. Many seem to have been based on inadequate observation, such as the double shell shown in the section of the western semidome in his plate IX, the backward bends of the crowns of the forward edges of the eastern and western barrel vaults shown in the same plate, and the non-existent course of stone above the cornice shown in the tympanum elevation.

[5] For earlier preliminary reports of the survey and the conclusions drawn form it, see W. Emerson and R. L. Van Nice, "Haghia Sophia, Istanbul: Preliminary report of a recent examination of the structure," *American Journal of Archaeology*, XLVII (1943), 403–436 (cited below as *A.J.A.*, 1943); idem, "Hagia Sophia: The collapse of the first dome and the construction of the second dome and its later repairs," *Archaeology*, 4 (1951), 94–103, 162–171 (cited below as *Archaeology*); R. L. Van Nice, "The structure of St. Sophia," *Architectural Forum*, 118 (1963), 131–139, and idem, "St. Sophia's structure: A new assessment of the half-domes," *Architectural Forum*, 121 (1964), 45–49.

able to go over much of the accumulated data with Mr. Van Nice and to assist in its interpretation. The complete study is in process of publication[1] and the present Paper is merely a preliminary note on some of the findings relating to the structural design and the sequence of construction and reconstruction. It is less definitive than was hoped because the necessary correlation and analysis of some of the data is still proceeding and some critical observations, whose significance was not originally recognised, remain to be made.[2]

Over the centuries the structure, frequently racked by earthquakes, has been consolidated or received additional buttressing on a number of occasions, notably at an unknown date prior to 1317 perhaps during the Latin occupation,[3] under the Emperor Andronicus Palaeologus in 1317,[4] by the architect Sinan in 1573,[5] and by the Fossati brothers as already mentioned. Only the last of these operations has significantly affected the original structure, in a manner which will be noted below, though all have altered the external aspect. The extent of each can be readily identified, and it is hardly necessary, for the present purpose, to consider them further. Some attention must be given to the reconstructions of the western and eastern semidomes and adjacent portions of the pendentives and dome which are recorded as having taken place in 989–94 and 1346–53 respectively, but attention is concentrated on the church as originally built and as partly reconstructed after two severe earthquakes in 553 and 557. In the light of evidence still visible in the structure and now recorded for the first time, an attempt is made to elucidate less speculatively than in the past the records left by contemporary historians of the initial construction and first reconstruction, to describe the sequence of events as far as the structure is concerned and to analyse in part the designers' intentions and reasoning.

None of this would be possible without Mr. Van Nice's patient accumulation and plotting of the relevant data. To him in particular, for allowing completely free access to it and for helping in every way possible, the writer is deeply indebted and would like here to express his thanks. He would like also to acknowledge the many courtesies in Istanbul of Bey Feridun Dirimtekim, Director of the Ayasofya Museum and Mr. Ernest Hawkins, and to thank Dumbarton Oaks for sponsoring his recent work on the church and sanctioning the publication of this Paper.

Structure and Construction

Divested of its later peripheral buttresssing and of the four ramps which have always projected from the four corners to give access to the galleries,[6] the church is almost square in plan to the level of the gallery roofs, with a narthex added across the west front and an apse projecting slightly at the east (Fig. 1). Above a lesser square in the centre, with sides averaging 30·975 m. measured between the skirtings, the dome is carried on the arches and pendentives spanning between four main piers

[1] It will consist of about 50 plates showing general and detail plans and sections reproduced at scales of 1:250 and 1:100 and a text volume (with additional drawings and photographs) by C. Mango, R. J. Mainstone and R. L. Van Nice. The first instalment of plates has now appeared: R. L. Van Nice, *Saint Sophia in Istanbul: An architectural survey* (Dumbarton Oaks, Washington, 1966).

[2] Between the reading of this paper and the preparation of the text for publication, two further seasons' work have been completed in Istanbul and analysis of the data already assembled plus that newly acquired has continued. As a result it is now possible to substantiate even further the principal conclusions about the extent and manner of the major reconstructions, but some doubts still remain about the precise sequence of the initial construction and it is possible that some of the conclusions more tentatively advanced about this may finally have to be modified. The Paper is, nevertheless, reproduced substantially as it was read, brief references to subsequent work having been added later in notes. For a definitive account the reader is referred to the forthcoming text mentioned in the previous note.

[3] E. H. Swift. "The Latins at Hagia Sophia," *American Journal of Archaeology*, XXXIX (1935), 458–474; idem, *H.S.*, 87–88, 117–119.

[4] E. H. Swift, *H.S.*, 14, 112–116.

[5] W. Emerson and R. L. Van Nice, *A.J.A.* (1950), 33–34.

[6] E. H. Swift, *H.S.*, 120–122.

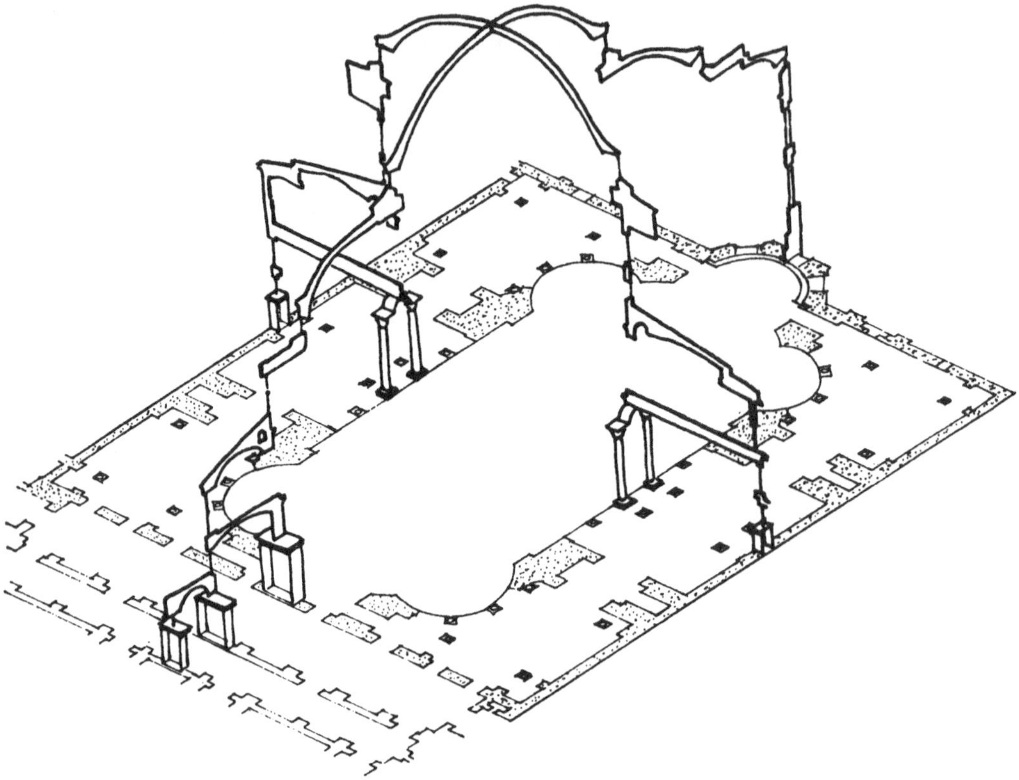

Fig. 1. St. Sophia, Istanbul; isometric plan, longitudinal section and cross section

(Fig. 2). At the east and west its outward thrusts are counterbalanced and carried downward and outward by the inward thrusts of the two great semidomes which abut the main eastern and western arches and are themselves carried on secondary arches spanning between the main piers and secondary piers lying further back on the extremities of the main square. Smaller semidomes, rising over four exedrae, abut the secondary arches which lie diagonally to the main longitudinal axis, while the secondary arches which lie transversely to this axis at the east and west ends are themselves deep barrel vaults inherently stiff in the directions of their own spans. At the north and south, however, the thrusts are transmitted wholly to the main piers by pairs of interconnected arches spanning between them on each side. The upper arch on each side, which is the only one visible on the interior, corresponds in span to the single main arches at east and west, though its width, as shown in the figure, is purely conjectural. This upper arch is stiffened and underpinned by a lower arch of greater width and lesser span[1] prominently visible on the exterior and springing from slightly recessed projections of the main piers. The north and south thrusts are then carried back by transverse arches spanning the aisles, galleries and gallery roofs and by transverse walls above the gallery roof arches to buttress piers which lie against the northern and southern extremities of the main square.

[1] These lower arches now have spans of about 22·60 m. as compared with about 31·60 m. for the upper arches at the north and south and 32·30 m. for the main arches at the east and west.

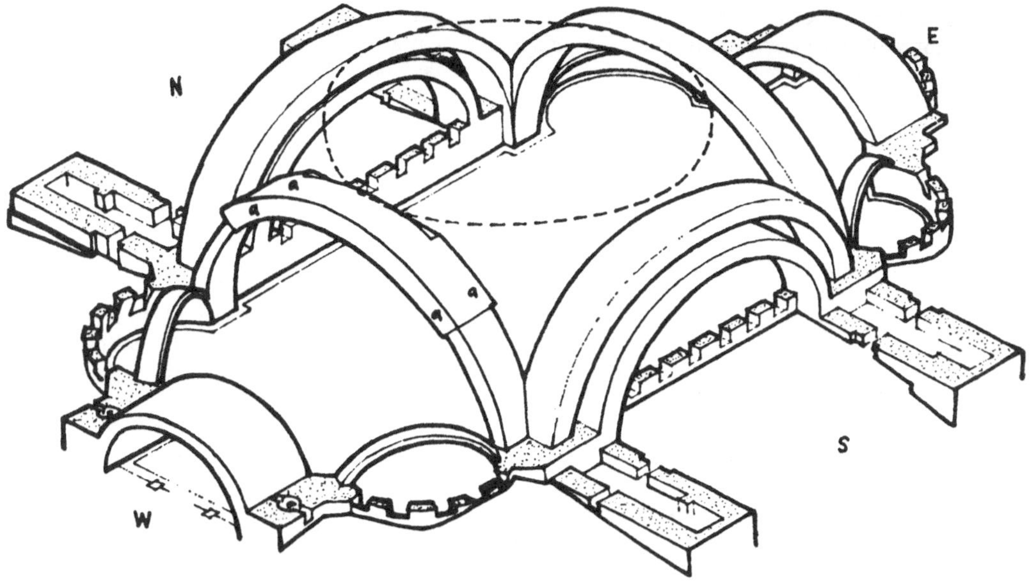

Fig. 2. Isometric view of the arches that carry the semidomes, pendentive and dome. The increased section of the main western arch *aaaa* is part of the tenth century reconstruction. (Upward extensions of buttress piers removed for clarity.)

The completely open nave thus created, extending for most of the length of the building but across a little less than half its width, is flanked by the open colonnades of the aisles and galleries whose domical, groined and barrel vaults have all the variety of form which became typical of later Byzantine construction.[1]

To liken this structure to that of the Pantheon's dome raised over the central bay of the Basilica of Maxentius and constructed without the benefit of Roman concrete ignores much, but it does perhaps give some indication of what was achieved, structurally, by Justinian's architects, Anthemius and the elder Isidorus. The architects of Imperial Rome had been the first to exploit architecturally the space-enclosing potentialities of the dome.[2] But even when its inner surface was scalloped it remained a solid concrete monolith immensely thick towards its base, and even when the walls below were hollowed into niches or exedrae they remained, with few exceptions, a continuous and similarly monolithic support. Problems of outward thrust and its containment hardly arose as a real constraint on design.[3] Nor had they ever arisen to the same extent in structures like the great basilicas and the similar tepidaria of some of the Imperial baths as they did when St. Sophia's dome was constructed on four isolated piers and ultimately stabilised only by others lying much further back.

[1] The vaults and minor arches of the aisles and galleries, though important for a complete assessment of the structure, have had to be largely omitted from the present paper, but will be considered in the final publication referred to above. For previous accounts see W. R. Lethaby and H. Swainson, *op. cit.*, 219–226; A. Choisy, *op. cit.*, 136–137 and plate XXV; E. H. Swift, *H.S.*, 104–106, 123–125.

[2] See, for instance, J. B. Ward Perkins, "The Italian element in late Roman and early Mediaeval architecture," *Proceedings of the British Academy*, XXXIII (1947), 163–194; *idem*, "Roman concrete and Roman palaces," *The Listener*, LVI (November 1, 1965), 701–703; E. H. Swift, *Roman Sources of Christian Art* (New York, 1951).

[3] This point is well made, though slightly exaggerated, in J. B. Ward Perkins, Notes on the Structure and Building Methods of Early Byzantine Architecture, *Second Report on excavations in the Great Palace of the Byzantine Emperors at Constantinople* (Edinburgh, 1958), 78–81.

Anthemius and Isidorus started, at least, with one great advantage without which the whole edifice might well have perished in the first earthquake or that of 557—"such a great and terrible shock as is not within the memory of man on earth in this day and generation."[1] This is that, as far as can be seen, it is founded on rock brought up to level, under the north-west buttress, by immense blocks of stone set in very tenacious mortar.[2]

On this base the main piers and the lower parts of the buttress and secondary piers are constructed of large blocks of limestone or greenstone (a local granite). These are carefully fitted together as described by Procopius,[3] though the running of the joints with molten lead which he also described[4] is only partly accurate. Most joints are of mortar, but sheets of lead are substituted, where it has been possible to inspect, at the springings of arches and vaults. Similarly constructed of large blocks of stone is a series of projections from the main and buttress piers at aisle and gallery levels, these projections being linked, across the aisles and galleries, by bracing arches, also of stone, beneath the main transverse aisle and gallery vaults. All freestanding columns are marble monoliths, some bound with brass collars to prevent splitting, and marble is also used structurally for the principal cornices.

Elsewhere in the building, with only minor exceptions, the construction is of brick, with courses of stone inserted at the springings of arches and vaults. The bricks vary in size from very large ones approximately 0·70 m. square, which seem to have been specially imported or manufactured for use in the main arches of the original structure but reappear as reused material in later reconstructions, to others averaging 0·375 m. square in the original construction or slightly less when they are of a later date.[5] Structurally more important than the size of the bricks, all are set with joints thicker than themselves so that the deformational characteristics of the brickwork are largely those of the mortar used. The brick sizes did, nevertheless, limit the choice of thickness for vaults and arches and it is of interest that the measured thickness for the former is often slightly less than would be expected if the bricks had all been set radially. This seems to indicate the adoption of a slightly flatter inclination in the lower parts of the minor vaults. With this slight qualification, all minor vaults and those of the exedrae are one of the normal bricks in thickness and the main semidones are about twice this thickness. The dome is more complex, having on its undersurface projecting ribs which die out near the crown. The webs between these ribs are 0·75 m. thick in the sixth century parts and the fourteenth century reconstruction and 0·90 m. thick in the tenth century reconstruction.[6] Neither the dome nor any of the semidomes is, however, a simple continuous shell. Each is ringed by arched window openings in its lowest part and becomes here, in effect, a series of buttress-like piers appearing externally as a vertical sided drum, but conforming internally to the curved form of the shell proper. Where, therefore, tension would develop in a continuous shell even without any spreading of the base, the structure is already open.

Across these window openings of the dome and semidomes, timber ties were built in, wrapped in lead for protection. This timber, being in relatively exposed positions, has long since rotted, but its positions can be seen as patches in the plaster and mosaic surfacing.[7] Other timber ties or struts, of which a considerable number remain in these more protected positions, spanned some of the minor

[1] Theophanes, *Chronographia*, quoted by E. H. Swift, *H.S.*, 13.

[2] A preliminary account of the foundations appears in W. Emerson and R. L. Van Nice, *A.J.A.* (1943), 407–413.

[3] Procopius, *De adeficiis*, I, i, 51–52, Loeb edition with translation by H. B. Dewing (London, 1940), 22–23.

[4] Procopius, *op. cit.*, I, i, 53, Dewing, pp. 22–25.

[5] A preliminary account of the brickwork appears in W. Emerson and R. L. Van Nice, *A.J.A.* (1943), 413–423.

[6] Inspection of the upper surface of the brickwork of the dome is extremely difficult, but, from the limited observation that has been possible during the present survey, it appears that Swift (*H.S.*, pp. 166–167) was mistaken in inferring from the little that he saw that the ribs projected externally. A jog in the upper surface at the edge of the tenth century reconstruction seems to correspond only to an overall thickening of this sector as compared with the remainder of the dome.

[7] W. Emerson and R. L. Van Nice, *Archaeology*, pp. 168–169.

arches of the aisles and galleries. There is also evidence of the incorporation of bonding timbers in the brickwork of the upper buttress walls. Numerous iron ties are found across the vaults of the aisles and galleries, cramping together the stones of the cornices or consolidating the masonry of several different periods in the upper parts of the north and south buttresses. There is evidence also of others embedded in the filling above the aisle vaults and below the gallery floor. These iron ties are obviously of many different dates, but a considerable number, even of the largest, seem without doubt to date from the original construction or the first reconstruction after the earthquake of 557. They vary in size but many are about 25 sq. cm. in cross section.

The interconnections between adjacent elements are also important structurally. In several places where the working masonry is accessible or has been briefly bared, an unexpected lack of bond has been observed. In some cases this almost certainly means that one mass of masonry has been added to strengthen or stabilise the adjoining mass. This applies most notably to several fillings of earlier voids in the main piers at gallery level (in relation to their greater bulk above and below), to numerous additions to the mass of the north and south buttresses above gallery roof level, to a heightening of the outer drum of the western semidome and to the fillings of several windows in the semidomes and the sixth century dome. In another case of considerable importance the meaning is less clear, partly because it is not yet known with complete certainty how far the bond is lacking. This is the case of the aisle and gallery projections from the main and buttress piers and the bracing arches which these carry which were mentioned above. If, as seems most likely, these are entirely unbonded at aisle as well as at gallery level to the adjacent piers, this could mean that they also were later additions. But it might equally mean that, though part of the original construction, they were deliberately left without positive interconnection to the main structure, even where stone abutted against stone, because some slight relative movement was anticipated and some articulation to accommodate it was considered desirable. In one place, on the contrary, bond appears to exist where, at first sight, none would be expected. The generally convincing (but in some respects suspect) drawing by Salzenberg of the inside of the south tympanum[1] shows the bottom half of the wider lower arch of lesser span between the main piers as being integral with the lower part of the infilling wall (Plate XV (*a*). The brickwork at this level is coursed horizontally throughout whereas, in the freestanding widths of the same arches at the same level outside the infilling walls, the bricks have been seen to be set radially. It appears therefore, if the horizontal brickwork on the inside face of the arch is not merely a shallow facing, that a significant part of each arch was separated from the rest of its total width and bonded to an infilling which was bound to be subjected to different constraints in the event of any spreading and settling of the arch.

Deformations and Dislocations

The structure now shows many significant variations in line, level and inclination from the regular forms hitherto shown in all conventional representations of it. The most important may be summarised as follows:

(1) The piers all tilt away from the centre of the church (Figs. 3 and 4). Within the margin of error that results from having to take measurements to facings rather than to the working masonry behind, they do so symmetrically in relation to both axes of the church, those of the main piers being greater in the transverse direction than longitudinally, and greater above the gallery cornice (at which level these piers are considerably reduced in bulk) than below it.

(2) Apart from two groups which will be referred to below, the monolithic columns also tilt from the vertical in a generally symmetrical manner.

[1] W. Salzenberg, *op. cit.*, p. 21.

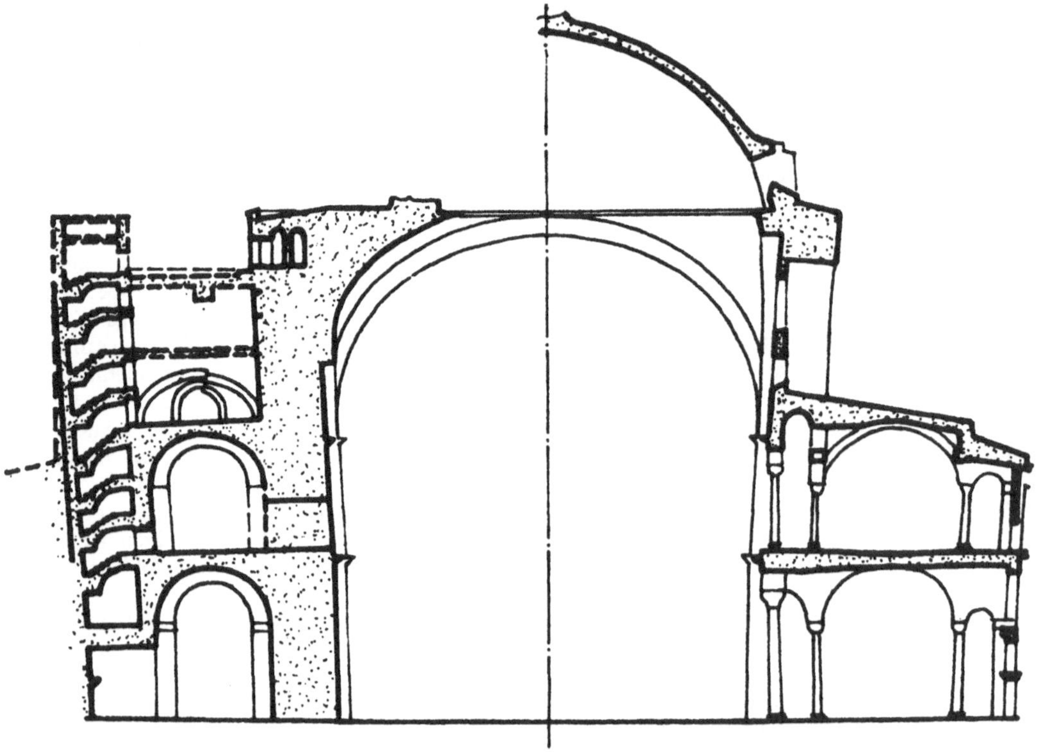

Fig. 3. Cross-section of the church looking East: distortions exaggerated two times.
Left: through the middle of the south-east pier and buttress (reflected).
Right: through the centre-line of the church
(Broken lines signify later additions to the sixth century structure, or later rectifications of the facings.)

(3) The horizontal cornices of the north and south galleries between the main piers show curves in plan, again symmetrical in relation to the axes of the church.

(4) The vertical profiles of the arches spanning between the main piers and between the main and secondary piers to carry the larger and smaller semidomes and the pendentives are consistently flattened at the crown and spread outwards at the haunches as compared with a true semi-circular profile (Fig. 5).

(5) The same arches all show slight curves in plan when projected vertically, these again being symmetrical in relation to the axes of the church, except that the forward bend of the main western arch is less than that of the main eastern one (Fig. 6 and Plate XV (*b*)).

(6) Cross sections of the dome and large and small semidomes also show deviations from the circular arc. These are systematic and symmetrical except for more marked irregularities in the upper part of the western semidome, and to a lesser extent in the western and southern sectors of the dome (Fig. 7).

Fig. 4. True cross-sectional detail on the transverse centre-line, looking East

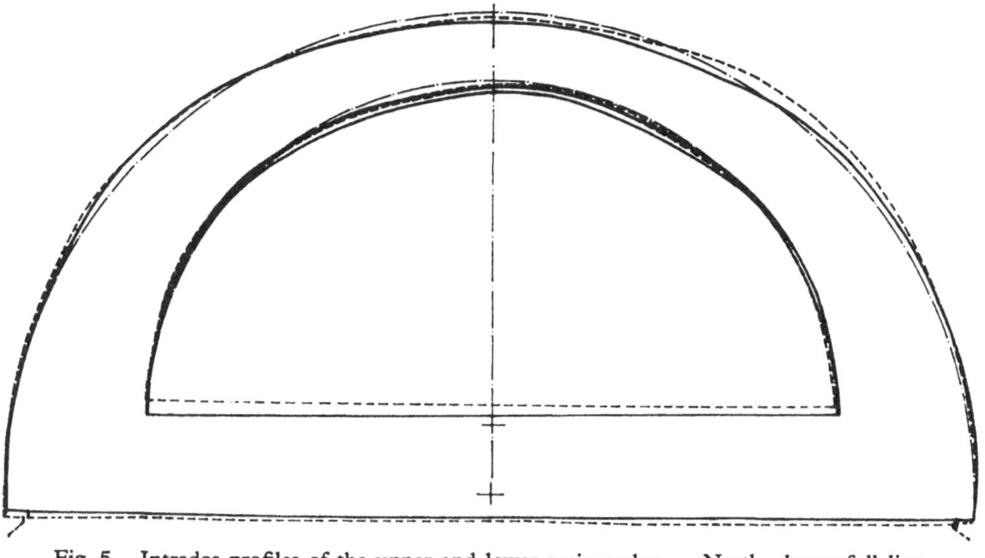

Fig. 5. Intrados profiles of the upper and lower main arches. North—heavy full line. South—heavy broken line. Presumable semi-circular setting-out curves—faint chain-dotted lines.

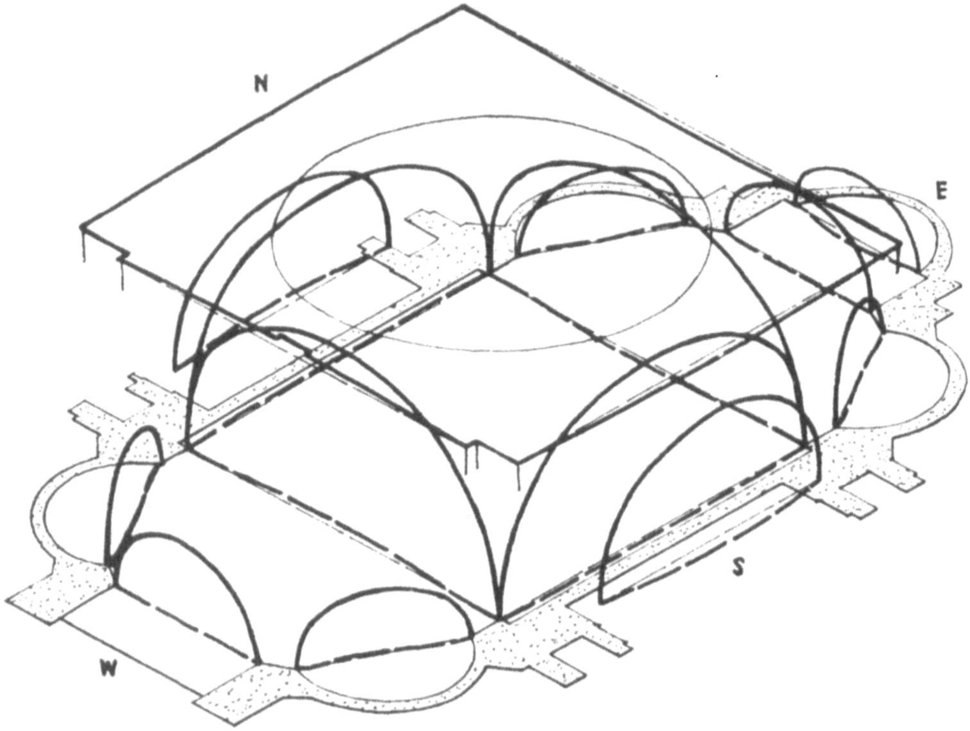

Fig. 6. Isometric representation of the edges of the dome base and of the main and secondary arches; deviations magnified two times. Distorted base—heavy full lines. Vertical projection of arches—heavy broken lines. Straight lines between corners, or springings—faint chain-dotted lines.

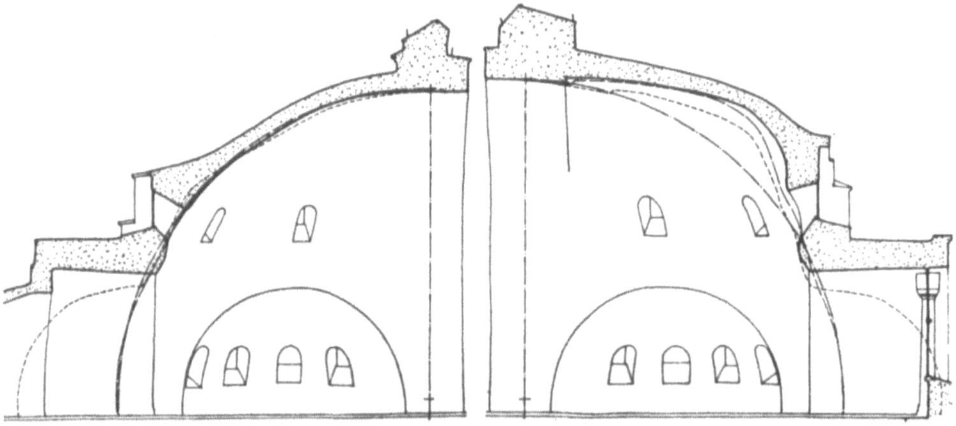

Fig. 7. Cross-sections of the main semidomes. Left—eastern. Right—western. Longitudinal sections—full lines, thicknesses shaded. Diagonal sections—broken lines, superimposed. Presumable setting-out curves for the original semidomes—chain-dotted lines.

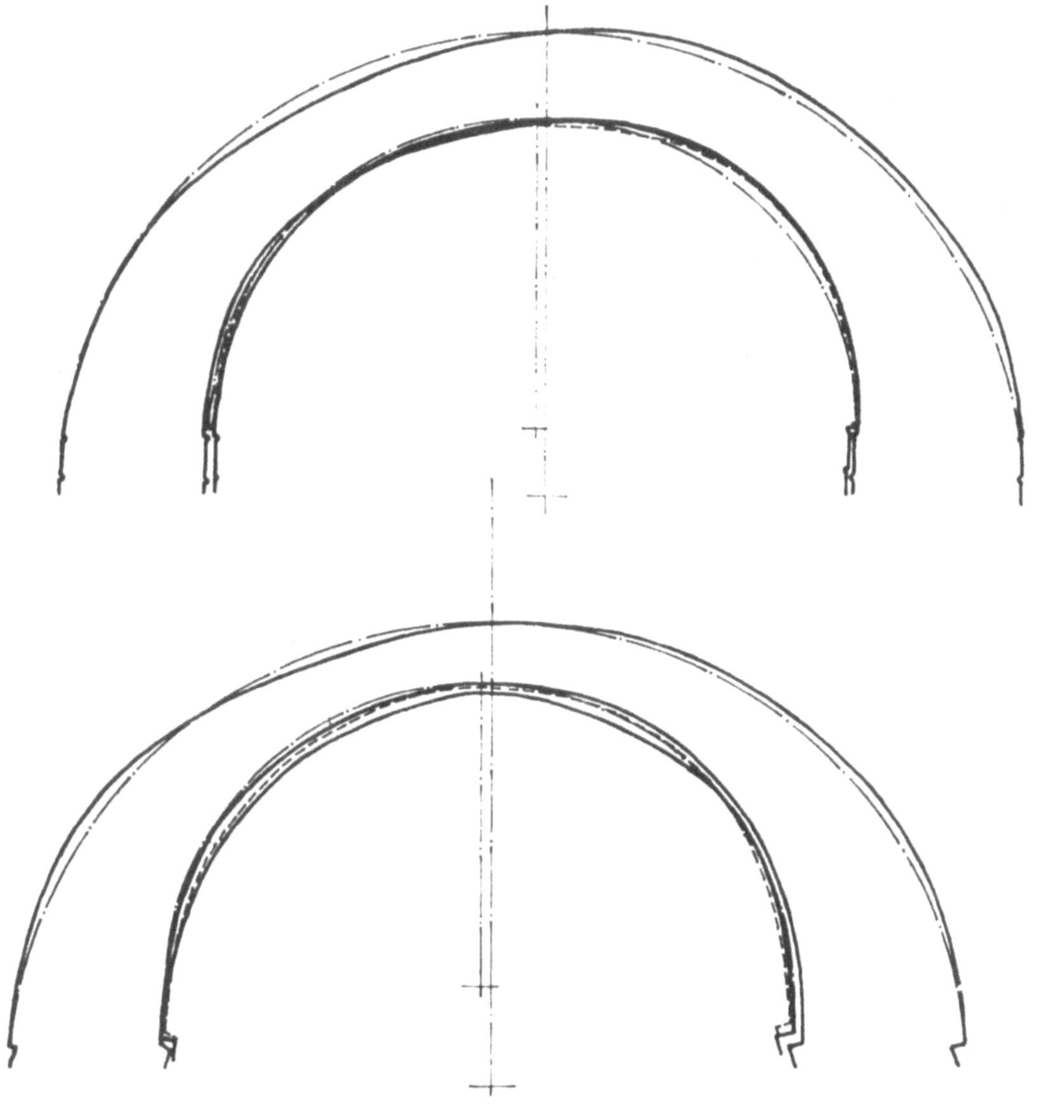

Fig. 8. Typical intrados profiles of the arches linking the main and buttress piers; the dome is on the left.

The outer curves represent the intradoses of the brick barrel vaults, the inner curves those of the stone bracing arches that span between the projections. (See Fig. 3.)

Presumable semicircular setting-out curves are shown by chain-dotted lines.

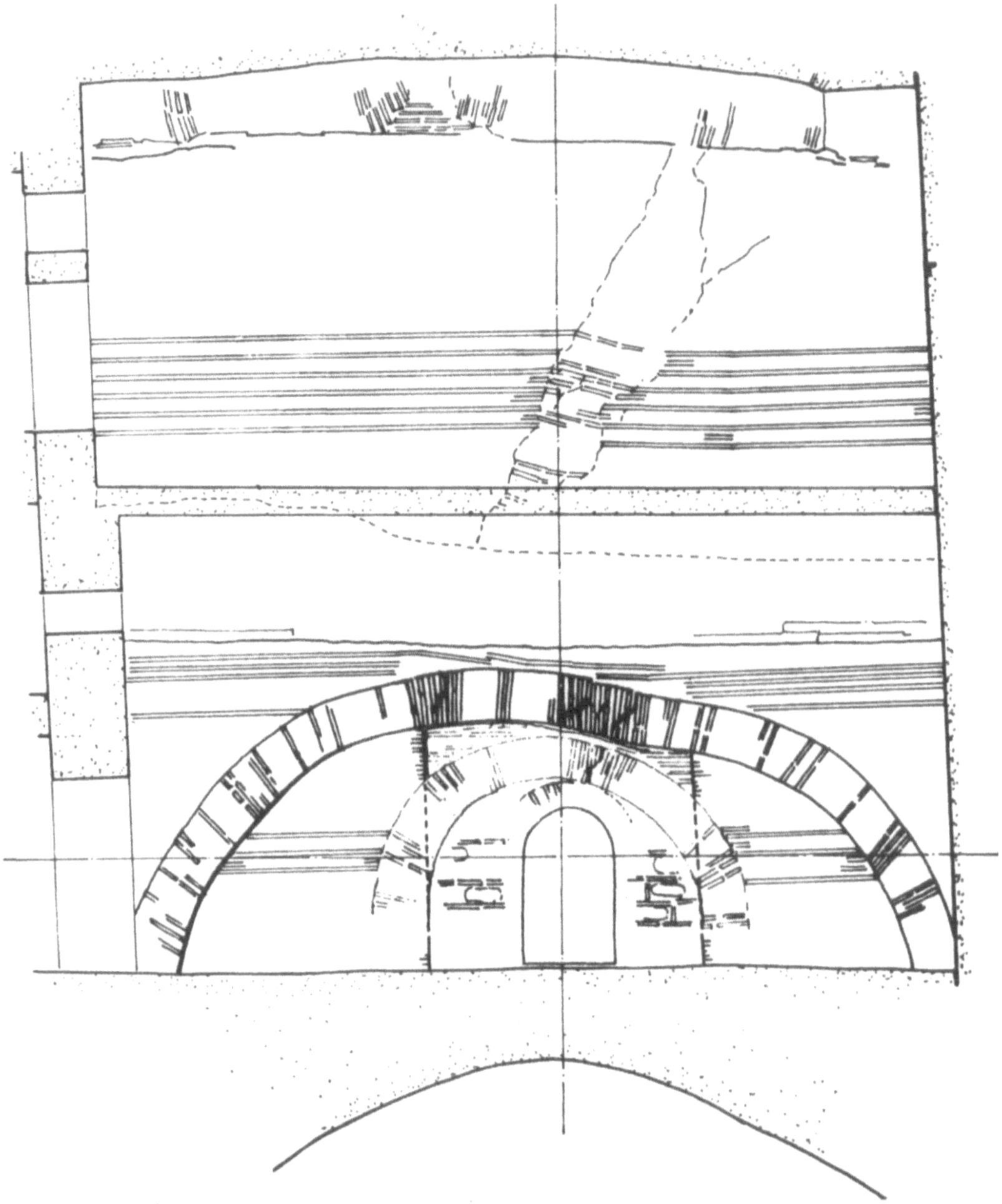

Fig. 9. Elevation (looking east) of the east wall of the north-east buttress above gallery level. The back of the north-east pendentive is on the right. The fillings below the dislocated crown of the segmental arch are successive later additions.

(7) All arches connecting the main piers to the buttress piers have profiles similar to those shown in Figs. 8 and 9, those above the gallery roof (whose brickwork alone is exposed) being dislocated at the crown in the manner shown in Fig. 9.

(8) None of the pendentives conforms to a spherical surface, the northeast one, in particular, showing a very pronounced bulge under the dome cornice in the centre (Plate XVI (*a*)).

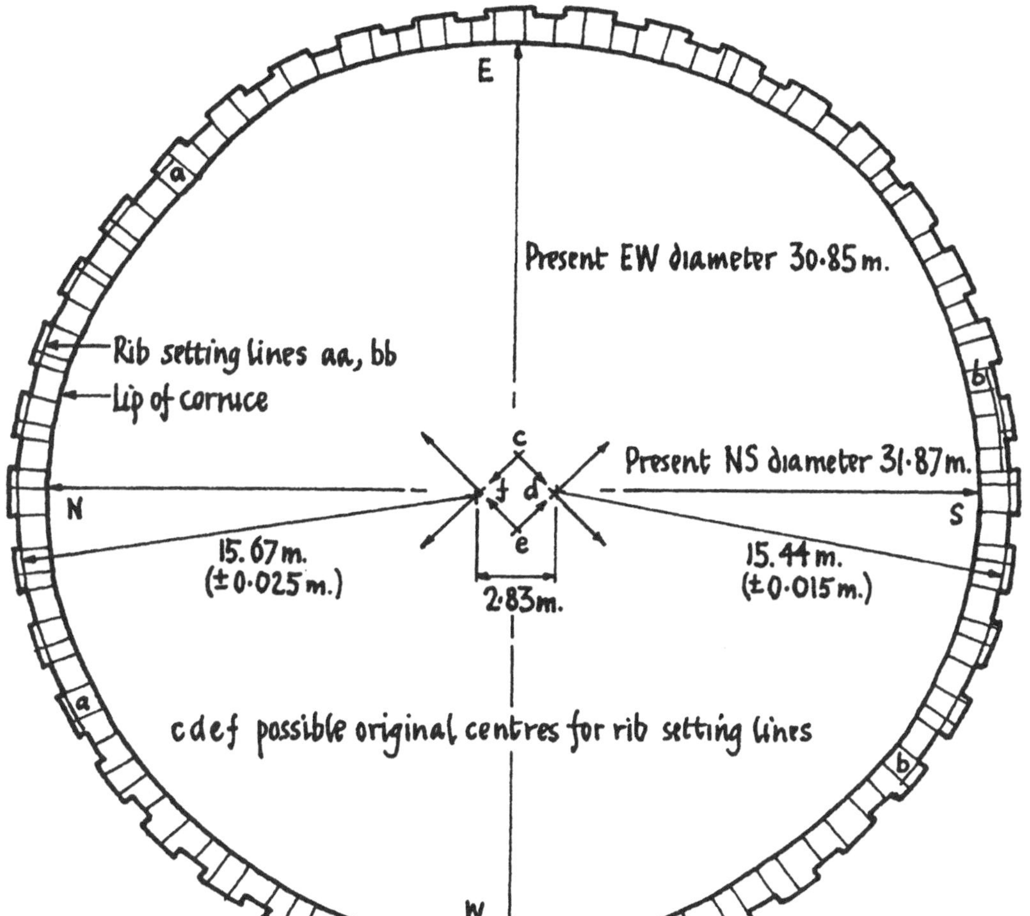

Fig. 10. Plan of the dome cornice at present, with a suggested basis for the setting out of the first reconstruction in the sixth century.

(9) The outer sides of the approximately square dome base are all curved in plan in the same sense as the main arches below (Plate XVI (*b*)).

(10) The dome itself is not circular in plan, being wider transversely than longitudinally, as shown in Fig. 10.

On the other hand the dispositions of the main elements in plan show a remarkably high degree of accuracy in setting out. The sides of the square formed at ground level by lines joining the projections from the main piers from which the four arches carrying the pendentives spring, differ at most by 2 cm. from their average length of 30·975 m. for instance.

Symmetrical vertical and horizontal curves and inclinations associated with very precise workmanship are, in a structure like the Parthenon, unequivocal evidence of intention: they must have been there from the start and must be deliberate.[1] Here this is not the case. One cannot reject out of hand all suggestions[2] that one or other of the groups of curves or inclinations just listed was an intentional feature of the original design which has remained unchanged since it was built. On the other hand, a known history of difficulties in construction and partial collapses, the evidence of successive additions of buttressing masses, infillings and ties, and the presence in many places where it is possible to observe of cracks, separations and other dislocations, all strongly hint at the likelihood of considerable deformations during and after construction of the kind that are to be expected, in the light of modern knowledge, under the action of large thrusts on masonry incorporating such large quantities of mortar.

Because the mortar used[3] will have set slowly in all the larger masses, considerable plastic flow is to be expected, resulting in deformations considerably exceeding the elastic ones necessary for final equilibrium. The dependence of even the latter on the sequence of construction will therefore have been greatly accentuated in the dependence of the total final deformation on this sequence and on the rate at which the mortar hardened in different places, as well as on the restraints offered by adjacent members. The last, in turn, besides depending on the original interconnections, will have been progressively modified by each crack or separation.

A definitive interpretation of the precise present forms must therefore consider and reconcile not merely the evidence of these forms and the associated dislocations, but also all other evidence of the sequence of construction visible in the structure itself (from the character of the masonry and applied decoration for instance) and recorded in documents, together with all that is known of the likely effects of the structural actions. Such an interpretation is not easily arrived at, but it offers, at the same time, knowledge of the whole sequence of construction and reconstruction, of the forms as originally intended and set out and of the deformations which they had undergone at different times. The last of these are, moreover, of interest not only in themselves but also as the evidence that presented itself to those responsible for remedial work.

[1] The extensive literature on Doric refinements is summarised in W. B. Dinsmoor, *The Architecture of Ancient Greece* (3rd edition, London, 1950), 164–170 and A. W. Lawrence, *Greek Architecture* (London, 1957), 169–175.

[2] W. H. Goodyear, "Vertical curves and other architectural refinements in . . . early Byzantine churches at Constantinople," *Memoirs of Art and Archaeology*, I, No. 4 (Museum of the Brooklyn Institute of Arts and Sciences, New York, 1904), suggested that, as in the Parthenon, both the main inclinations from the vertical and some of the horizontal curves mentioned above were intentional aesthetic refinements. K. J. Conant, "The first dome of St. Sophia and its rebuilding," *The Bulletin of the Byzantine Institute*, 1 (1945), 71–78, suggested (p. 72) that the dome, the main semidome and the main arches were laid out as non-circular "functional" curves. W. Emerson and R. L. Van Nice, *Archaeology*, pp. 167, 170, originally suggested that the forward horizontal curves of the eastern and western main arches were deliberate camberings of these arches into the expected thrusts from the dome, but Mr. Van Nice has since agreed that this interpretation is now untenable.

[3] Specimens of the mortar hitherto analysed (by Professor F. K. Morris) have all been lime mortars to which had been added various amounts of crushed brick. A brief summary of Professor Morris's findings is given in W. Emerson and R. L. Van Nice, *Archaeology*, pp. 101–102.

THE STRUCTURE OF THE CHURCH OF ST. SOPHIA, ISTANBUL

The Church in the Sixth Century

The most useful contemporary accounts of the original sixth century church and its first partial reconstruction are those of Procopius,[1] Paul the Silentiary[2] and Agathius.[3]

Procopius, writing at about the time of the first partial collapse, but apparently describing the church before this collapse (since he makes no mention of it or of the earthquake which preceded it), gives a remarkably clear layman's account of its form without, however, giving any precise clues to the forms of those parts which were subsequently altered. His singling out of the four pendentives for a detailed description before mentioning the dome, and his reference to these as "triangles"[4] suggests, at least, that these were always separated by a cornice from the dome and not continuous with it, as has been suggested previously.[5] His description of the infilling of the main north and south arches by "certain structural elements including a number of rather small columns"[6] may also be contrasted with Paul's reference to "a wall, mighty in strength"[7] rising to the underside of each arch. His account is chiefly valuable, though, for the mention of two difficulties experienced during the initial construction.[8] We learn first that before the eastern main arch was quite finished "the piers, above which the structure was being built, unable to carry the mass which bore down upon them, somehow or other suddenly began to crack, and they seemed on the point of collapsing." The Emperor is said to have ordered the completion of the arch, "For when it rests upon itself it will no longer need the props beneath it." This is ambiguous because, in the original, the same word is used for "piers" and "props." If it should be rendered "props" in both cases, the advice was reasonable. But the rendering given seems on other grounds more likely, in which case the advice given, if correctly reported, may be significant elsewhere, as mentioned below. Secondly, we are told that when the northern and southern arches were first completed "everything underneath was labouring under the load, making the columns which stood there throw off tiny flakes, as if they had been planed." This necessitated some reconstruction "as soon as the dampness of the masonry had abated enough." Again the account is ambiguous, but, like the first, it indicates the very early onset of deformations in the brick masonry of the arches.

The extent of the first collapse, precipitated by the earthquake of December 557, though delayed, in part at least, until the following May, is not very adequately described. It certainly involved a part of the main eastern arch with which trouble had already been experienced and parts of the dome and semidome dependent on it. The documentary sources agree, however, that in the reconstruction the whole dome was rebuilt to a greater height, different figures being quoted for the extent to which it was raised.[9]

Paul gives the most complete account of the church after this reconstruction, an account agreeing in nearly all respects with what we see today, as far as the basic structure is concerned. In describing the collapse he adds the information that in the collapse "the very piers were shaken to their foundations."[10]

---

[1] Procopius, *op. cit.*, I, i, 20–78, Dewing, pp. 8–33.

[2] Paul the Silentiary, *Descriptio Sanctae Sophiae*, abridged translation by H. Swainson in W. R. Lethaby and H. Swainson, *op cit.*, pp. 35–52.

[3] Agathius, *Historiae*, quoted by W. R. Lethaby and H. Swainson, *op. cit.*, p. 30, by G. Millet: "La coupole primitive de Sainte-Sophie," *Revue belge de Philologie et d'Histoire*, II (1923), 599–617 (which includes a collation of other accounts of the first reconstruction by Malalas, Theophanes, and an anonymous writer); and by K. J. Conant, *op. cit.*, pp. 75–76.

[4] Procopius, *op. cit.*, I, i, 43–44; Dewing, pp. 20–21.

[5] R. Traquair, "The origin of the pendentive," *Journal of the Royal Institute of British Architects*, XXXV (1927–28), 185–187.

[6] Procopius, *op. cit.*, I, i, 40; Dewing, pp. 18–21.

[7] W. R. Lethaby and H. Swainson, *op. cit.*, p. 43.

[8] Procopius, *op. cit.*, I, i, 67–78; Dewing, pp. 28–33.

[9] Malalas and Theophanes give the increase as 20 ft. or more, whereas Zonaras gives it as 25 ft. (G. Millet, *op. cit.*, pp. 602–607).

[10] W. R. Lethaby and H. Swainson, *op. cit.*, p. 36.

Agathius gives the fullest account of the actual reconstruction, though it is unfortunately made obscure by his probable inability to understand completely what was done and the lack of a sufficiently precise vocabulary to refer explicitly to different parts of the structure. One passage, which had been variously rendered by different scholars, is worth quoting in full.[1] "Through the earthquake the church had lost the most central portion of its roof, that part which was higher than the rest. The Emperor restored it, strengthened it and gave it greater height. Although Anthemius was then dead, Isidorus the younger and the other architects considered the previous design, and by that which remained they judged of what had fallen, made out its structure and its weakness. At the east and west they left the great arches in place as they had been. At the north and south they took the construction on the curve and caused it to project inward more, making it gently wider to correspond better to the others and made the equilateral harmony more perfect, thus reducing the distortion of the void, stealing off a little of its extent to make the plan rectangular. Then they rebuilt the crowning middle part—orb, hemisphere or what you will—and this became more nearly perfect, well turned, and everywhere true to line, narrower and steep in profile. It did not astonish the observers as before, but it was much more securely set."

Two subsequent collapses involved, in the tenth century, a partial rebuilding of the western main arch and semidome, together with parts of the adjacent pendentives and part of the dome, and, in the fourteenth century, a similar partial rebuilding of their eastern counterparts. Again the sources are not very explicit about the extents of the damage or of the reconstructions, but we learn that the immediate or proximate cause was, in each case, another earthquake and that, while the first reconstruction took six years,[2] the second was completed in two short campaigns of a few months separated by about seven years.[3]

Finally it is recorded[4] that the Fossati brothers rectified twelve inclined columns in the exedrae of the galleries, and they, and no doubt others as well, also rectified a number of the facings and otherwise obscured some of the evidences of previous deformations.

Fossati's rectifications can easily be discounted and the extents of the reconstructions of the tenth and fourteenth centuries can now be defined with considerable confidence as embracing, in each case, the part of the semidome and main arch above a line passing through the drum containing the windows above the exedrae arches, plus that part of the dome indicated in Fig. 11 and the intervening sections of the pendentives. The break behind each pendentive can actually be seen in the truncated stairways now occupying the corners of the external dome base and was, for a short time, visible externally at the east when the cement rendering was removed from this face. The manner in which the unbonded junctions between the earlier structure and the repair have opened here, as shown in Fig. 12, confirms, together with other observations on the dome cornice, that the inward bends of the main east arch and the east face of the dome base are solely the result of deformations occurring during and after construction, rather than deliberate as suggested previously on the basis of too limited data. Below the repair, on both east and west, the main arches have, instead, a backward inclination which is the same as that of the upper stage of the pier below. The identification of the repairs is further confirmed by observations on the character and bonding of the brickwork where this has been visible, by a detailed study of the dome cornice[5] and of the surface irregularities of the dome itself, and by a comparison of the forms of the repairs with those of the earlier structure. The tenth century repair shows throughout an exaggerated stiffening and strengthening of the earlier form by

---

[1] The version given follows closely that by Conant cited above, but with a few small changes.

[2] The sources are summarised and discussed in C. Mango, *op. cit.*, pp. 76–78.

[3] The sources are summarised and discussed in C. Mango, *op. cit.*, pp. 66–68, and, in more detail, in I. Sevcenko, "Notes on Stephen, the Novgorodian pilgrim to Constantinople in the XIV century," *Südost-Forschungen*, XII (1953), 165–175.

[4] C. Mango, *op. cit.*, pp. 12 and (quoting Fossati), 110.

[5] Reported in part in W. Emerson and R. L. Van Nice, *A.J.A.* (1943), 423–436.

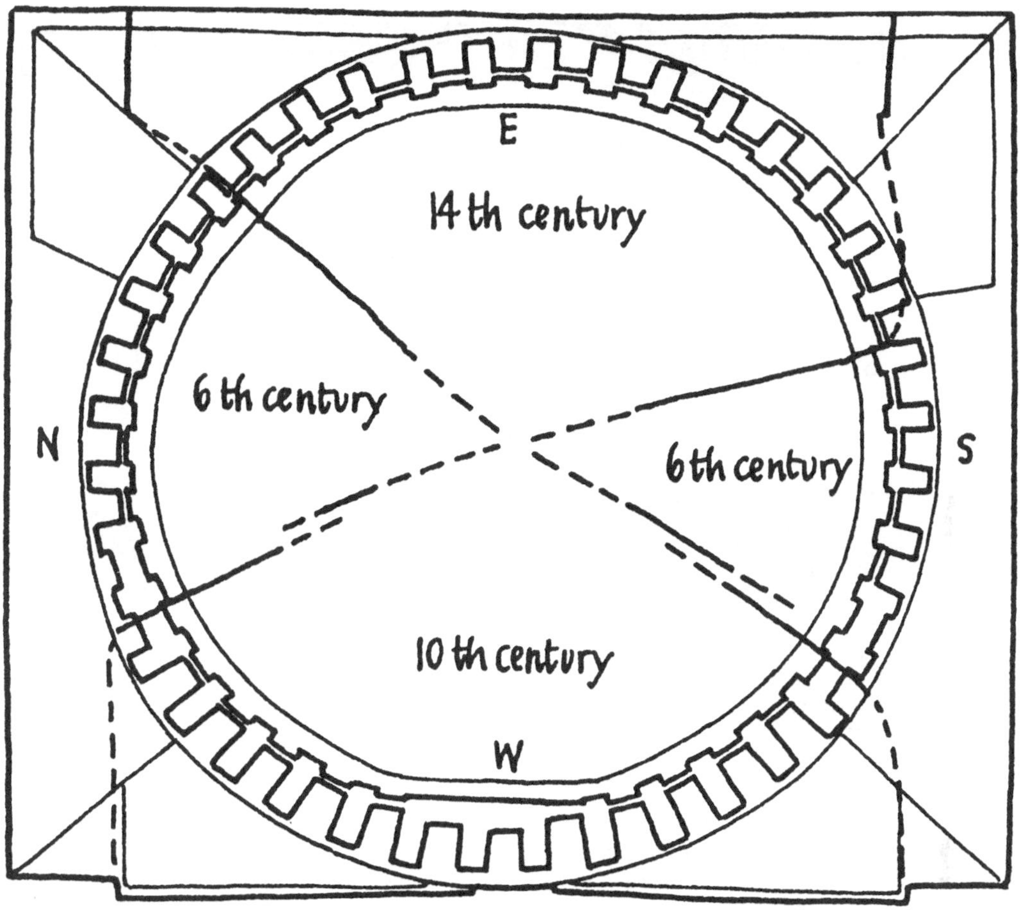

Fig. 11. Plan of the base of the dome showing the extents of the reconstructions in the tenth and fourteenth centuries.

increases in sectional dimensions and the addition of buttressing masses and fillings. In the peculiar profile given to the semidome there is also a strong suggestion that the work advanced very cautiously in limited stages. The fourteenth century repair, on the other hand, shows none of this. It appears to have been undertaken with great confidence and to have followed very closely the previous form. In fairness here one must note, however, that in terms of the prior experience of those responsible for the two repairs, the first must have been a formidable challenge, whereas the second was undertaken after the first peak of Gothic structural invention and one at least of the architects, an Italian, Giovanni Peralta, was very likely familiar with the structural expertise then acquired.

The clue to the identification of successive phases in the sixth century construction and partial reconstruction lies in the recognition that, in nearly all that remains of these, deviations from the vertical, from the straight line or from the circular arc are not only of the kind that would be expected

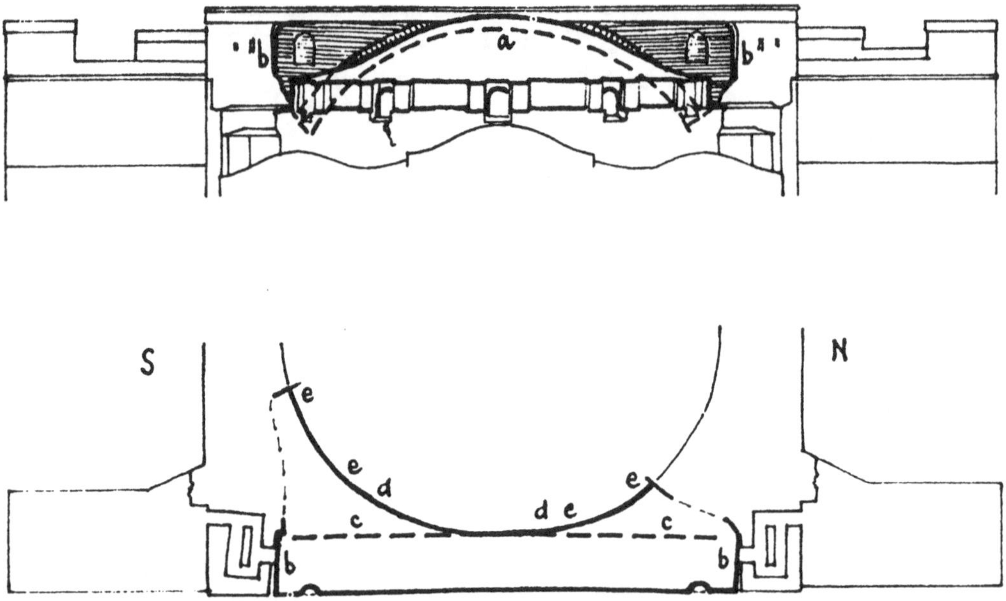

Fig. 12. The fourteenth century reconstruction of the eastern great arch.

Above. Elevation from the East showing the depressed crown *a;* courses of the brick spandrel filling dip at about 1 in 30 towards the crown; breaks at *bb* between sixth and fourteenth century work (originally butted close) now widen considerably towards the top.

Below. Plan at a level slightly below the cornice; breaks at *bb* widen towards the outside; the dome base and the inner face of the arch *cc* are bowed inwards. The splayed gaps between the stones of the dome cornice suggest that, after the cornice was constructed, its curvature diminished at *dd* and increased at *ee.*

to result, as permanent deformations, from the likely structural actions, but also of a magnitude consistent with those actions, with what is known of the behaviour of the materials used, with the observable crack widths and other separations, and with a probable sequence of construction as indicated by the sources. The working hypothesis has, therefore, been that originally all the major forms of what remains of the sixth century structure were intended to conform to the vertical, the horizontal, the straight line or the circular arc and were so set out with a high degree of accuracy similar to that still apparent where there has been no possibility of deformation. Within the limits of the investigation so far completed, it is believed that this hypothesis is fully substantiated by the consistency of the conclusions drawn from it with other available evidence. On such a basis it becomes possible, by correlating the deformations of different members of the structure, to estimate the sequence in which they became part of it and, with the help of the documents to fix certain points in the sequence, to estimate the progress of the deformations.

It can now be confirmed that Conant's interpretation[1] of the description given by Agathius of the first phase of the first reconstruction is in principle correct: the strong lower arches spanning between the inward projections of the main piers on the north and south sides were left as they were, but the lighter upper arches rising to the dome cornice were extended forward in their upper parts. As seen

[1] K. J. Conant, *op. cit.*, pp. 74–76 and plate XXXVII.

in Figs. 3 and 4, the present inclinations of the vertical planes of the springings of the lower arches are the same as those of the piers from which they spring but the crowns lie further back because the arches also bend backward horizontally, as a result of the dome thrusts, in curves which are closely parallelled by those of the outer faces of the dome base (Fig. 6). The exposed soffits of the upper arches, on the other hand, are progressively widened in a somewhat irregular manner above the haunches, so that the planes containing their springings and crowns incline backwards at angles which are only about one third of the pier inclinations. The irregularity of the "piecing out" makes it unwise to assume that these planes were ever exactly vertical. But, assuming that they were approximately so at the time of the reconstruction, we have here one indirect indication of the amount of pier tilt at the time of the first collapse.

A more recent suggestion by Underwood and Hawkins,[1] on the basis initially of evidence of renewal of the mosaic decoration in the soffits of the gallery arcades under these north and south arches, that, at about the same time, both the tympana which fill the arches and the arcades themselves were also reconstructed can also be confirmed. The necessity for such a reconstruction is apparent if at this time, in the east-west direction, a similar proportion of the present total inclinations of the piers to that indirectly indicated in the north-south direction by the evidence discussed above is assumed to have occurred. It would have resulted in the piers separating by something of the order of 0·2 m. at the level of the springings of the arches and, even if the arcades and tympana were not fully completed until after some of the separation had taken place, they would have been badly cracked open. The present tympana are also cracked, but no more than is consistent with the subsequent spreading of the arches and separation of the piers. The inner faces of the present tympana necessarily follow the inclinations and horizontal curvatures of the lower main arches where they ajoin the inner faces of these arches. But, as will be seen from Fig. 4, the outer faces are much more nearly vertical and they parallel closely the inclinations of the "pieced out" upper arches. The arcades, moreover, are now almost straight and their columns upright in sharp contrast to the deformations of the gallery floor and the vaulting lying behind; but there is none of the evidence of later rectification as seen where the Fossati brothers moved the columns in the exedrae. To the alterations first suggested by Conant we must, therefore, add a reconstruction, roughly contemporary, of the tympana, the arcades immediately below them and the narrow tunnel vaults carried by the arcades.[2] The iron ties between the heads of the columns must have been inserted at this time, but it is not yet clear whether there were ties there originally.

Even after the supports of the dome had been brought forward on the north and south, the transverse diameter of its base must have considerably exceeded the longitudinal one, so that Isidorus the younger was not able to reconstruct it on a circular plan. In the portions of it which remain, his actual setting lines, scribed in the stones of the cornice, can still be seen (Plate XVI (*c*)) and they prove, as a result of a very precise survey in 1965, to approximate very closely indeed to two arcs of circles with radii considerably less than half the transverse diameter, as shown in Fig. 10. Careful examination of all available evidence, particularly the present condition of the closely fitted joints of the cornice stones, shows that these curves can have distorted little, so that it is particularly significant that the present separation of their centres is considerably greater than the present difference in diameters. Originally it would have been less than this by about 0·4 m., but the original longitudinal diameter is not precisely known and there is no direct evidence to show how the curves were completed at the

[1] P. A. Underwood and E. J. W. Hawkins, "The mosaics of Hagia Sophia at Istanbul: The portrait of the Emperor Alexander," *Dumbarton Oaks Papers*, No. 15 (1961), 189–217.

[2] This leaves open the question of whether this reconstruction was the one referred to by Procopius in the passage quoted, in part, above (I, i, 74–78) or a subsequent and more extensive one presumably associated with the major reconstruction of the dome and the preparatory "piecing out" of the upper arches. Further evidence confirming conclusively that it did take place and defining more precisely its manner and extent was uncovered in 1967, and it is intended to present this, with a fuller discussion, in *Dumbarton Oaks Papers*, No. 23 (1969).

east and west. It was probably as shown in Fig. 10, with two more centres similarly placed on the longitudinal axis from which arcs of larger radius were struck, the common radii on which the larger and smaller arcs met lying at about 45° to the axes. The present transverse cross-section of the dome is consistent with its having been originally a circular arc somewhat less than a full semicircle. Since the same arc could not be rotated to generate the complete surface, it is probable that successive ribs were formed as circular arcs of decreasing radius, to give a constant height at the crown.

Nothing remains in the present structure to indicate the form of the original dome except the evidence of the bending and spreading of the arches and tipping of the piers. Part of these deformations, including a forward bend of the eastern and western main arches analogous to but probably greater than that which is seen now, but no outward bend of the northern and southern arches, must have occurred during construction of the main arches and the pendentives, so that this original dome must also have been slightly elliptical from the start. But further inference can be based only on the documents already mentioned.[1]

The main piers through most of their height, and the buttress piers up to the level of the springings of the main arches are, without doubt, those of the original structure. But the upward extensions of the buttress piers, the important bracing arches under the wider arches linking the main and buttress piers, and numerous less important fillings all show features, such as lack of bond, which could indicate that they were added later. The analysis of these features is complicated by the inaccessibility of critical details, and by the fact that some measured inclinations are misleading because the facings were obviously altered in the nineteenth century and made more nearly vertical than the masonry behind. Tentatively it is concluded from the evidence now available that the bracing arches were constructed with the arches that they fill, and indeed served as centering for these, and that the two upper stages of the buttress piers, though added as successive afterthoughts, were constructed as soon as the piers began to deform and tilt under the thrusts of the main arches and pendentives and that they are, therefore, roughly contemporary with these.[2]

From this necessarily abbreviated account, from which much of importance has been omitted, it is clear that, though the basic structural concept was sound, Anthemius and Isidorus the elder either failed to appreciate quantitatively the effects of the thrusts of the main arches and dome on the fresh masonry of the piers or were spurred on to undue haste by Justinian's enthusiasm and against their better judgement. Remedial measures were taken as the need became apparent, but too late to forestall the serious weakening that resulted from the virtual disruption of the connections between the main piers and the buttress piers. This disruption, which allowed the piers to yield too independently of one another, sharing the outward thrusts but no more, was assisted, moreover, by a fatal flaw in the design: by a reliance on full semicircular and segmental arches, subject at their crowns to large shear and bending forces they were unable to withstand, to unite the two. Distortions were further aggravated by the excessive reduction in the bulk of the main piers at gallery level, as is seen in the large increases in inclination which occur there.

Earthquakes precipitated or hastened the first collapse as they did later ones, but they were not the main cause. Because the structure is founded on a rock base which shows no evidence of dislocation,

[1] For previous attempted reconstructions of the form of the original dome see: W. R. Lethaby and H. Swainson, *op. cit.*, pp. 209–210 and Fig. 4; E. M. Antoniades, *Hagia Sophia*, III (Athens, 1909), 101–106; R. Traquair, *op. cit.*; K. J. Conant, *op. cit.*, p. 76 and plates XXVI and XXXVII.

[2] Limited tests behind the facings have now provided more reliable measurements of the inclinations of the abutting unbonded faces of the main and buttress piers and the projections from them at gallery level and of their relative displacements. Though not yet conclusive, they do now suggest that these projections, and the bracing arches which they carry, may all have been later additions made as soon as the thrusts of the main east and west arches began to be effective. More detailed study of the transverse arches and walls of the upper stages of the north and south buttresses has confirmed that these also were added at a very early date. The pronounced fractures, typified by those seen in the more recent drawing reproduced in Fig. 9, and the associated dislocations, are consistent with most of the pier tilts having taken place after both upper stages were in place.

the amplitudes of the imposed movements at ground level would be small. The partial disruption of the structure by previous cracking may have led to buffeting, but will have led also to many different natural frequencies and modes of vibration so that there would be no large resonant amplifications of the ground movements. It would also increase the already large damping characteristics of the masonry, and the consequent rapid absorption of the energy of the earthquake would further limit the resultant movements. The worst that is likely to have happened is that some existing cracks would have extended and widened and been held apart by debris, and that some non-reversible slips would have occurred as the frictional resistances normally opposing them were momentarily reduced.

THE SIXTH CENTURY ARCHITECTS AND THE GENESIS OF THE DESIGN

A final assessment of the intentions and methods of Anthemius and Isidorus the elder, should start from a knowledge of the constraints under which they worked. All that we know, unfortunately, is that Justinian, their master, obviously knew what he wanted and did not hesitate to impose his will. He may well have been responsible for the crucial decision not only to vault the new church but to make it a domed basilica and not a fully centralised structure like the later cross-in-square church, the churches of the Italian Renaissance best exemplified today by S. Maria della Consolazione at Todi, or many of the great mosques built after the Turkish conquest. Certainly the decision to combine a buttressing system of large and smaller semidomes on two sides with a different and more conventional one (as in the basilica of Maxentius) on the other two was not made on purely structural grounds. Structurally it would have been both more logical and more obvious to adopt the same system on all sides. Non-structural considerations are equally apparent in the relationships to one another of the many monolithic marble columns structurally employed. Those standing some way back from the main arcades of the nave answer directly to one another at gallery and ground levels. But those of the main arcades do so in a manner which is structurally quite arbitrary. Indeed, there are many details which cannot be explained in purely structural terms.

This cannot indicate a failure to recognise the structural implications and give full weight to them; it shows rather a high degree of confidence in refusing to allow them to become dominant. For, apart from the remarkable structural success of the design, we know something of the backgrounds of both designers. They were dignified by Procopius with the profession of *mechanicos* and it has been shown by Downey[1] that this implied a complete mastery of all the academic disciplines relevant to architecture including, in particular, a knowledge of the contemporary science of mechanics. Isidorus seems to have been, in fact, a professor of geometry or mechanics and wrote a commentary on the lost treatise "On Vaulting" of Heron, while Anthemius also appears to have been a teacher and wrote treatises on mathematics and mechanical devices.[2] It is less certain whether this theoretic knowledge was supplemented by previous practical experience, but it is possible that one or both had been associated with the slightly earlier church of SS. Sergius and Bacchus which may, in some respects, be considered a much smaller prototype for St. Sophia.

On the theoretic level their adoption throughout the building wherever possible of the semi-circular arch, including its use to link the main piers and the buttress piers, the associated adoption of a spherical or near-spherical surface for the dome and semidomes, and the relative sections given to different arches and vaults are of particular interest.

These all, with slight qualification in the last case, suggest a primarily geometric approach to the structure. They suggest, in other words, that, rather than seeing it as an abstract network of forces

[1] G. Downey, "Byzantine architects: their training and methods," *Byzantion*, XVIII (1948), 99–118; Procopius, *op. cit.*, I, i, 24, 50, 76.

[2] G. Downey, *op. cit.*, pp. 112–114; T. L. Heath, *A Manual of Greek Mathematics* (London, 1931, Dover edition, New York, 1963), p. 519; G. L. Huxley, *Anthemius of Tralles, a Study in Later Greek Geometry* (Cambridge, Mass., 1959).

in equilibrium with one another and calling into play the characteristic strengths and resistances to deformation of the constituent elements, both designers saw it primarily in terms of the forms of the elements, the geometric characters of these forms and the simple geometric conditions which distinguish stable static structures from mechanisms. The almost exclusive use of the semi-circular arch form, for instance, recalls Alberti's much later insistence on its superior strength,[1] and there are many similar echoes in structural practice up to the seventeenth century.[2] The relative thicknesses of the dome and greater and lesser semidomes (assuming that the present eastern semidome closely reproduces the original large semidomes) could indicate a fairly accurate appreciation of their respective structural roles as understood today. But these thicknesses are roughly proportional to the diameters which could, equally, indicate a purely geometrical proportioning. Similarly the obvious realisation of the inherent strengths and stiffnesses of these shells, with their uniform and equal double curvatures, could indicate an early anticipation of modern understanding. But again there is an echo of Alberti,[3] and a little reflection will show that, even today, these strengths and stiffnesses can most readily be intuited in purely geometrical terms by trying—in vain—to imagine possible modes of deformation that do not involve considerable extensions or contractions of the surface. Paradoxically, the most obvious structural innovation of the whole design—the use of the semidomes in the dual role of vault and buttress—is the one most readily explicable in terms of the known background of knowledge at the time. It is unlikely, moreover, on the analogy of the writings of Guarini[4] and Wren,[5] that we should find anything in Heron's treatise on vaulting or the commentary by Isidorus which would significantly enlarge that known background. They almost certainly contained little more than geometrical rules for setting out and some simple Archimedean statics. For ease in setting out, of course, the spherical surface had the great advantage of giving interpenetrations of the simplest kind and of the preferred semicircular curve.[6]

Questions arise chiefly in relation to the dome and the arches linking main and buttress piers. The first dome was apparently much less than the hemisphere and the second still slightly less. Does that indicate a correct realisation that in the full hemisphere tension would inevitably develop in the lower part, but a sufficiently shallow dome would be wholly in compression given adequate support at the base? The channelling of the thrusts of the second dome into ribs which became free-standing buttress-like piers at about the level where, with a firm base, tensions would first appear and were linked there by continuous ties lends colour to such a suggestion. On the other hand, it must already have been apparent, even when the first dome was set out, that the base was less rigid than had been hoped. The total depth given to the arches linking the main and buttress piers at ground and gallery levels (including the bracing arches) was much greater, in relation to the span and vertical load, than that of any of the other arches except perhaps the lower main arches on the north and south which also had to carry more than a pure vertical load. This can only indicate a clearer recognition of their task of uniting the piers than might otherwise be inferred. It may also be relevant that in the aisles and galleries any arch profile but the semicircular would have been

[1] L. B. Alberti, *Ten Books on Architecture* (translated by J. Leoni) (fascimile edition, London, 1955), Book III, Ch. XIII, p. 57.

[2] R. J. Mainstone, "Structural theory and design: Part 1, Early proportional theories," *Architecture and Building*, XXXIV (1959), 106–113.

[3] L. B. Alberti, *op. cit.*, Book III, Ch. XIV, p. 59.

[4] G. Guarini, *Architettura Civile* (Turin, 1737).

[5] C. Wren, Reports on Old St. Paul's, Westminster Abbey and Salisbury Cathedral and Tract II on architecture, reproduced in S. Wren *Parentalia of the Wren family* (London, 1750). See also S. B. Hamilton, "The place of Sir Christopher Wren in the history of structural engineering," *Trans. Newc. Soc.*, XIV, 27–42.

[6] In fact, owing to the manner in which the main eastern and western arches and those spanning the exedrae are incorporated in the greater and lesser semidomes (with internally exposed soffits and faces continuous with their surfaces), the semicircular arcs of the former and the quarter-spherical surfaces of the latter reciprocally imply one another, assuming the surfaces to be regular geometrical forms.

aesthetically discordant, and that in the stage just above the roofs a segmental arch of slightly reduced rise was substituted. There still remains, however, a possibility that most of the arches were somehow thought of as binding together the piers from which they sprang. Procopius' reference to Justinian's intervention when the main eastern arch threatened failure is one hint of this; the use here of *loros* to signify "arch," with its connotation of binding is another.[1] But neither is conclusive and neither, even if the translation of Procopius given above is the correct one, necessarily implies more than a laymen's imperfect understanding.

In the heightening of the buttresses after the piers began to yield and in the rebuilding of the dome after the first collapse, it was possible to learn much from the observation of what had already happened, though what was learnt would still be limited by the observer's ability to interpret what he saw. In the first case, Anthemius and Isidorus were doing little more than adding a further obstacle to an impending movement. In the second case one would give much to have a fuller record of the conference of architects to which Agathius refers in the passage quoted above. They seem to have concentrated their attention on the dome rather than its supports, intent partly on giving it a more regular curve of slightly reduced maximum diameter and partly on reducing its thrust as compared with that of the previous dome, which was by now amply manifest. By bringing forward slightly the north and south upper arches and the pendentives they would have increased the inward tipping moments of the latter about the piers and this might have been intended, on the basis of Archimedean statics, to help counteract to dome thrusts.[2] Or it may have been no more than "reducing the distortion of the void."

An intuitive understanding of natural law may be given to some before it is possible to formulate the law. Usually such understanding is very incomplete and not easily generalised or directed to problems outside the limited range of direct experience from which it sprang, but it may transcend these limits, as indeed it must when new scientific insights are born. How far, on the evidence of their work on St. Sophia, can Anthemius and Isidorus be credited with such insights into the as yet unformulated laws of static equilibrium for forces acting in any direction and of structural strength and deformation? In the writer's present opinion the answer must be—to a very limited extent only: perhaps as far as Wren but not as far as Leonardo or Guarini. They must have learnt much in the hard practical experience of putting their ideas to the most searching test of all. But their work is to be acclaimed first as one of the supreme architectural masterpieces, secondly as a great structural *tour de force* conceived largely in geometric terms.

Discussion

Mr. Mainstone, in reply to a query by Mr. Earl, said that architectural plans had survived from the Roman period and earlier but, as far as he knew, no working drawings. For these they could look back only as far as the early Gothic period. Setting out from them on the site seems then to have been based on the use of very simple geometric grids. Indeed what used to be called the secret of the masons seems now to have been no more than a few simple geometric tricks employed in establishing these grids and, for instance, halving the square by joining the mid points of its sides. If we

[1] I am indebted to Professor Cyril Mango for helpful discussions on the possible ranges of meaning of certain words and passages in the sixth century texts, though he must not be held responsible for the suggestions made in this paper.

[2] This possibility may be compared with Wren's analysis of vaults and their abutments in the unfinished Tract II (S. Wren, *op. cit.*, pp. 356–358). Wren's analysis implies that the whole section of vault carried by a pier constitutes, with the pier, a single rigid body. It then ignores completely the horizontal thrust at the crown and leads to a precise but quite erroneous conclusion. What is now suggested implies at least a qualitative recognition of the horizontal thrust and a clearer recognition of the static role of the pendentives. But it is still far from a complete appreciation of the conditions of equilibrium and unlikely that it could have been the basis of any precise calculations.

could legitimately argue back from the thirteenth century, the setting out of a building like St. Sophia was possibly done from relatively few drawings of a very simple kind, but he could not say precisely what form those drawings took.

The Chairman, Mr. ANDRÉ KENNY, said that the treatise of Frontinus on Roman aqueducts in the first century A.D. referred to drawings "which are kept in the office." There seemed to be little doubt that at that time there were people not only drawing things they wanted to make, but also keeping records on bits of paper in the drawing office.

Professor CYRIL MANGO said that plans on paper certainly existed at the time. There were references to them in tax statements and literature. Concerning the destruction of the church of Gaza in the year 400, we were told that a plan was actually sent from Constantinople to Gaza. A special term for it occurred a number of times in Byzantine literature. Apparently, after the town was besieged, they somehow set out the walls by marking furrows or lines on the ground, but that was as far as it was possible to go on the basis of literary evidence. Was there any connection between St. Sophia and San Vitale in Ravenna (A.D. 526–547), designed, he believed, by one Julianus Argentarius and roughly contemporary?

Mr. MAINSTONE replied that both churches belonged to the general class of structures in which a central space, usually vaulted, was surrounded by another for circulation. Nearly, if not all, the others were more fully centralised than St. Sophia, but they all had, at least, a common ancestry. If one was looking for direct connections with St. Sophia, he thought that SS. Sergius and Bacchus was more important than the closely contemporary and related church of San Vitale, but the latter certainly ought also to be considered.

Mr. R. J. M. SUTHERLAND asked how thick were the bricks in St. Sophia; why were the joints much thicker than the bricks; was this due to inaccuracies in firing a two-foot square brick; and how were the bricks made?

Mr. MAINSTONE replied that the sixth century bricks averaged about 4·5 cm. thick: those of later date were mostly a little thinner. They appear to have been formed in open moulds and levelled by hand, leading to slight variations in the thickness before firing. There was, however, no evidence of appreciable warping due to firing in the one fairly large area of sixth century wall that could now be seen in cross section, so that it was unlikely that such warping had necessitated the thick joints. Perhaps it was primarily a matter of economy—mortar being cheaper than these large bricks. Even in the major curved members of the structure—the principal arches and vaults—it seemed that the joint thickness could easily have been reduced if desired to something closer to that which was seen, for instance, in the Red Basilica in Pergamon. In the smaller vaults and arches, though, the wide joints were possibly necessary to give the required flexibility in the choice of form.

Mr. FRANK NEWBY asked if there was any information on the method of construction of the dome, the sequence of erection, and on problems associated with this.

Mr. MAINSTONE replied that there was now no evidence on the manner of construction of the first dome and only indirect evidence of the procedures adopted in the first complete reconstruction and the subsequent partial ones. Considering first the sixth century reconstruction, he thought that each rib at least had been fully centred. The scribed setting line that had been mentioned implied, for instance, the use of a chord fastened at the centre of each arc to a fairly rigid central platform and carrying, probably, a chisel at the other end, and it seemed unlikely that such a platform would be constructed for this purpose alone. Where the brickwork of the sixth century ribs had been bared alongside the window openings it could be seen, moreover, that the bricks were set almost radially, whereas, if they had not been constructed throughout on centering, one would expect the inclination of the bricks here to be a good deal flatter and to increase only slowly from that of the cornice slabs. More generally, both the transverse section of the dome and the profiles of the surviving main arches of the sixth century structure were so close to what one would now expect to find as deformations of circular arcs, precisely set out in the first place, that it seemed, bearing in mind the thickness of the mortar joints and the character of the mortar, that they must all have been constructed on centering. It was less clear though whether the webs had been constructed integrally with the ribs throughout their height or added subsequently as infillings, but the former seemed at present to be more likely. There must have been problems of other kinds in the subsequent partial reconstructions when about threequarters of the dome remained and the new work had to be fitted to this. There is always some surface irregularity along the join, from which it appears that the edges of that part of the structure which remained tipped in slightly. In the tenth century reconstruction, but not in the fourteenth century one, the surviving masonry seems to have been carefully trimmed back to the nearest complete rib which now has, at both north and south, an almost vertical edge instead of the splayed edge seen elsewhere. In both reconstructions the bricks of the ribs have, for a considerable height, an inclination appreciably less than the radial one seen in the sixth century sections. This seems to indicate a more limited use of centering in these reconstructions. A further indication of this is a greater irregularity of surface and of the rib profiles, particularly at the east, where it is probable that the reconstruction was completed in two distinct phases.

Mr. R. W. M. Clouston was surprised that the semidomes collapsed, in both cases, while the supports on the north and south sides of the dome remained standing, still in their sixth century form. Was it true that the bricks used in the centre of the dome were made of some lava material of a light density? What evidence was there of damage to the brickwork by corrosion of the iron tie rods?

Mr. Mainstone said that it was, indeed, a little paradoxical, at first sight, that the semidomes should have collapsed and the supports to the north and south should, each time, have remained standing when the semidomes were undoubtedly the more efficient members in resisting the outward dome thrusts. The best index of that greater efficiency was the fact that each main pier was now pushed outwards, by the thrusts of the dome and the principal arches, about three times as far to north or south as it was to east or west. The collapses of the east and west arches and semidomes were largely the results of the associated north-south spreads of their bases. These spreads were sufficient to allow the crowns of the arches to drop out, whereas the lesser spreads of the bases of the north and south arches (which are also in the aggregate considerably deeper) resulted only in the depressions of the crowns that had been mentioned. It should perhaps be added that it is most unlikely that there would ever have been any collapses without these spreads of the bases of the main arches under the action of the principal outward thrusts. The east and west arches and semidomes were amply proportioned to resist even the worst earthquakes if their supports had remained immovable.

It was difficult to give a final answer to the question about the possible use of material of light density in the upper part of the dome. Precedents for this were, of course, known elsewhere, but from the little that had been seen of the actual masonry of the dome of St. Sophia it did appear that the normal bricks had been used throughout, apart from a springing course of marble for the window arches. He had seen no evidence of damage due to corrosion of the iron tie rods.

Mr. Ernest Hawkins said that, though he had not had any opportunity of seeing the interior of the masonry of St. Sophia, he, likewise, was not aware of any serious damage due to corrosion of the early iron ties there. On the other hand, at St. Theodosia, a monastery church in Constantinople of the first decade of the tenth century, an iron chain (lengths of iron rod with circular eyes at each end through which the adjacent lengths were pinned together) had done considerable damage to the masonry by corrosion salt. But in that building the weather had been coming through the masonry for generations.

Mr. MAINSTONE added that most of the known early iron ties in St. Sophia were inside the building, either spanning between the capitals of the columns at aisle and gallery levels, across the arches that these carried, or cramping together the stones of the cornices. Little damage was to be expected from these.

In reply to a reference made by a visitor to the great arch of the ruined Sassanian Palace of Ctesiphon (A.D. 550), over 80 ft. in span and reputed to have been built without centering, Mr. MAINSTONE doubted whether this great vault could have been built without any centering, but at most it possibly called for a light centering for the lowest voussoir ring. It was perhaps the finest remaining early example of a very ancient technique of pitched brick vaulting which could be seen very well in the store rooms of the Ramesseum at Thebes, and which had been described by Dr. Norman Davey in *A History of Building Materials* (1961), Chapter 17. Examples of the same technique could be seen in many of the minor vaults of St. Sophia, both barrel vaults and groin vaults. They could be seen also in the neighbouring church of St. Irene and in the earlier rotunda of St. George in Salonika. Not only was the technique, in origin, a very ancient one: it was still practiced in the old way from Spain to Iran and a variant of it had been carried over to the United States in the late nineteenth century by the Spanish Guastavino brothers. They were responsible for many centreless tile vaults constructed during the following decades in the principal cities of the eastern seaboard. The most notable was probably the temporary dome, with merging pendentives, constructed in less than four months over the crossing of the cathedral of St. John the Divine in New York in 1909. The dome itself, with a span only slightly less than that of St. Sophia, had a maximum thickness of only 19 cm.—an interesting contrast. It was, however, set on a substantial base and was, in addition, reinforced there by six circumferential tie rods.

Replying to a question by Dr. C. St. C. B. DAVISON, on the origin of certain forms of arched and vaulted construction, Mr. MAINSTONE said that St. Sophia could be regarded both as one culmination of the development of the double-shell domical form that had already been referred to in connection with San Vitale in Ravenna, and as the finest example of the transposition of Roman techniques of heavy large scale vaulting in concrete into less massive brick-and-mortar construction. The latter transposition had a considerable history behind it and one could even say that it had taken place to some extent in Rome itself in the later days of the empire. The greatest leap forward in the design of St. Sophia had been in the organisation of its forms and the spaces they enclosed. If one wished to trace the basic double-shell form back beyond San Vitale and SS. Sergius and Bacchus, the direct line of descent would possibly pass through structures like Santa Constanza and the vestibule at the north end of the Piazza d'Oro of Hadrian's Villa to the octagon of the Domus Aurea, but one would also have to take note of a number of structures outside Italy such as Constantine's Golden Octagon in Antioch and his original rotunda of the Holy Sepulchre in Jerusalem. Recently there had been tantalising glimpses of other possible links in the chain in Istanbul itself, but it was still too early to say more about these.

Challenged by Mr. J. G. JAMES on his early dating of many of the iron ties in St. Sophia, Mr. MAINSTONE acknowledged that the great complexity of the structure presented many problems of dating and interpretation. By piecing all the evidence together, it was possible, nevertheless, to date quite certainly much of the working masonry and much of the decorative marble and mosaic at ground and gallery levels either to the initial construction or to the first reconstruction only a few decades later. The evidence for also assigning a sixth century date to many of the iron ties was the manner in which they were embedded in or accommodated by this undoubtedly early and undisturbed masonry or decoration. One found, for instance, sixth century mosaic in the soffit of an arch which had clearly been set around the tie which spanned the arch. On the cornices one even found chases cut in the stones in a number of places, obviously to accommodate ties in the original construction but no longer doing so as a result of subsequent but very early displacements of the stones. It was of particular interest, therefore, that some of the early ties were made up of individual forgings as large as 4 cm. × 5 cm. × 3·7 m. It was probable that those passing through the aisle vaults under the gallery floor were even larger.

Dr. W. I. CHALONER, referring to the brickwork used for constructing the arches and vaults asked whether this was not the local equivalent in Constantinople of the concrete used in Rome in the older Pantheon dome.

Mr. MAINSTONE agreed, but thought that it was important to bear in mind both that there had been changes even in Roman concrete during the centuries that had elapsed between the construction of the Pantheon dome and that of St. Sophia, and that there were both similarities and differences if these equivalents were compared. The later tendency in Rome had been increasingly to use roughly coursed brick fragments as the concrete aggregate and to incorporate quite complex systems of brick ribs in the vaults, as seen, for instance, in the so-called temple of Minerva Medica. When embedded in the concrete that filled the intervening spaces and provided an unbroken outer surface, these ribs would have no independent structural identity, but, in terms of erectional procedure, they might serve a purpose very similar to that served by the ribs of St. Sophia's dome. In terms of the structural action of the completed dome, he thought that the chief difference between St. Sophia's dome and nearly all Roman ones was the great thickness given to the latter up to about half their height, giving the characteristic shallow stepped external profile usually associated with the Pantheon, as compared with the almost uniform thickness of the former and its punctuation all round the base by windows. The substitution of bricks and mortar for the concrete of the Pantheon dome was unlikely, once the mortar had matured, to have made much difference to the characteristic strengths, though it is possible that sliding deformations would occur more readily (along the bed joints) in the former than in the latter.

Mr. J. FOSTER PETREE, in proposing the vote of thanks, said that the Paper in many ways represented a beautiful example of what a Paper given to this Society should be. The Author had not only described with amazing clarity a structure of great complexity—he had also given Members an insight into some of the thinking that went into that structure, and, further back still, an idea of what the architects were after. He had told them about the relation between the conception of an architectural structure and its execution. He had also hinted at the difficulties which confront every architect who is working for a master: the boss says that something has to be done; he does not say how. The building was erected at a particularly interesting time in history. The city of Constantine, the New Rome, had by then got going and had asserted its influence as the capital of that part of the world: the name of the City of Constantine is still used in the encyclical letters of the Orthodox Patriarchs. It is of particular interest to us that a building of that period should have been described in such a fashion, because this is a time when many of the stores of knowledge, particularly technical knowledge, of the ancient world were being collected together and prepared, as it were, for our use. Many times that evening they had heard the name of Anthemius of Tralles, architect or *mechanicos*. His work would today entitle him also to be called a technical author. He wrote a great collection of mechanical knowledge, which has unfortunately largely been lost. He also transmitted a great deal of our knowledge of the static and hydrostatic theories of Archimedes. This had not only been a most enjoyable Paper to listen to; it was one which caused us to think a good deal about what we were trying to do in this Society, and about how our thoughts should be organised when we were writing Papers for presentation to it.

PLATE X

(*a*) From the South

(*b*) From the West:
St. Sophia, Istanbul, exterior

PLATE XI

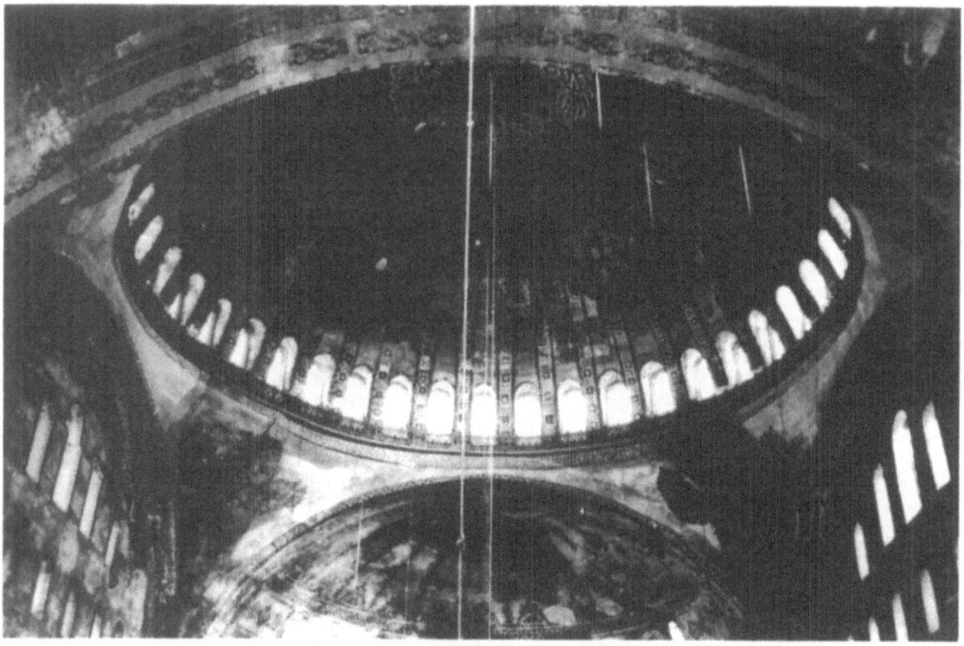

(*a*) The dome from the West Gallery

(*b*) Looking towards the North-west exedra from the dome cornice
St. Sophia, Istanbul; Interior

PLATE XII

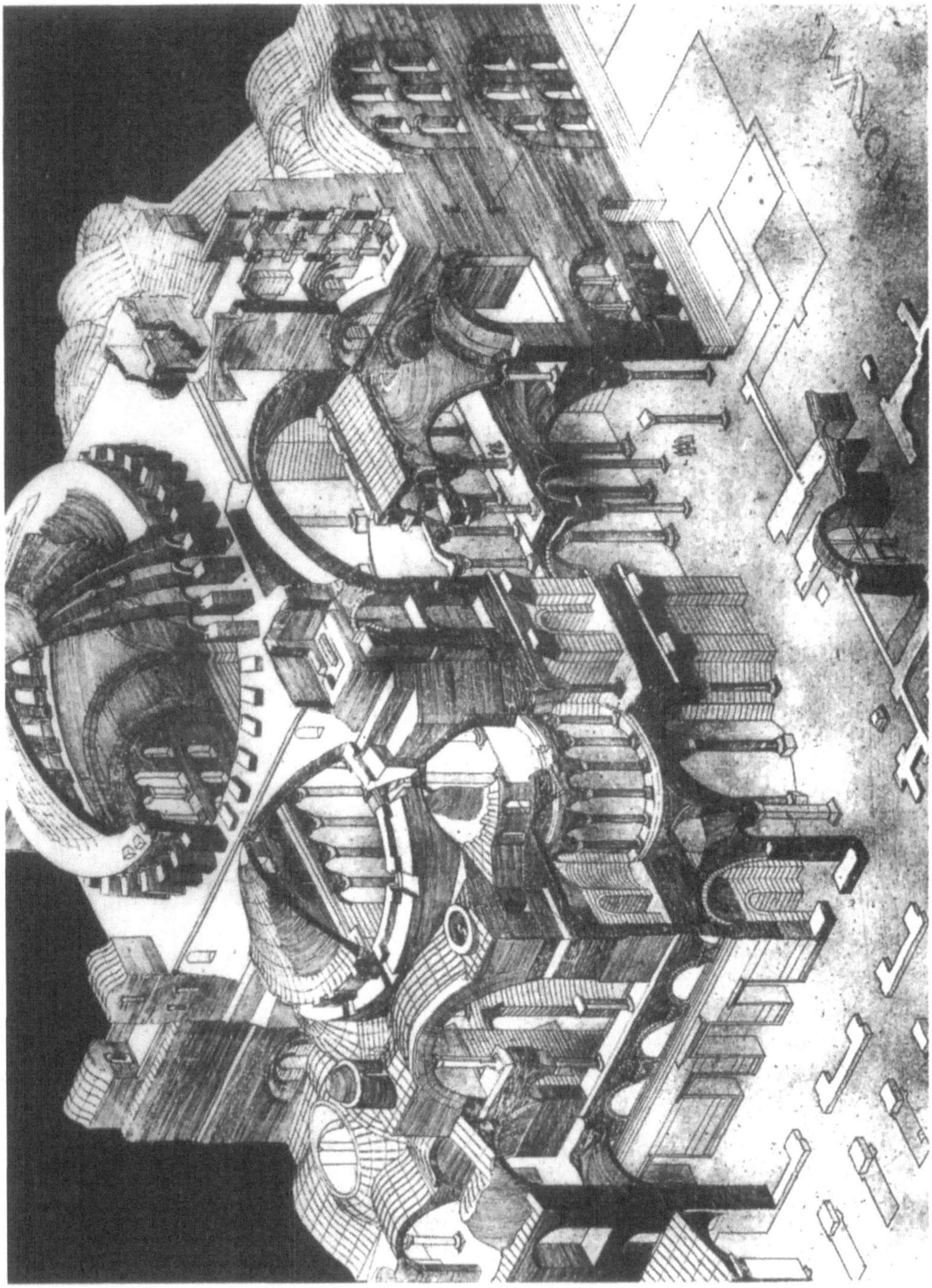

(Henri Prost, 1906)

St. Sophia, Istanbul; a cut-away isometric view from the South-west

PLATE XIII

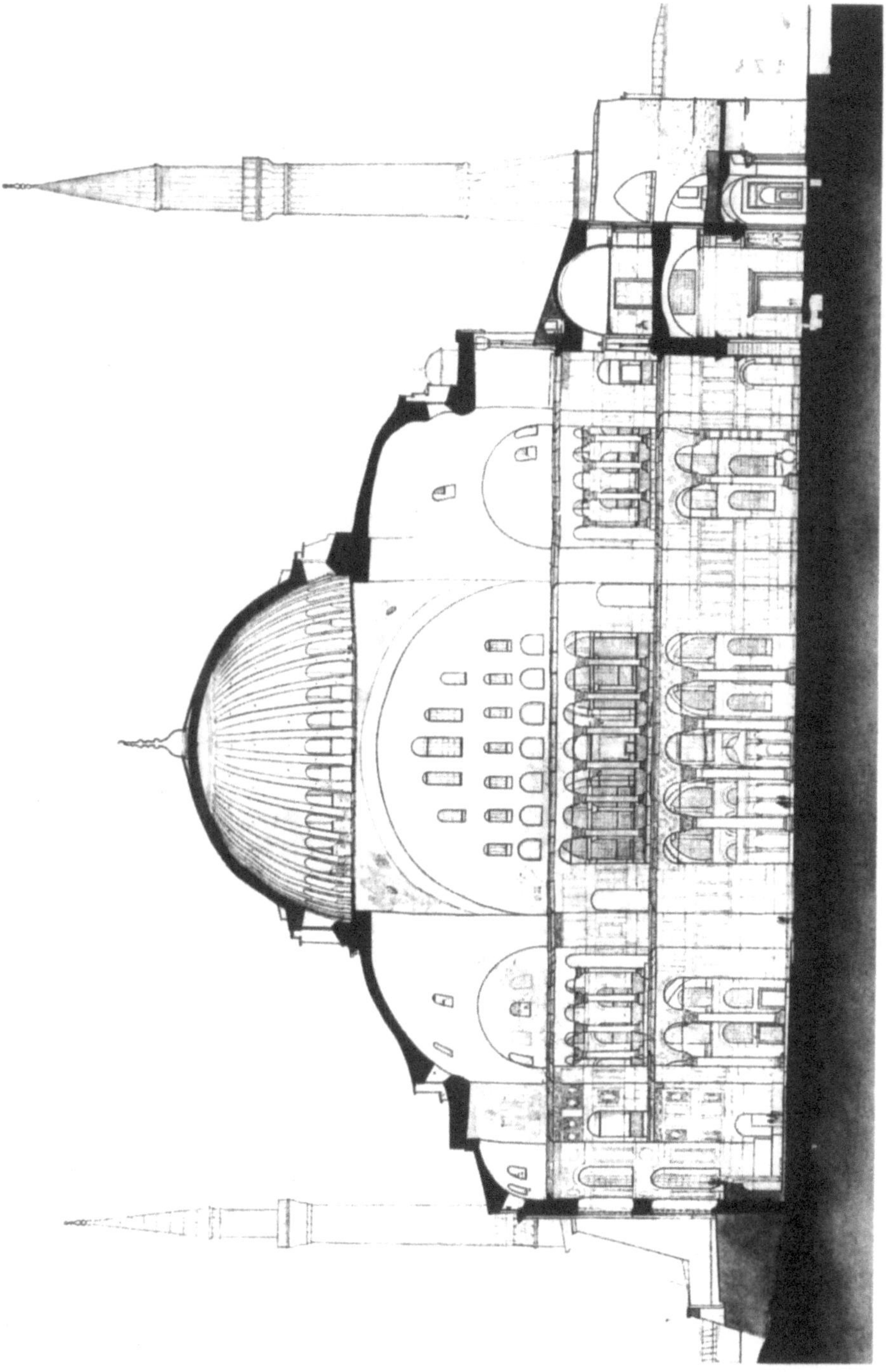

(*Courtesy R. L. van Nice and Dumbarton Oaks*)

St. Sophia, Istanbul; longitudinal Section, looking South

PLATE XIV

St. Sophia, Istanbul; transverse section, looking East

(Courtesy R. L. van Nice and Dumbarton Oaks

PLATE XV

(*Salzenberg*, 1854)

(*a*) Plan and interior elevation (partly conjectural) of the South tympanum with facings removed

(*Courtesy Dumbarton Oaks*)

(*b*) A vertical photograph of the forward edge of the apse vault showing its present pronounced forward bend in plan.

St. Sophia, Istanbul

PLATE XVI

(*a*) North-east pendentive and the dome, showing the surface irregularities which mark the limit of the 14th century reconstruction

(*b*) Eastern edge of the dome base showing its horizontal curve

(*c*) Sixth century dome cornice showing the scribed setting line for the forward edges of inwardly projecting ribs

St. Sophia: Details

IX

The Reconstruction of the Tympana of St Sophia at Istanbul

Introduction

THE known contemporary records give only tantalizing glimpses of the progress of the initial construction of Justinian's church and of subsequent structural changes prior to the mid-nineteenth century. The diversity of interpretation which these records will bear has already been amply demonstrated by scholars, particularly in their attempts to establish the original forms of the great north and south tympana.[1] A major obstacle to the writing of a definitive structural history has been the inaccessibility of most of the working masonry behind marble revetments, mosaics, and painted stucco. Lacks of bond, changes in the character of the masonry, and similar features which would otherwise furnish independent evidence of probable structural changes are thus largely hidden from view, and the great size of the building would alone preclude a general stripping merely in search of such indications. Further less direct indications are, however, provided by the distortions—tilts of columns and piers, spreads and bows of arches and vaults—produced progressively throughout the building by the large and changing forces acting on all principal elements. Given an adequate knowledge of the forces that would act at successive stages for different sequences of construction and of the distortions that would be likely to result, it is, in principle, possible to establish the actual sequence simply from a correlation of precise measures of the actual present distortions. The texts may then be used to fix in time certain points in the sequence. Unfortunately the present distortions can be inferred from the present forms only by first making assumptions about the manners in which these forms were initially set out. The only internal test of the validity of these assumptions is their reasonableness (in terms of what is known about the practices of the time) and their self-consistency coupled with the reasonableness and self-consistency of the conclusions drawn. It is necessary finally, therefore, to fall back on the more direct evidence of the masonry itself, of other related constructional features, and of the surface decoration. Since the evidence can now be sought much more purposefully to test specific conclusions, it becomes possible to substitute very limited and more practicable tests for the much more extensive exploration that would otherwise be called for.

A comprehensive study, on these lines, of the sequences of construction and subsequent reconstruction and of the associated principal structural actions throughout the building was begun in 1964. It was made possible by Mr.

[1] See, for instance, William R. Lethaby and Harold Swainson, *The church of Sancta Sophia, Constantinople* (London, 1894); Eugenios M. Antoniades, *Ekphrasis tes Hagias Sophias* (Athens, 1907–09); Cornelius Gurlitt, *Die Baukunst Konstantinopels* (Berlin, 1907–13); Emerson H. Swift, *Hagia Sophia* (New York, 1940); Kenneth J. Conant, "The First Dome of Haghia Sophia and its rebuilding," *The Bulletin of the Byzantine Institute*, I (1946), 71–78; and Panagiotes A. Michelis, "L'esthetique d'Haghia-Sophia," *Corsi di cultura sull'arte ravennate e bizantina* (Ravenna, March–April 1963), 207–76.

Robert Van Nice's precise measurements and delineation of the building as it is today and by his intimate knowledge of all aspects of its construction, and it has been undertaken in close collaboration with Mr. Van Nice and with his invaluable assistance in taking additional measurements where necessary. Since the tympana and their supporting arcades are only infilling screens set between the main piers and the main north and south arches, they were not a major focus of attention. They are, nevertheless, of great importance architecturally. It was therefore considered worthwhile in 1966 to make a very limited test to confirm the conclusion then reached that they must have been entirely reconstructed at a date well after the completion of the initial construction. The failure of this test to disclose the limits of the reconstruction —while seeming to leave no doubt that it must have taken place—led to further tests being made in 1967 and 1968. These not only disclosed the limits at all points where they were sought, but shed considerable further light on the probable date of the reconstruction and the probable original form of the tympana. Together with tests made on some elements of the main structure, they also provided the fullest confirmation of the validity of the initial approach to the interpretation of the precise geometry of the present forms.

This paper concentrates on the tests and on the conclusions drawn from them. In presenting this account of them it is a great pleasure to acknowledge Mr. Van Nice's assistance and to thank the Turkish authorities for permitting the tests to be made. We are especially grateful to Bay Feridun Dirimtekin, Director of the Ayasofya Museum, for his unfailing cooperation and sympathetic interest. I also wish to thank Mr. Van Nice for preparing the drawings from which the folding plate and figure 1 have been reproduced.

Previous Investigations

For the present purpose it seems unnecessary to review the long history of scholarly speculation about the tympana prior to the discussion by Underwood and Hawkins[2] of the dates of the ornamental mosaics alongside the portrait of the Emperor Alexander (fig. 1). In seeking a date for the later (phase 2) mosaic, they were led to suggest that the entire tympana plus the supporting gallery arcades were reconstructed at some time after completion of the original (phase 1) mosaic. It then appeared that this reconstruction was probably part of that known to have taken place immediately after the first collapse of the dome in 558, though there is no reference to the tympana in the fairly detailed account of that reconstruction left by Agathias.[3]

The principal evidence on which this suggestion was based came from a close study of the mosaic itself and of other related details. It was noted that the original mosaic had been replaced by that of phase 2 in all the soffits of the arcades carrying the tympana, although it survived almost intact in all

[2] Paul A. Underwood and Ernest J. W. Hawkins, "The Mosaics of Hagia Sophia at Istanbul: The Portrait of the Emperor Alexander," *Dumbarton Oaks Papers*, 15 (1961), 189–217, esp. 210–15.

[3] *Historiae*, bk. V, Patrologia Graeca, 88, cols. 1556C–1557A.

the corresponding soffits of the exedrae arcades; that the mosaics of the two phases alongside the Alexander portrait were differently related to the marble revetment and *opus sectile* panels on the south face of the pier and arcade and to the tie bar across the first arch of the arcade; and that the seemingly original *opus sectile* panels were a poor fit to the spandrels of the present arches.

Supporting evidence was found in three observations relating to the present alignments and forms of the tympana and arcades. The first was the setting back of all the columns of the arcades well behind the original setting marks incised in their stylobates. The second was the relative straightness of the upper cornices immediately above the arcades as compared with the more pronounced inward bends (toward the nave) of the lower cornices. The third was the greater verticality of the outer faces of both tympana above the gallery roof as compared with the pronounced outward leans of their inner faces. To the first of these observations could be added two related facts: most of the columns are almost vertical in contrast to the backward leans of the main piers which terminate the arcades, and they stand in almost straight lines although the original setting marks follow the inward bends of the lower cornices.

Conclusions from the Present Forms

Some of these characteristics of the tympana and arcades are illustrated in more detail in relation to the surrounding structure in figure 3.[4] The outward leans of the main piers commence well below gallery level but increase there about threefold to about 3 cm. for every 1 m. rise. Above the upper cornice two arches at each side carry the pendentives and the dome. The lower arches (referred to here as the main arches) span only the clear distance between the piers and are the full width of the piers, excluding the added width of the projections from which the great east and west arches spring. Above the gallery roof they are prominently revealed on the exterior. Since, however, the tympana are constructed with their inner faces continuous with those of the arches, they show today on the interior only as discolorations in the stucco which betray the pattern of the underlying brickwork, as shown more clearly in Salzenberg's partly conjectural engraving reproduced in figure 4.[5] The upper arches span the considerably greater distance between the projections from the piers just referred to and, like the east and west arches, directly carry the pendentives. Their outer faces are now hidden within the dome base so that their width is indeterminate, but their soffits are partly revealed on the interior.

[4] This figure is taken from plate 4 of Robert L. Van Nice, *Saint Sophia in Istanbul: An Architectural Survey*, Installment 1 (Dumbarton Oaks, Washington, D.C., 1966). Other relevant details may be found on pls. 5, 18, 22, 25, and 27.

[5] From W. Salzenberg, *Alt-christliche Baudenkmäle von Constantinopel vom V. bis XII. jahrhundert...* (Berlin, 1854), 21. Apart from its failure to show the joins between the original structure and the reconstruction and to represent more faithfully the major cracks and the precise geometry of the present forms, the chief faults of that part of the engraving which is reproduced are the representation of a non-existent course of stone above the upper cornice and some incorrect or implausible details in the representation of the brickwork.

The exposed faces of both the main and upper arches lean outward at their springings at the same angle as the piers from which they rise. There is a marked difference, however, between the outward displacements of their crowns, and this difference is both greater than and of opposite sense to the small one which would result merely from the greater rise of the upper arches. The difference corresponds to a progressive increase in the width of the exposed soffit of each upper arch as it rises, the main arches being of uniform width throughout. Thus, the crowns of the upper arches are displaced outward some 35 cm. less than would be expected as a result of the outward leans at the springings, while the crowns of the main arches are, on the contrary, displaced some 15 cm. further than would be expected. The inner faces of the tympana are, of necessity, displaced outward at their heads to the same extent as the crowns of the main arches. Instead of a gradual warping outward of these faces toward the center, though, the slope is almost uniform for most of their height throughout their length and there is a conspicuous adjustment in the last 3 m. The outer faces, if some local slight irregularities are ignored, have a uniform slope throughout with no adjustment near the head and, as noted by Underwood and Hawkins, this slope is much less than that of the inner faces. As the section shows, it is very similar to that of an imaginary plane drawn through points on the face of either upper arch at its springings and its crown.

The comprehensive study of the sequences of construction and reconstruction showed, at an early stage, that the present forms of all main elements of the structure could be best interpreted as distortions of forms initially set out with all unengaged upright faces vertical and with the profiles of all arches semicircular (with one relatively minor exception where this was impracticable).[6] The distortions clearly resulted from the large thrusts generated by the dome and the principal arches and semidomes. In magnitude they are now much greater than the recoverable elastic distortions inseparable from the bearing of load by any structural member. In part they could be attributed to plastic distortions in the slowly setting mortar used in large quantities in the upper parts of the building. To this extent they would clearly be dependent on the interval between construction and the time when a particular load became effective. Even bearing in mind the known speed of the initial construction and the references by Procopius[7] to some of the difficulties that ensued, it was evident that major distortions such as the outward leans of the main piers could not be accounted for solely by such plastic distortions. Tensile cracking accompanied by marked separations or shearing dislocations must also have played a part. The inevitable progressive increase in these separations and dislocations, accelerated from time to time by earthquakes, would be the main contributory factor to the successive collapses of the east and west arches and the dome.

[6] Rowland J. Mainstone, "The Structure of the Church of St. Sophia, Istanbul," *Transactions of the Newcomen Society*, XXXVIII (1965–66), 23–49.

[7] *De aedificiis*, VII (Loeb ed.), 28–31.

Confirmation that this interpretation was correct was initially provided by the remarkable consistency with which the distortions, so derived, conformed to those that would be expected on the basis of the already firmly established structural history. One example, directly relevant to the present discussion of the tympana, must suffice. Attention has been drawn to the greater outward displacement of the crowns of the main arches framing the tympana as compared with that which would be expected if the faces of these arches were inclined throughout at the same angle as the piers from which they rise. Other measurements show that this further outward movement was a consequence of an outward horizontal bending of each arch and that it is accompanied, in the vertical plane of the presumed semicircular setting-out curve, by a marked depression of the crown and outward rotation of the haunches. This rather complex three-dimensional movement, involving also some twisting of the cross sections, is precisely what would be expected to result from the dual structural role of the arch and the movements of its supports under the action of its own and other thrusts. It is called upon not only to carry the vertical weights of the structure above to the main piers but also, by virtue of its considerable transverse width, to span horizontally in transmitting to the main buttresses the horizontal thrusts of the dome that act to north or south. Quantitatively, moreover, the movements are in all respects consistent with the arches being, like the main piers, part of the original construction. They may be contrasted with the present distortions of the surviving sixth-century segments of the dome cornice immediately above. These segments are known to have been reconstructed in 558–63[8] and, though they do show some evidence of subsequent relative displacement of the individual marble blocks, both the horizontal bends and the depressions over the crowns of the arches are much less than those of the arches themselves.

Any other element of the main structure whose present distortions are similarly inconsistent with those of elements that were undoubtedly original was also likely to have been reconstructed either wholly or in part. Thus, as first pointed out by Conant,[9] the upper arches above the tympana must have been partly reconstructed, their exposed faces being brought forward progressively as they rise by amounts corresponding to the progressive increases in the widths of exposed soffits. This reconstruction must have immediately preceded that of the dome cornice just referred to. If the exposed faces were then made approximately vertical again, their present mean outward inclinations should give good indications of the extents of the outward leans of the main piers to north and south subsequent to the reconstruction. On this basis, about 60 per cent of the present total movement occurred up to 558 and the remaining 40 per cent subsequently.[10]

[8] William Emerson and Robert L. Van Nice, "Haghia Sophia, Istanbul: Preliminary Report of a Recent Examination of the Structure," *American Journal of Archaeology*, XLVII, No. 4 (1943), 403–436, esp. 423–36.

[9] *Op. cit.*, 75.

[10] This and similar estimates have been rounded off, for the purpose of the present discussion, in terms of the degree of reliance which it is thought can be placed upon them. The basic measurements have been checked to a high degree of accuracy and most estimates have been arrived at indepen-

If the same argument could be applied directly to the tympana, it could be concluded, from the similarity between the present inclinations of their outer faces and those of the inner faces of the upper arches, that they also must have been constructed at about the same time. This, however, would fail to explain the noticeable inward bends at the level of the upper cornice (which, though less than those of the lower cornice, cannot be ignored). It would also lead to the improbable conclusion that the columns of the arcades below must have been rectified much more recently since most of them are still much more nearly vertical.

Greater care is clearly called for in seeking the right interpretation of the evidence here. Both the tympana and the arcades are merely infillings of the main structure. They must fit the openings which they fill and must, in any reconstruction, have been adapted to any existing distortions of these openings. Some, at least, of the present inclinations, for instance, must be due partly to these adaptations, and the proportion which can be attributed to subsequent distortion must be somewhat indeterminate because of the options open to whoever was responsible for the reconstruction. Nevertheless, it is possibly correct to assume that the columns of the arcades were set vertically when moved to their present positions and that the outer faces of the tympana were rebuilt approximately vertically in their central portions. It is also reasonable to assume that the reconstructed lengths of the upper cornices and of the bases of the tympana were initially straight. It must then be remembered that, after construction, both the tympana and the arcades, as infillings, will have been subject to much more complex patterns of loading than the main elements of the structure.

If all the probable indications of distortion in the present forms of both tympana and arcades are considered together, they are much more consistent with the assumption that both were reconstructed at the same time. Further backward inclinations of the main piers after the reconstruction would inevitably carry the tympana back to the same extent where they are in contact with the main arches, but there would be no such direct action on the columns of the arcades. These, above the capitals, and the lower central portions of the tympana would be acted upon more directly by the gallery vaults thrusting inward toward the nave. The inclinations of the columns and the horizontal bends of the upper cornice, as well as of the tympana and spandrels of the arcades immediately above and below it, all indicate, if examined in detail, that this more direct action was sufficient largely to neutralize the indirect one of the backward movement of the piers, except in the area close to the piers. A better estimate of the extent to which the leans of the main piers increased after this reconstruction is given, therefore, either by correcting the present inclinations of the outer faces of the tympana to allow for the bends at the level of the upper cornice or by averaging the present outward inclination of

---

dently from several different sets of measurements. Nevertheless, certain possibilities of error are inherent in the allowances that must be made for probably built-in irregularities and for the variable thicknesses of stucco etc. where the working masonry is inaccessible.

both tympana and arcades over their full height. On this basis, the increase has been probably not more than 30 per cent of the present total. This figure, in relation to the 40 per cent estimated as having occurred after 558, suggests that the reconstruction was undertaken well after that of 558–63, though it must obviously have been before the initiation of the post-iconoclastic scheme of mosaic decoration.

First Series of Tests: June 1966

Close inspections, in June 1966, of the main piers and buttresses, particularly in the spaces above the gallery roof where the working masonry can be studied with little difficulty, further confirmed the validity of the general approach to the interpretation of the precise present geometry of the forms that has just been outlined.

A direct independent test of the conclusions relating to the tympana and arcades was, nevertheless, considered desirable, but was more difficult because here none of the working masonry could be inspected without laboriously uncovering it first. As a preliminary to a possible more extensive exploration, one very limited test was proposed and undertaken just before my departure from Istanbul. This was the cutting of a narrow chase in the stucco to expose the brickwork immediately above the upper cornice along the length of the south tympanum. A similar chase was cut on the north side, but this had to be interrupted where gaps in the wooden railing alongside three of the niches would have made it too conspicuous. It was intended primarily to show the aggregate width of open cracks between the main piers. It could be assumed that as the piers inclined further to north and south they would also incline further to east and west and that the latter movements would be a fairly constant proportion of the former. The movements to east and west would, of course, increase the gaps between the piers that are filled by the tympana. Since the present total increases in the gaps are about 30 cm., the increases since 558, for instance, could be assumed to have been about 12 cm. The elasticity of the brickwork could be ignored, since it was unlikely even to have accomodated the full drying shrinkage of the newly constructed tympana, and the aggregate crack widths could thus be taken as fairly direct measures of the actual increases in the gaps.

Cracks totalling 8.8 cm. were, in fact, found at the foot of the south tympanum between the piers and there were two further cracks totalling 2.2 cm. slightly further west. Within the limits of accuracy to be expected here the previous conclusions were, therefore, confirmed. Paradoxically, however, a first inspection showed no signs of the expected joins between the original brickwork and the reconstruction. Nor were any seen in the more limited areas exposed at the north, though the extent of the cracking there was generally similar.

Subsequently Van Nice made two small test holes at the eastern extremities of the barrel vaults behind the tympana. On the east or west face of each main pier is a pilaster which receives the ends of the arcade. It is wider, at gallery level, than the arcade itself, and the excess width is terminated just above the upper cornice (fig. 1). The openings were made immediately above these truncated pilasters and the only significant observation was that of a completely unbonded join between the brickwork of the pier and that of the springing of the barrel vault at the southeast.

Second Series of Tests: September 1967

Re-examination, in 1967, of the chases cut at the feet of the tympana likewise failed at first to disclose the joins in construction. It was clear, though, that a possible reason for this was the minimal width of the chases and the damage done to the original surfaces of the bricks and mortar in removing the very hard cement stucco. Since it was desired to restrict any further cutting to what was most usefully revealing, attention was directed next to the cornices. It was noted that almost the entire length between the piers could have been dismantled with ease at the north. At the south, however, a block firmly embedded in the pier projected beyond it for a considerable distance at each end. At the southwest, moreover, this block had a chase cut in its top surface, obviously to receive a cramp, and this chase was now about 4 cm. forward of its other half on the adjacent block (F on folding plate, and fig. 7). The insertion of the cramp, intended to halt the separation of the piers, was obviously a repair before the reconstruction, and the misalignment of the chases now indicated equally clearly that the block to the east had been reset. Other cramps, still in place across this joint and across the joint almost in line with the edge of the northwest pier (fig. 5) and each set partly behind the face of the brickwork, must have been inserted in the reconstruction. The most likely locations of the construction joins seemed, therefore, to be almost in line with the edges of the piers at the north and somewhere nearer to the ends of the embedded blocks of the cornice at the south.

Permission was sought and readily granted to extend the previous narrow openings in these positions. Proceeding with care to hold damage to the brickwork to a minimum, the openings were enlarged to varying extents to disclose, in all cases, the expected joins (folding plate and figs. 5, 6, 7—the joins being marked by broken lines in these figures and in figs. 2, 8, 9). On first inspection there was no appreciable separation at any join, but its presence was unmistakably betrayed in several ways:

1) The regular coursing of the brickwork was interrupted to a greater or lesser degree. This interruption was most noticeable at the northwest, where the bottom two courses of the reconstruction corresponded to three courses of the original.

2) The pointing of the original brickwork was, like that of most sixth-century brickwork elsewhere in the building, finished to a concave profile leaving the full width of each brick exposed. The pointing of the reconstruction, on the other hand, was finished almost flush with the bricks and, by overlapping slightly their rounded edges, tended to diminish their apparent width and to exaggerate that of the joints.

3) The mortar used in the reconstruction differed also from the characteristic, almost white, sixth-century mortar. It was definitely pinkish, due to the incorporation of brick dust, but less pink than that used in the tenth-century reconstruction of the dome. This pinkish mortar had previously been noted by Van Nice at a higher level on the exterior of the south tympanum when some areas had been exposed by falls of stucco.

Careful probing in the joins with a fine tool showed that each contained, and was clearly distinguishable by, a fine black deposit that was not found where cracks occurred within previously continuous brickwork. The joins also differed from hair cracks in that, where original bricks had been cut in the trimming that preceded the rebuilding, no matching half was to be found on the other side of the join. On this side all brick ends, whether finished or cut, were mortared.

At the south it was possible to extend both openings sufficiently to follow the joins into the shallow niches. As expected, the original brickwork had been cut back obliquely at each end. More significant was the discovery that only in the reconstruction had the niches been formed from the start. In the two places where they now extend into the original brickwork they had been cut out of a mass that was originally of uniform thickness (figs. 6, 7).

All subsequent exploration was directed to establishing more completely the limits of the reconstruction and to finding further evidence of its probable date and of the probable original form of the tympana. The knowledge acquired on the upper cornice of the character of the reconstructed brickwork assisted greatly in identifying the joins where they were not immediately recognizable because of the small size of the initial test hole or because of damage to the original surfaces.

It was now apparent that the unbonded join observed by Van Nice in 1966 above the truncated pilaster attached to the southeast pier was part of the join between the original brickwork of the pier and the reconstruction. Some further uncovering above the corresponding pilaster at the northeast, and fresh holes above those at the southwest and northwest, laid bare the joins there also (D on folding plate, and fig. 2).

The most important feature noted in each opening was the unmistakable evidence of a fully bonded connection, originally, between the pier and the brickwork above the arcade. Ignoring a slight lateral dislocation at the northeast and northwest (accompanied by separations of about 2 to 3 cm.), this original bonded return was in line with the present springing of the barrel vault. In general, it had been cut back to the face of the pier alongside. One brick only, bonded into both the pier and the original brickwork above the arcade, projected over the stone block which terminated the pilaster at the southeast and northeast, but there was no indication that an arch or any other construction originally sprang from this.

No test was possible on the spandrels or soffits of the arcades below the upper cornice, but there was one indication that the arcades were reconstructed at the same time as the tympana. In place of the separations just referred to above the northeast and northwest pilasters, a wide crack was found running into the original brickwork above the pier alongside the join at the southwest. At the southeast there was no such crack but there was clear evidence, in nineteenth-century stucco repairs, of a major crack running from the soffit of the barrel vault to the soffit of the arcade, roughly in line with the join in the cornice about 1.2 m. west of the pier. A continuous crack of this nature would not be expected if the arcades and tympana had not been reconstructed at the same time.

Third Series of Tests: July 1968

To supplement the picture obtained from the tests above the truncated pilasters, four more smaller openings were made in 1968 at the outer ends of the barrel vaults (E on folding plate). Here the brickwork of the vaults belonged entirely to the reconstruction and no original bonded returns were found. In every case the finished face of the brickwork of the pier continued uninterrupted behind the springing of the vault. This did not necessarily indicate that the original vaults had a different form. They would almost certainly have been constructed after the piers, and normal practice would not demand a bond in such situations where relative movement might be anticipated but need not be regarded as serious.

In parallel with other investigations of the original form of the main piers at gallery level and of the dates of subsequent additions to them, attention was then transferred to the exterior of the tympana. Here the gallery roof gave easy access to the lower range of windows. One earlier test by Van Nice and Hawkins at the west end of the south tympanum had briefly exposed the crack shown in Salzenberg's engraving (fig. 4) and a closely related Fossati sketch, but it had shown a fully bonded connection between the tympanum and the main arch and no apparent intermediate break in construction between this and the late (probably Turkish) narrowing of the first window opening. It was now considered that joins between what remained of the original tympana and the reconstruction must exist here and in the three corresponding positions at the southeast, northeast, and northwest.

Permission to look for the joins having been granted, the stucco was removed first in the area of the previous test just above the roof at the southwest, then in the corresponding positions at the other three extremities, and, finally, at the level of the springing of the window arch at each extremity. Since this removal proved to be particularly revealing at the northwest and the stucco was relatively loose, a larger area was subsequently uncovered there (folding plate and figs. 8, 9).

The join just above the roof at the southwest, which had previously escaped notice, proved to have been obscured on the surface by a later patch. A few of the other joins were also not immediately apparent, but all were traced. In every case part of the original tympanum was found. Because it was fully bonded to and appeared to have been constructed simultaneously with the main arches up to the springings of the window arches, it was clearly original and not part of the rebuilding that followed the partial dismantling during initial construction that is referred to by Procopius.[11] Here a normal bond became impossible, and the few bricks that remained of each window arch extended as one unit into the mass of the main arch. The pointing on their faces also extended into this mass, indicating that these window arches had been constructed first and the main arches then built up around them (fig. 9).

The original form was best displayed at the northwest. Here the finished (pointed) jamb and part of the finished soffit of one window opening had been left untouched, with the new work simply butted up against them (AA on folding plate, and fig. 8). Elsewhere the original jambs had been cut away to a varying extent. At the southeast this seemed to have been done deliberately to bond in the new work. The irregular joins at the southwest and northeast may, however, simply have followed the lines of earlier cracks.

In relation to the earlier conclusion that the greater verticality of the outer faces of the tympana, as compared with the inclinations of their inner faces and the main piers, denoted a rectification at the time of reconstruction, the direct evidence of rectification that was discovered was particularly interesting. Enough remained of the original outer faces to show that they did indeed incline outward to the same extent as the main piers. To avoid surface discontinuities of several centimeters at the joins it was necessary, therefore, to adjust the inclination of the new work locally to match that of the original or to trim the face of the original back alongside the join. The latter alternative was adopted at the northwest and southeast, where the trimming back of the original faces could be very clearly seen at the level of the window arches (fig. 9).

Extent and Date of the Reconstruction

The extent of the reconstruction of the upper cornice, the tympana, and the barrel vaults behind the tympana was well defined by these tests (folding plate), but no tests were possible on the faces or soffits of the arcades below the cornice, on the arches that carry the outer springings of the barrel vaults, or on the gallery vaults beyond these arches. In considering the full extent of the reconstruction it is necessary, therefore, to rely on less direct evidence.

The present alignments of the arcades, the distortions they have undergone at the ends, and the surface indications of underlying cracks all show that they also have been reconstructed, almost certainly at the same time as the tympana.[12] They would, in any case, have been cracked open to the same extent as the cornice and the bases of the tympana; so it would be surprising if the reconstructions had not been undertaken together.

[11] *Op. cit.*, 30–31.

[12] Measurements, available only after this paper was written, of the inclinations of all the columns provide more conclusive confirmation that the arcades were reconstructed at the same time as the tympana. At both north and south, these inclinations diminish progressively toward the center of the arcade as the inward thrusts of the gallery vaults increasingly neutralize the contrary pulls of the backward leaning piers. The inclinations of the columns nearest the centers average less than 30 per cent of the estimated increases in the pier tilts subsequent to the reconstruction of the tympana, whereas those of the columns nearest the piers average more than 80 per cent.

A fuller examination than has yet been possible of the precise forms of all the arches and vaults of the central bays of the gallery may indicate whether the reconstruction extended beyond the barrel vaults behind the tympana and, if so, how far, but a definite conclusion will be more difficult to reach here. One reason is the probable lesser degree of accuracy achieved in setting out these relatively minor elements. Another is the likelihood of local patching here rather than of complete reconstruction. Such patching has already been confirmed in the main domical vault over the northwest bay of the gallery.

Further evidence for the date of the reconstruction has been furnished by the observations on the character of the mortar and pointing and by C14 measurements on three specimens of wood. It points, like the precise present geometry of the forms, to a date well after the reconstruction of 558–63 but before the initiation of the post-iconoclastic scheme of mosaic decoration.

Both the mortar and the pointing of the reconstructed brickwork are, as has been noted, intermediate in character between those typical, elsewhere in the building, of the sixth and tenth centuries,[13] and they differ sufficiently from both the earlier and later types to indicate considerable changes in practice. Only one close parallel to both mortar and pointing has been found. This is in the reconstruction of the upper parts of the main structure of St. Irene which George[14] dates immediately after the earthquake of 740 and which cannot have been earlier than this. Both the mortar and the pointing closely resemble those of the reconstruction of the tympana. Similar pointing has also been found inside the present upper storey of the Treasury of St. Sophia and inside one of the towers of the land walls just north of the Silivri Gate. The Treasury cannot yet be dated independently, but the tower bears an inscription of Leo III and Constantine V and is, therefore, roughly contemporary with the reconstruction of St. Irene.

Three specimens of wood were dated from their C14 content by the University of Michigan.[15] One came from a rough length of timber built into a hole in the easternmost niche of the south tympanum, possibly to support scaffolding during the reconstruction. The second came from a rough beam spanning a small arch at the level of the upper cornice above the pair of columns that stands just behind the south gallery arcade. The third came from the stub of a tie above the east face of the impost block of the eastern column of this pair. Thus the first and second could, almost certainly, be associated with the reconstruction, while the third, if its age proved to be similar, probably was. The ages were estimated as 1260 ± 130, 1340 ± 130, and 1160 ± 120 respectively. Solely in terms of the possibilities of error inherent in the C14 counts, there is a two in three chance of the true ages of the specimens falling within the limits stated. If the limits are doubled there is still a one in twenty

[13] W. Emerson and R. L. Van Nice, *op. cit.*, 216–21, discuss and illustrate the principal characteristics of sixth- and tenth-century brickwork as seen in the stairways of the southeast buttress. Tenth-century brickwork is distinguished by slanted weathered pointing of the horizontal joints. The characteristics of the mortars of the joints themselves have been referred to above.

[14] Walter S. George, *The Church of Saint Eirene at Constantinople* (London, 1913), 69–71.

[15] Specimen nos. One=M-2209; Two=M-2210; Three=M-1942.

chance that the true ages will fall outside. Since the ages relate to the short periods of growth of the specimens as parts of the living cambiums of the trees from which they came, the dates of incorporation of the timbers in the structure will, of course, always be later, and may be substantially later, than those derived directly from them. Bearing this in mind, these latter dates—690, 610, and 790 with the same ranges of probable errors—are approximate termini post quem for the construction of those parts of the structure from which the specimens were taken. In relation to the probable errors, the differences between these dates are not very significant and there is certainly no indication that the third specimen came from an earlier phase of construction than did the other two. The mean of all three—700—is quite consistent with all three timbers having been set in place soon after 740. There is, however, about a one in four chance that the true age of the second specimen was 1420 or more, in which case it could mean that the timber was taken from the original construction and reused. The mean age of the other two would then make a rather later reconstruction most probable, and this possibility cannot, on any interpretation, be ruled out.

To reach a more definite conclusion it is necessary, in the absence of any explicit textual reference to the reconstruction, to consider such textual evidence as does exist and some further evidence bearing on this.

Accounts of the earthquake of 740 make no mention of St. Sophia. The first mention of damage after the collapse of 558 occurs in references to the earthquake of January 9, 869. The earliest reference speaks of damage in many places.[16] It adds that the Emperor had the damage repaired, but gives no further detail. Since the reconstruction of the tympana must have preceded work on the mosaics and cannot, therefore, have been completed much after 869, there is thus some textual authority for assigning it to 869 or immediately afterward and none for any likely alternative. This, in itself, is inconclusive, but there are two further reasons for preferring this date to 740 or some intermediate date.

The first is the siting of the inscription which ran in several lines across the upper parts of both tympana. Two fragments were uncovered by the Fossati and, apart from the fragment of the uppermost line at the south, these have been identified with an epigram which has been quoted and discussed by Mango in his preliminary study of the mosaics.[17] The epigram does not specifically refer to the tympana, though it has been suggested that the uppermost line at the south might have done. It does, however, commemorate a repair by the Emperor, and there is good reason for believing this to be the repair undertaken by Basil I in about 869. Not only the prominent placing of the inscription on the tympana but also the nature of the reference to the repair strongly suggest that the reconstruction of the tympana constituted a major part of it. The repair was said to have been necessary because the church "had

[16] *Vita Ignatii*, PG, 105, col. 549A.

[17] Cyril Mango, *Materials for the Study of the Mosaics of St. Sophia at Istanbul*, Dumbarton Oaks Studies, VIII (Washington, D.C., 1962), 63–66.

been 'suffering' or 'afflicted' by the effects of time and time had 'threatened' to destroy it,"[18] and these words would exactly describe the progressive cracking apart and distortion of the tympana and arcades which would inevitably have occurred from a very early date as a result of the progressive separations of the main piers and distortions of the main arches. The second is the fact that the shallow niches below the windows appear to have been introduced for the first time in the reconstruction. Since they so effectively frame the mosaic figures of the bishops, it seems probable that they were introduced for this purpose, which would preclude any date much before 869.

I therefore date the reconstruction very shortly after the earthquake of 869, leaving open, however, the possibility that it may have been planned and even commenced shortly before. This possibility cannot be excluded because, as has just been stated, the damage which led to its being undertaken would, in large measure, have been long standing and it is doubtful whether the brief textual references can be relied upon to distinguish between repairs undertaken only after the earthquake and repairs already in progress.

Nothing was found that could be associated with the partial rebuilding that is referred to by Procopius, and this was probably much less extensive.

Original Forms of the Tympana and Arcades

Enough evidence now exists to show that the original tympana and arcades were probably almost identical with the present ones except for the absence of the shallow niches below the lower ranges of windows. Above these windows, however, no detailed examination was possible, and it is highly unlikely that anything now remains from the original construction. One can only speculate, on the basis of the Silentiary's description,[19] that in place of the present groups of three windows there were single, larger windows resembling the main west window.

There is no indication that the pilasters which receive the ends of the arcades (fig. 1) ever carried anything to justify their present excess width. This excess appears always to have been abruptly terminated, as now, at the level of the upper cornice. It seems to be explicable only in terms of a change in the design when the main piers had already reached some height above the gallery floor but work had not yet started on the arcades. It has been tentatively suggested[20] that the intention may have been to make all the arcades at gallery level similar in design to those below, as was done at SS. Sergius and Bacchus.

[18] *Ibid.*, 65.

[19] Paulus Silentiarius, *Descriptio Sanctae Sophiae*, line 536 (in P. Friedländer, *Johannes von Gaza und Paulus Silentiarius* [Leipzig-Berlin, 1912], 242), refers to eight windows in all. Assuming that there were, as now, seven in the lower range answering directly to the bays of the arcade below, this means that there can only have been one above. A possible form of this is illustrated in K. J. Conant, *op. cit.*, pl. XXXVIII. The present groups of three windows referred to are those shown in the Salzenberg engraving (fig. 4) without the later narrowings of the openings.

[20] R. J. Mainstone, "Justinian's church of St. Sophia, Istanbul: Recent studies of its construction and first partial reconstruction," *Architectural History*, 12 (1969), 39–49.

Alexander Mosaic and Northeast Pier and Pilaster at Gallery Level

2. Springing of Barrel Vault above Northeast Truncated Pilaster.
Detail, looking South

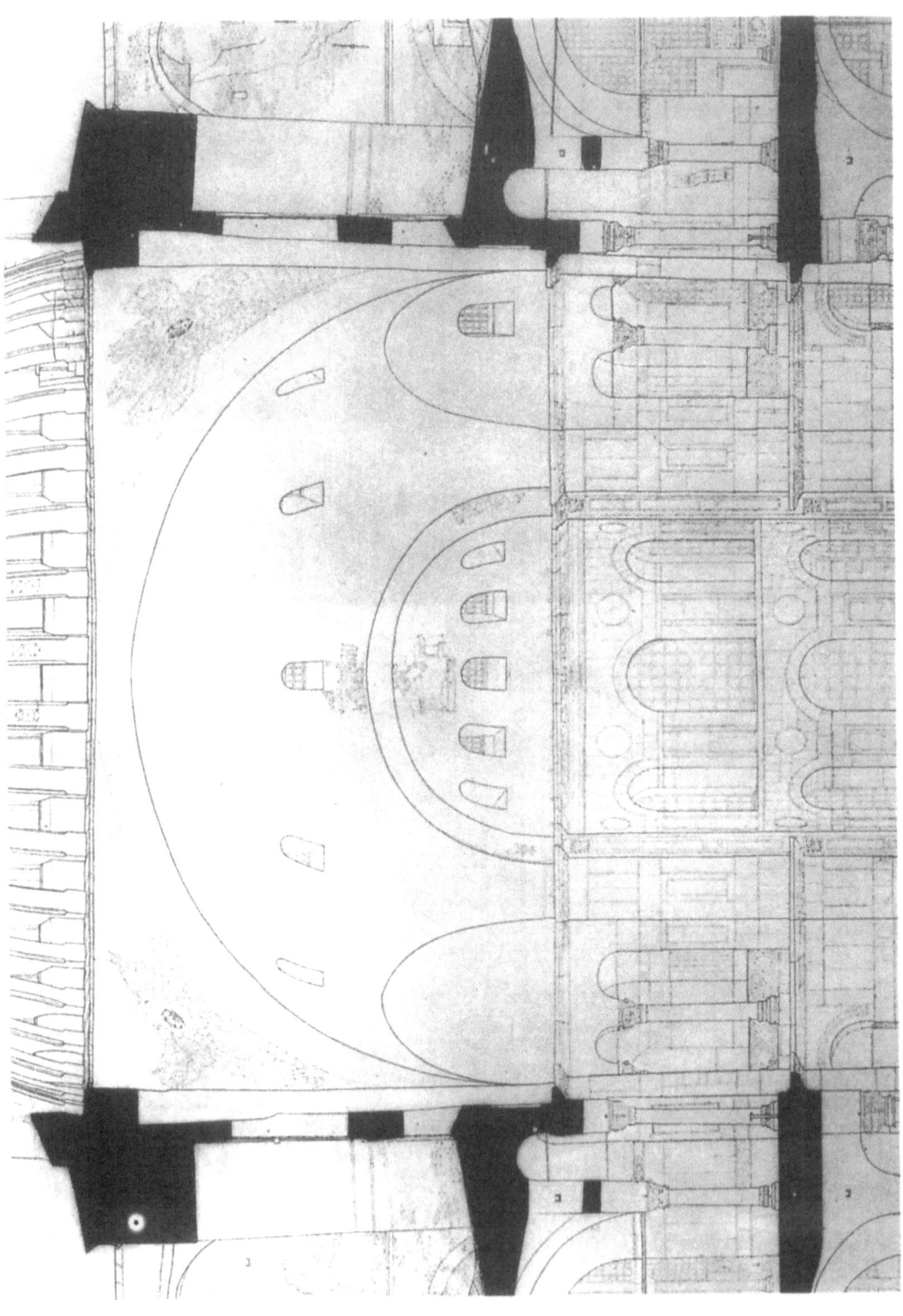

3. Istanbul, St. Sophia. Transverse Section (1:200)

4. South Tympanum, Interior Elevation, according to Salzenberg

5. North Tympanum, West End. Detail of Test on Upper Cornice

6. South Tympanum, Easternmost Niche

7. South Tympanum, Westernmost Niche

8. North Tympanum, West End, Exterior

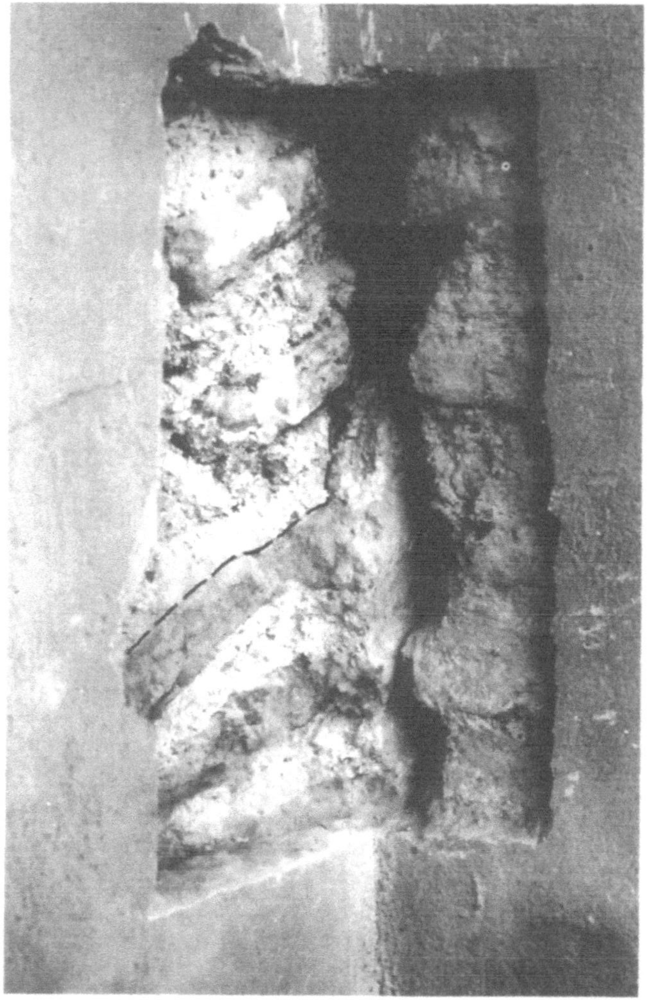

9. Upper Test seen in Figure 8. Oblique Detail

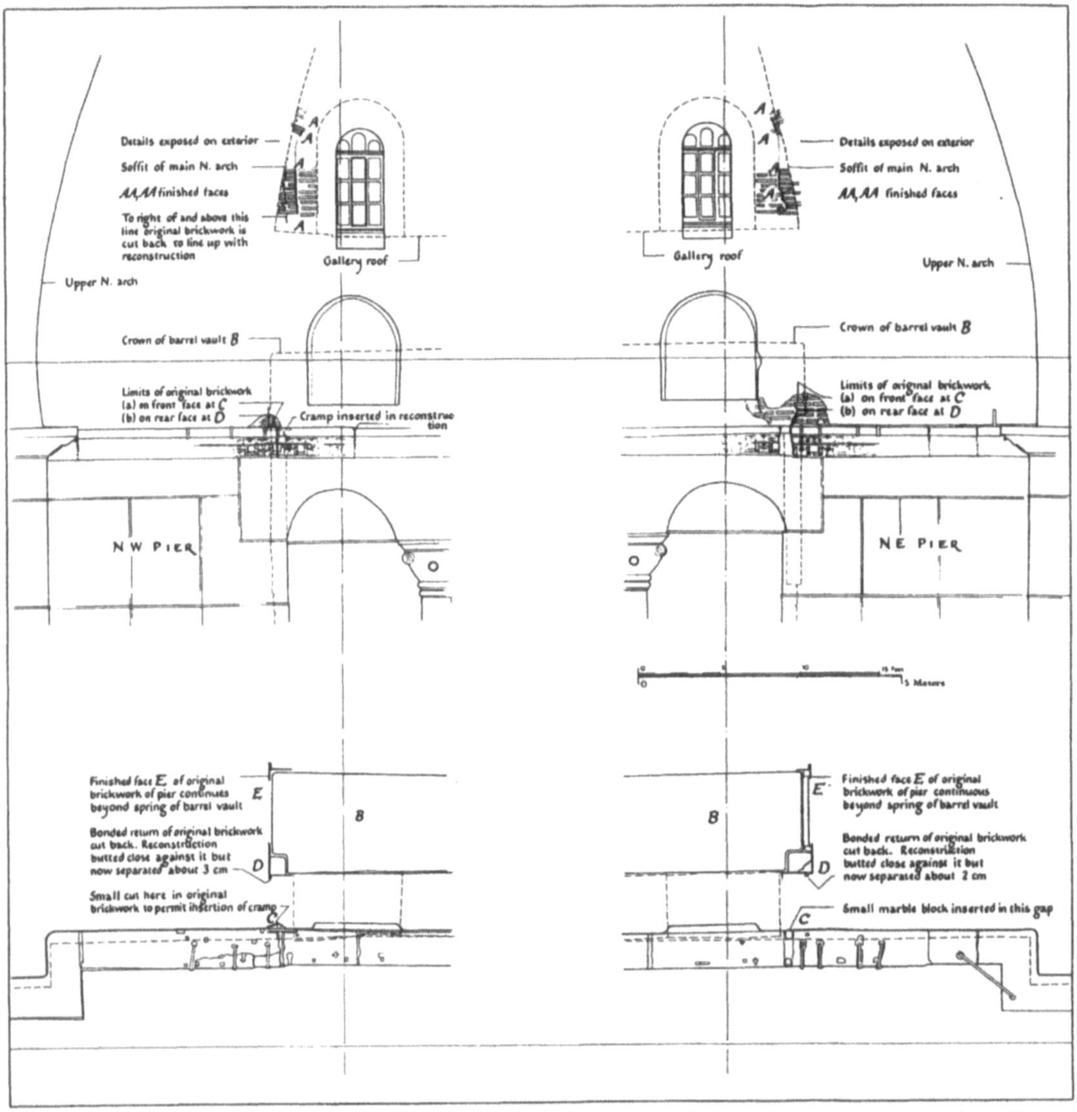

Istanbul, St. Sophia. Interior Elevations and Plans of Ends of Tympana, showing (in heaviest line) observed Limits of Reconstruction (1:100). Details exposed on Exterior are depicted in Reverse, as if seen from Interior. Plans are drawn just above Level of Upper Cornice.

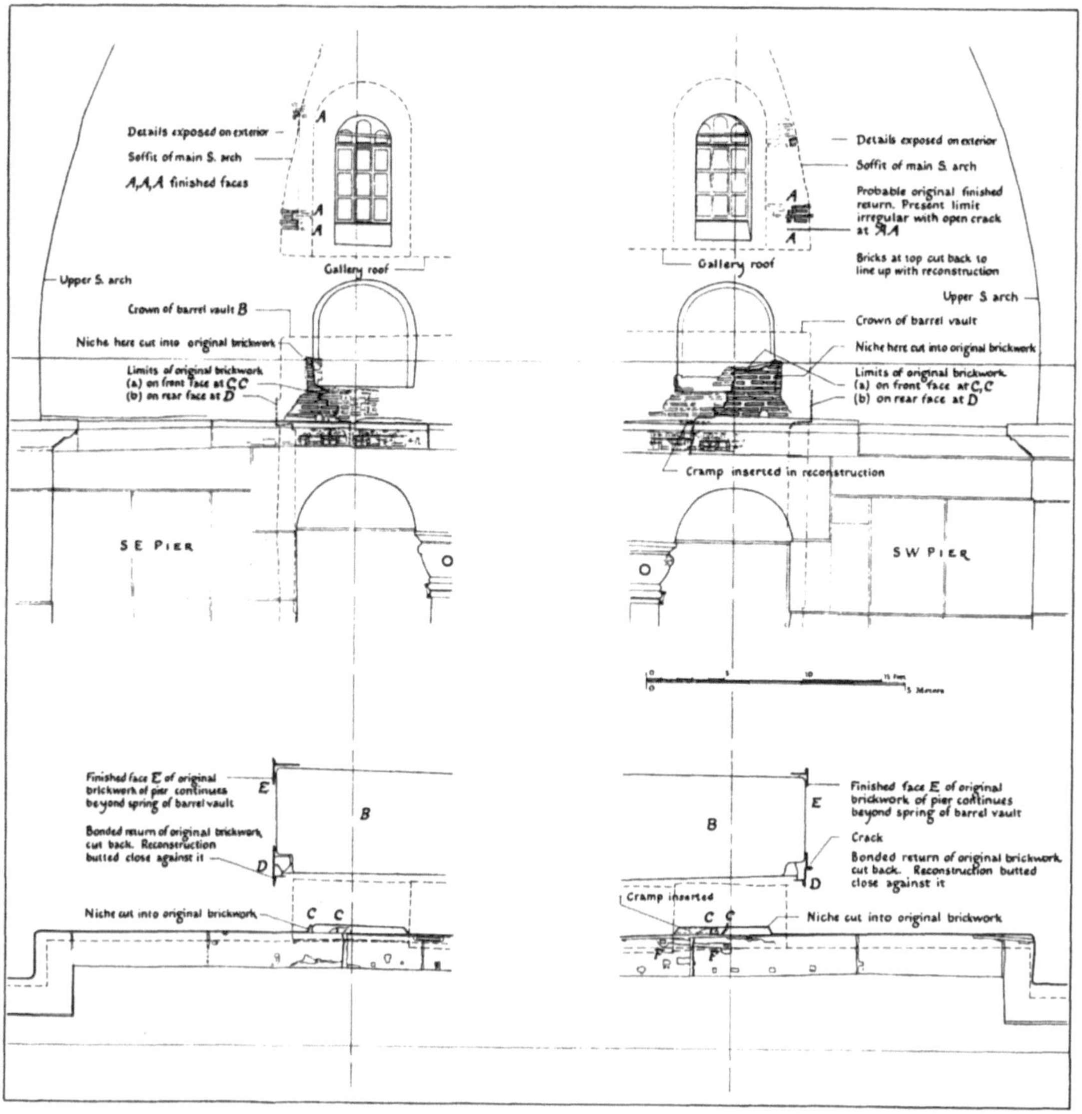
Details exposed on exterior
Soffit of main S. arch
A,A,A finished faces
A
A
A
Gallery roof
Upper S. arch
Crown of barrel vault B
Niche here cut into original brickwork
Limits of original brickwork
(a) on front face at C,C
(b) on rear face at D
S E Pier
Details exposed on exterior
Soffit of main S. arch
Probable original finished return. Present limit irregular with open crack at AA
A
A
Gallery roof
Bricks at top cut back to line up with reconstruction
Upper S arch
Crown of barrel vault
Niche here cut into original brickwork
Limits of original brickwork
(a) on front face at C,C
(b) on rear face at D
Cramp inserted in reconstruction
S W Pier
0
5
10
15 Feet
0
5 Meters
Finished face E of original brickwork of pier continues beyond spring of barrel vault
E
B
Bonded return of original brickwork cut back. Reconstruction butted close against it
D
Niche cut into original brickwork
C
C
B
E
Finished face E of original brickwork of pier continues beyond spring of barrel vault
Crack
Bonded return of original brickwork cut back. Reconstruction butted close against it
D
Cramp inserted
C
C
Niche cut into original brickwork
F
F

X

Hagia Sophia: Justinian's Church of Divine Wisdom, later the Mosque of Ayasofya, in Istanbul

Introduction

No engineer stepping into Justinian's Great Church of Hagia Sophia could fail to marvel at the structural daring with which a clear space 80 m long and more than 30 m wide—comparable with that of the Albert Hall—was covered more than 1450 years ago by small and large semidomes rising to a vast central dome whose crown is some 56 m above the floor (Fig 1).

It is difficult to accept that it was designed and built in a mere 6 years between 532 and 537 when nothing remotely comparable had been built for some 200 years, and that it went far beyond even the closest of the earlier precedents. It was not, like the Gothic cathedrals, the product of a relatively cautious step-by-step evolution. And, in spite of the fact that it now shares the Istanbul skyline with a number of imperial mosques built partly in emulation of it after its own conversion to use as a mosque in 1453, none of these quite equals it. Nor does anything comparable exist elsewhere.

Equally unique and remarkable is the character of the space that the high vaults enfold.

Today, with more complex requirements, the design of a structure of similar size, importance and originality would be more demanding, and would call for collaboration between architects, engineers, and numerous other consultants.

In the late Roman Empire—to which the Byzantine Empire ruled by Justinian was the successor—these professions had not yet been born as we now know them, though a distinction, unknown later in the West, was made between *architektones* and *mechanopoioi*[1]. As *mechanopoioi*, Hagia Sophia's principal designers, Anthemius of Tralles and the elder Isidorus of Miletus, appear to have been as close to the profession to which we belong as anything that existed subsequently up to the 16th century. But, since their theoretic background was limited to the geometry and mechanics of the day (which they both taught), they cannot have exercised their design skills in quite the same way as we do, and they filled the present role of

Fig 1. The interior today from the west gallery

architect also.

The way in which they approached and developed the design is the principal focus of this paper[2]. But some further description of the present structure, of the changes that it has undergone and of its original character, are necessary preliminaries, and the discussion of the changes will be prefaced by a brief review of the available evidence and its interpretation.

The church today: form and structure

Externally (Fig 2) one notices first the great lead-sheathed dome on a square

Fig 2. The exterior today from the south

base dominating the relatively squat bulk of the lower levels and, beside it, the contrasting forms of four tall minarets[3]. Above the gallery roofs the swelling vaults of the small and large semidomes are visible at east and west, while at north and south their place is taken by broad arches set well back between massive projecting buttresses. Below the gallery roofs, large deceptively solid-looking masses abut the lower walls on three sides, while in the corresponding position at the west, relatively slender flying buttresses straddle an outer narthex.

Internally (Fig 1) the nave is a far more complex and subtle space than anything known previously. Though it is almost as strongly centred on the dome as the Roman Pantheon is on its dome, one is equally conscious, on entering, of an east- west axis, and the true boundaries are elusive. As they alternately advance and recede, they are sometimes solid and impenetrable and sometimes no more than open colonnades through which, as one moves around, there are ever-changing glimpses of the half-revealed peripheral spaces of aisles and galleries. Expectations, moreover, are confounded by such features as the lack of correspondence between the colonnades at ground and gallery levels. A seemingly effortless unity is nevertheless achieved by a pervading geometrical discipline and by details like the continuous horizontal cornices.

The aisles and galleries flanking the nave (Fig 3) are twice narrowed along their lengths by the intrusive masses of the piers that directly support the dome and of four answering piers (the bases of the outer parts of the projecting buttresses at north and south just referred to) that stand against

Fig 3. The south east bay of the aisle looking towards the southeast pier (originally, the uppermost arch soffit at the left continued unbroken to the far end of the pier, the two projections from the pier and the arches they carry being later additions, as shown at A in Fig 11)

the outer walls. On the sides towards the nave they otherwise necessarily follow the undulating boundaries of the nave colonnades. At the west are inner and outer narthexes and, above the first, a transverse gallery.

Structurally, the narthexes and those major parts of the aisles and galleries that are open to the nave are virtually independent infills, with outer walls constructed almost entirely of brick and with brick vaults—barrel, groined, and domical—largely carried on monolithic marble columns. These columns are linked to one another above the capitals by iron tie rods or, when standing near to a pier, they are connected to it by a short length of timber or a projecting block of marble that serves also as an impost block.

The main structure (Fig 4) is of considerable complexity. Essentially, the supports for the high vaults consist of the four main piers, four more towards the east and west, the four others referred to above, and all the

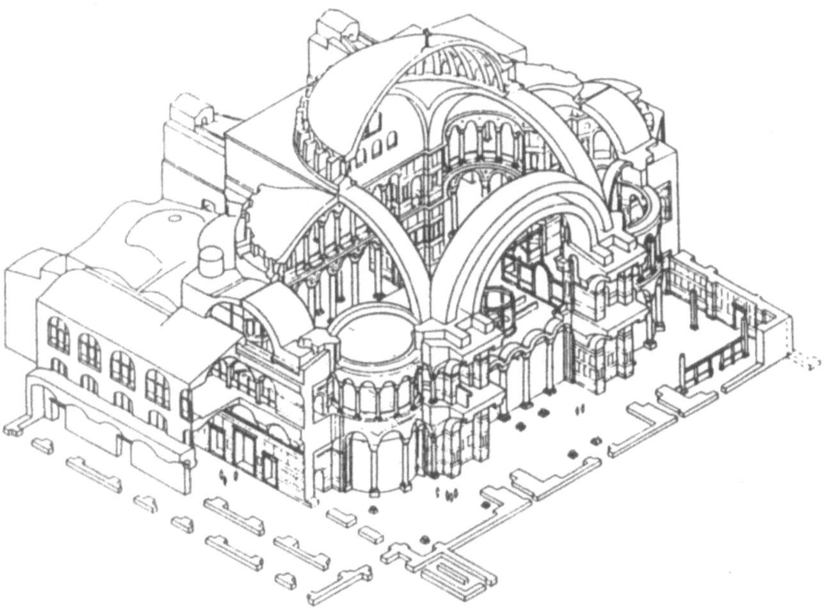

Fig 4. Partially cut-away isometric from the south west showing part of the main structural system and the principal internal furnishings at the time of the second dedication in 562

interconnecting arches. The principal arches span between the eight piers which stand around the nave while, over the aisles and galleries, arches of shorter span (shown sliced through their crowns) link the main piers with the four others that stand against the north and south walls.

The eight piers around the nave are constructed of large squared blocks of limestone and a local granite up to the springing levels of the main arches, and the outer piers to north and south are similarly constructed up to the springings of the aisle vaults. For their upward extensions and for all arches and vaults, brick was used, escept that there are continuous marble cornices at three levels and marble springers for some smaller arches. The stone masonry is closely jointed. But elsewhere there is more mortar than brick, so that premature loading resulting from the rapidity of construction must have led to large early plastic deformations.

The smaller semidomes are carried partly by the colonnades of the exedrae (the semicircular apse-like expansions of the spaces below the large semidomes) and abut the diagonally-aligned smaller principal arches which partly carry the large semidomes. These large semidomes, in turn, abut the transverse principal arches at east and west which, together with the longitudinal principal arches at north and south (actually pairs of arches of different span), carry the pendentives and dome.

Thrusts of the dome along the longitudinal axis are thus carried down fairly directly through the large and smaller semidomes to their supports, while transverse thrusts reach the main piers only indirectly via the pairs

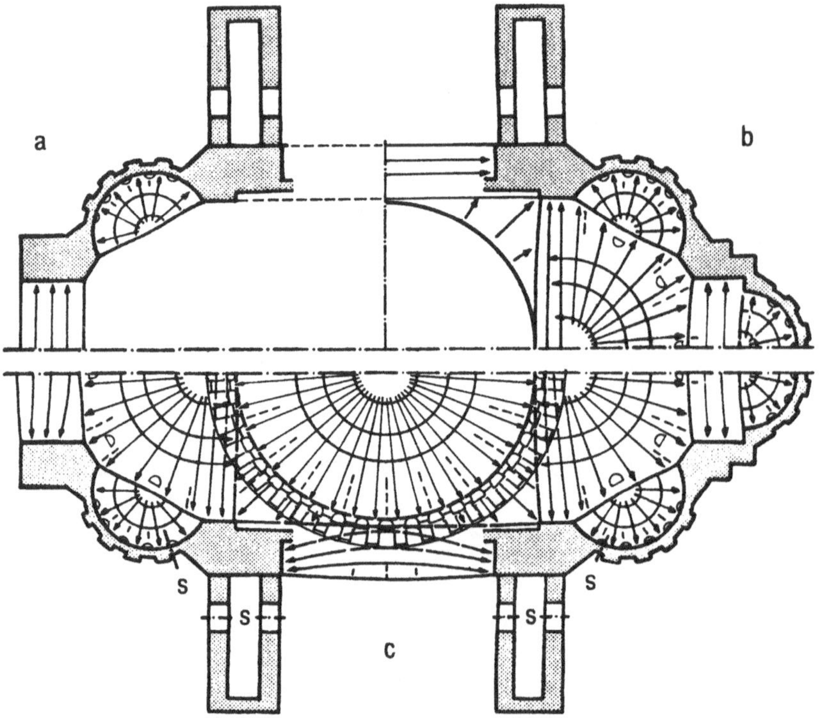

Fig 5. Thrust lines (arrowed) in the main vaulting system and the general pattern of associated cracking (broken line):
(a) after completion of the small semidomes (assuming that they were completed before the large semidomes);
(b) after completion of the large semidomes and pendentives;
(c) after completion of the dome

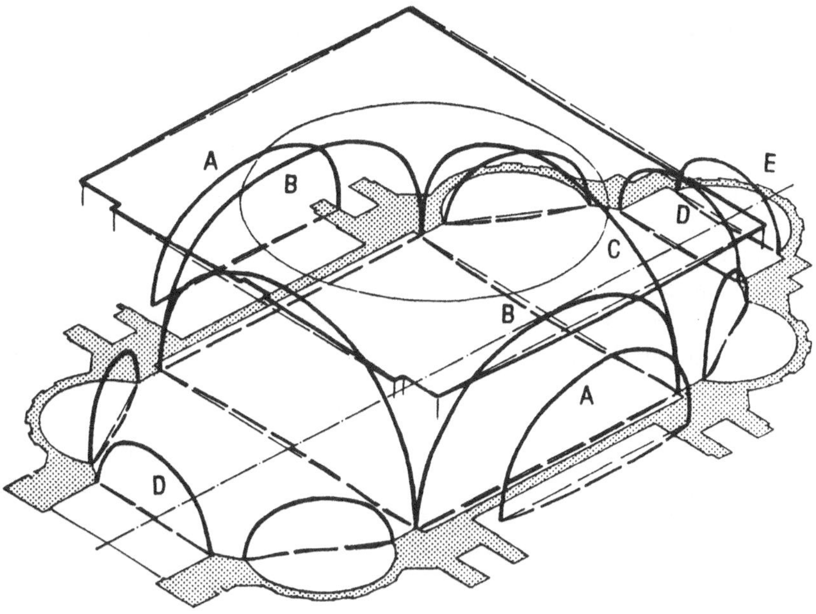

Fig 6. Isometric representation of the edges of the dome base (top) and the forward edges of all the principal arches (Each edge is shown in heavy full line with its present horizontal deformations magnified two times. It is shown similarly in heavy broken line projected to the horizontal plane through the springings, projection to the right of the centreline being vertical and that to the left of it being parallel to the inclination of the supporting piers to show only the extent of lateral bowing).

of arches at north and south. These arches are thus required to arch horizontally as well as vertically, which adds to the thrusts they exert on the piers in the longitudinal direction (Fig 5).

Evidence of the action of these thrusts is clearly visible in outward leans of all the piers, most notably in outward leans of the upper stages of the main piers in the transverse direction, and in lateral bowing of the main arches (Fig 6). There is also widepsread cracking of the type to be expected in these circumstances (Fig 5). Though it is mostly hidden behind the present surface renderings, its presence has been confirmed by exploratory probes. It includes both radial cracking of the vaults (shown by dashed lines) and cracks extending into the piers and associated with major shear displacements (in the regions indicated by short chain lines and marked 's').

Past changes and original form

The building we see today has, however, suffered changes from several partial reconstructions, from numerous external and some internal additions, and from additions and subtractions of other kinds to fit it to different uses—first as the principal church of the Byzantine Empire for some 9 centuries and then, for another 5, as one of the principal Ottoman mosques. Establishing the nature and extent of these changes, and ultimately the original forms, was not always easy, especially where these forms are now hidden beneath later additions or there is little direct evidence of what has been lost.

Documentary evidence helps. Because of the supreme importance of the church its history is unusually fully recorded, though the records are often obscure or admit of more than one interpretation unless there is something else to set alongside them, or they are tantalisingly silent when one would like to know more.

Three partial collapses are recorded, involving, in turn, the main eastern arch, the main western arch, and the main eastern arch again.

Each time, part of the dome supported by the collapsed part of the arch also fell, as did part of the semidome. These collapses occurred in 558, 989, and 1346 and each was preceded by one or more earthquakes, though none occurred during an earthquake. Only in the reconstruction after the first of these collapses was the whole dome rebuilt and its profile changed—being raised by some 20 Byzantine ft (about 6.5 m). The later reconstructions kept essentially to the earlier forms, though less so at the west where there is evidence of some uncertainty and seemingly excessive caution.

Other recorded additions and repairs include some of the present outer buttresses (early 14th century), the minarets (late 15th to late 16th centuries), and various repairs and consolidations, mostly unspecified, in the late 9th, late 16th and mid-19th centuries.

For many structures it is possible to fill gaps in a surviving record with reasonable confidence on the basis of close examinations of structural discontinuities, the character of the masonry, and such details as arch profiles, mouldings, etc.

Here, evidence of structural discontinuities and changes in the character of the masonry during the Byzantine period is largely hidden from view, except at the highest level where the extents of the reconstructions after the partial collapses of the dome and main semidomes can easily be traced. Evidence of the latter kinds serves chiefly to identify the more recent changes, especially the added outer buttresses and other additions and consolidations after the Ottoman conversion of the church to a mosque. Removing these additions, and adding an atrium at the west on the basis of what still survives and what was found in an excavation carried out in 1935, it is possible to reconstruct more of the exterior as it would have appeared in 562 after the first partial reconstruction (Fig 7).

To learn more, it was necessary to draw also on the evidence provided by the present deformations. Deformations that are inconsistent with the pattern of movements over the whole life of the structure were *prima facie* indications of later work.

Of course, deformations cannot be measured directly—only deviations from an assumed setting-out. So the procedure was less straightforward than may, at first sight, appear. It called for a clear understanding of the forces that would have been acting at any time and an ability to visualise, always in three dimensions, the resulting movements. So as not to be misled, it was also necessary to adopt a rigorously consistent approach checked against all other available evidence, as has been described elsewhere[2,4].

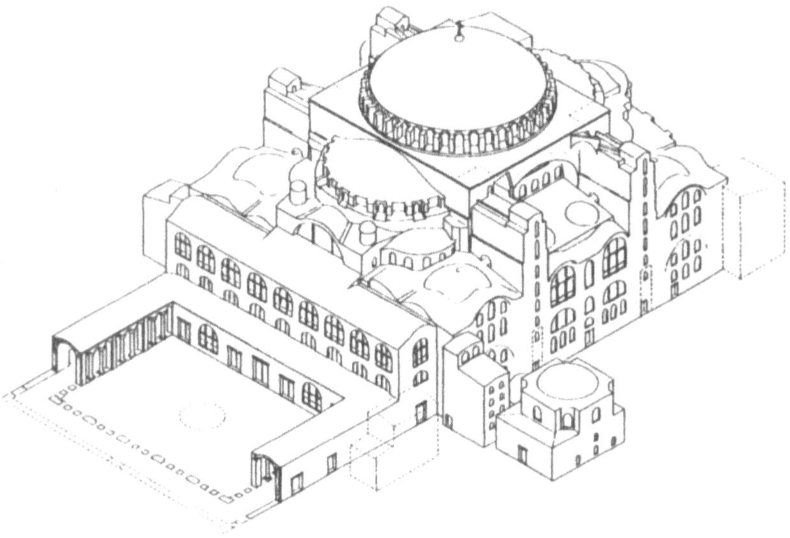

Fig 7. Reconstructed isometric of the exterior as it might have appeared from the south west on completion of the first partial reconstruction in 562 (broken lines denote features whose precise form is uncertain.)

It was possible to show that all upright faces were indeed built with negligible deviation from the vertical, except where they had to conform to an adjacent face that was already leaning, and that all major arches were set out in vertical planes as semicircular arcs and not as the 'functional curves' that some previous observers have seen in them on account of the present deformations. It was then possible, on this basis, to follow through the successive increases in inclinations and changes in arch profiles on the principal transverse and longitudinal cross-sections from the time of initial construction onwards.

Figs 8 and 9 show examples of successive states. Fig 8 shows, at (a), the original flatter dome and the initial inclinations of the main piers in the

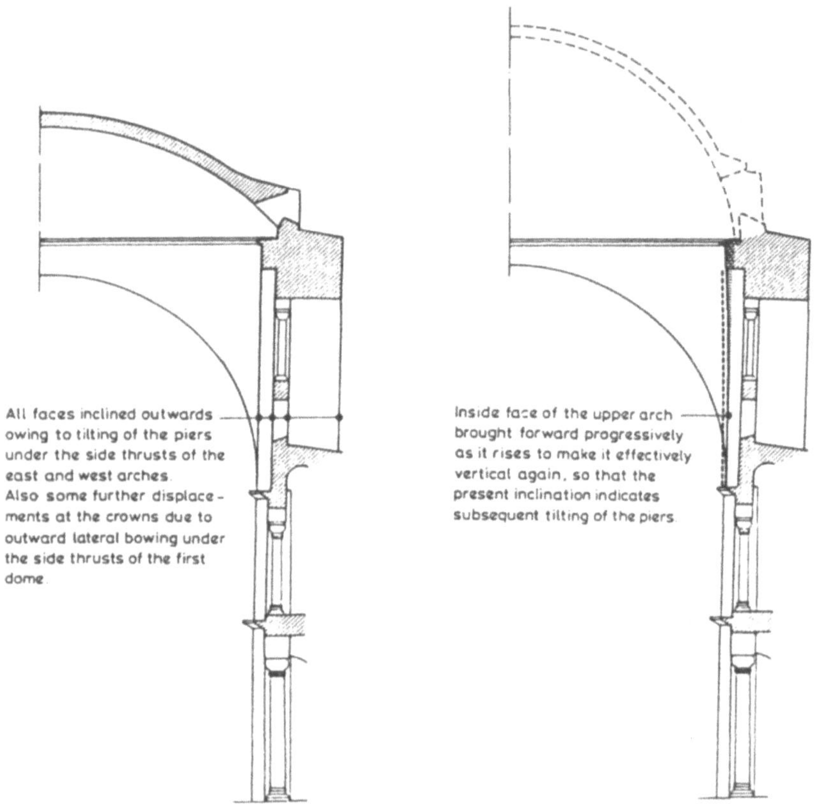

Fig 8. Partial cross-sections of the church looking east on its transverse centreline: (a) in 557 shortly before the first partial collapse; (b) in about 560 after preparatory work for rebuilding the dome; (c) (facing page) after the reconstruction of the tympana and the gallery colonnades below them in about 869

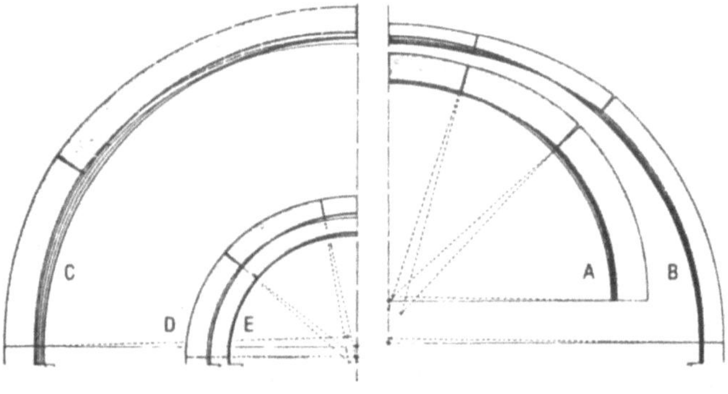

Fig 9. Successive semi-profiles of the principal arches, with the full depths shown only for the present state (Rotations and displacements of the springings and displacements of the centres of curvature are also indicated. The arches are lettered as in Fig 6. For the eastern arch, C, the broken line denotes the reconstructed part.)

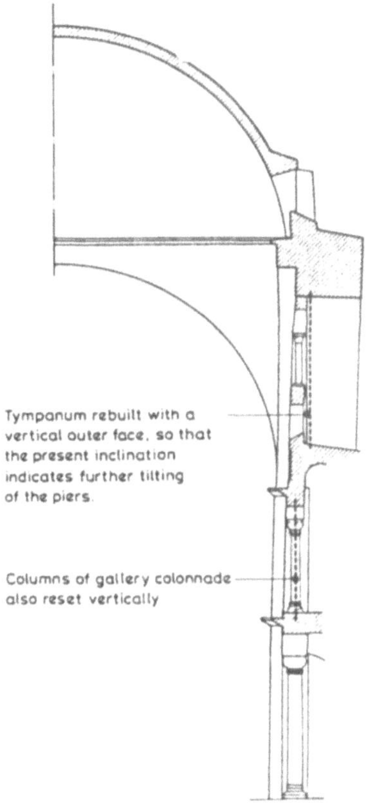

Fig 11. Part-sectional elevation of a typical main pier and its associated staircase tower showing, at AA, BB, C (underpinning arch only), and D (the whole uppermost stage), additions made in an attempt to halt the tilting of the piers as the main arches were constructed (see also Figs 3 and 4)

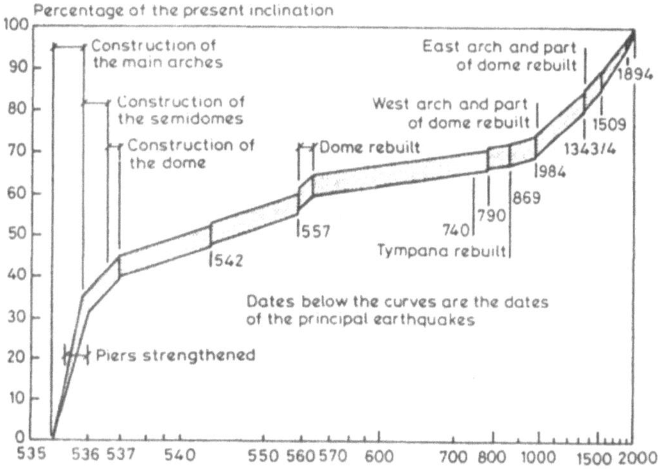

Fig 10. Average north-south inclinations of the upper stages of the four main piers expressed as percentages of the present average inclination, showing the progressive increase during and after construction (the width of the band denotes the margin of uncertainty)

transverse direction. Before the rebuilding to a higher profile, there was some widening towards the crowns of the upper arches at north and south, as shown at (b). At a later date, after the inclinations had increased further, the infilling tympana and gallery colonnades below the arches were rebuilt as shown at (c), the present relative inclinations of the rebuilt outer faces of the tympana and the original faces of the arches themselves being part of the evidence that first led to the identification of this undocumented reconstruction and to its approximate dating. Limited explorations beneath the 19th century plaster subsequently furnished confirmation and gave useful indications of the more open original form of the tympana.

Approximate dating of the latter reconstruction on the basis of this evidence was possible because the historical record gave the date for state (b). A plot against time of average transverse inclination of the upper stage of a main pier, drawing on all the relevant evidence, is reproduced in Fig 10. It will be seen that roughly half the present total occurred during construction, while load was increasing and the fresh mortar was gaining strength, and the remainder at a diminishing rate thereafter. Though there is no direct evidence for the size of the increases due to earthquakes, they must have been relatively small.

Fig 11 is particularly revealing in relation to the process of design and construction. It shows early underpinning additions (AA, BB, C) to the interconnections between the four main piers and the piers that stand to north and south of them and serve partly as staircase towers, and (at D) a heightening of these towers and the connecting walls to heighten the projecting buttresses seen in Figs 2 and 7. Because of the rapid initial rate of tilting, the differences between their present inclinations and other deformations and those of the adjacent piers and original interconnecting arches showed that they must have been made during the final stage of construction of the main east and west arches or immediately afterwards.

Design and construction

Looking back to the antecedents and probable genesis of the design, Justinian's Hagia Sophia replaced another which was already the second on the site. Both seem to have been conventional basilicas—long, axially planned structures like the present St. Demetrius, Thessaloniki, with a central nave roofed by timber trusses, probably double aisles on each side with galleries above, and clerestory lighting of the nave. Each, in turn, suffered serious damage by fire on account of the timber roof. From the very limited excavation hitherto undertaken beneath the floor of the present church, it seems probable that the overall dimensions of its immediate predecessor were enlarged by some 25 % in the rebuilding, with little change in the ratio of length to width.

Although there is only indirect evidence for the objectives underlying the new design, there was clearly a wish to make a more forceful statement without altogether loosing the previous basilican character.

The basic inspiration for this new statement must have been a parallel architectural tradition initially associated chiefly with audience halls in Roman imperial palaces. After the official adoption of Christianity, this tradition was taken over for some major churches, especially in the Eastern Empire. One of the first such churches was San Lorenzo in Milan, which still survives under a Baroque remodelling. By Justinian's time, there were at least a dozen further east. Today, these have mostly been badly damaged or much changed, or they are known only from descriptions or excavations.

Fig 12. Interior of Sts. Sergius and Bacchus

But the form is well seen in a small church commissioned by Justinian himself and probably built shortly before the rebuilding of Hagia Sophia (Fig 12).

All these churches have centralised plans, focused on a central space covered by a dome or other high vault. Characteristically also, the central space was expanded by four semicircular exedrae and surrounded by aisles or an ambulatory with a matching gallery above.

To stretch this centralised plan to give an elongated nave as in a conventional basilica, while also removing the risk of future fire damage, called for a completely new vaulting system. Whatever the earlier ideas leading to it (for which suggestions have been made elsewhere[2]), the answer was the masterly combination of semidomes, part-spherical pendentives and dome referred to briefly at the start. Individually these elements, or approximations to them, were also to be found in earlier buildings. But they had never before been used on this scale and combined in this way. Most of the reduction of overall width that would otherwise have occurred as the semidomes closed in at the east and west was avoided by opening them out into the smaller semidomes over the diagonally placed exedrae.

The planning of the piers that were to serve as supports (initially without the projections AA in Fig 11) leaves no doubt that this whole vaulting system had been decided on before construction began (Fig 13). They were clearly intended to carry the arches shown hatched in the upper part of the figure, and there would have been no need for such broad arches at north and south if a groined vault had been intended over the centre.

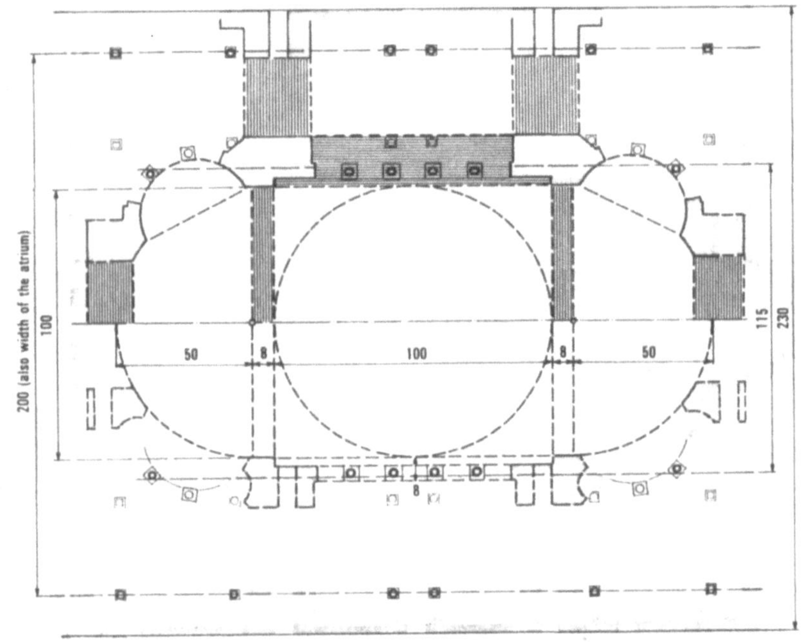

Fig 13. Ground-level setting-out indications of intentions for vaulting the nave, showing also other generators of the plan and (superimposed in broken line in the lower half) the reduced original outlines of the piers around the nave at gallery level (Dimensions in Bvzantine feet.)

Thanks to a most unusual (and probably unmatched) accuracy in setting out, the leading plan dimensions are also revealing. The 100 ft (1 Byzantine ft = 0.312 m) diameter of the dome and main semidomes clearly marks them out as the generators of the plan. This dimension is doubled for the overall width measured between the outer row of columns in the aisles. Most other principal dimensions are multiples of 3 ft, 5 ft, and 8 ft, modules—10 ft, 15 ft, 30 ft; 6 ft, 18 ft, 24 ft, 60 ft, and 72 ft. There also seems to have been a deliberate alignment of the easternmost and westernmost columns of the exedrae with the columns lining the nave between the main piers. The overall width between these alignments (115 ft) is given in a contemporary description as the width of the nave, and it likewise is precisely doubled for the overall external width. Further echoes of the basilican origins of the plan?

The chief structural concerns must have been the choice of dimensions for the piers and their connecting arches and the linked choices of materials. The tendency of all arched forms to thrust outwards on their supports would have been well known. But, in the virtually certain absence of a quantitative understanding of the balance of forces acting in any direction to match the Archimedian understanding of the balance of weights acting vertically, there would inevitably have been a greater need to rely on tradition and past experience in making these choices, without even the guidance of all our present insights into the effects of increases in scale[5].

The choice of well-constructed ashlar for the piers, at least for part of their height, did indeed follow the best earlier practice. And the dimensions of the piers could have been derived from those of the most nearly comparable earlier structure—the Basilica Nova (Fig 14)—by enlarging them roughly in proportion to the increase in vault spans. Similarly, the choice of brickwork for the arches and vaults followed earlier practice in the Balkans and western Asia Minor.

It is less easy to establish the original intentions for the elevations and vertical cross-sections. The choice of height of the colonnades around the nave was probably influenced strongly by the availability of suitable monolithic columns. A more difficult problem would then have been the vaulting of the aisles. It was made no easier by the highly irregular boundaries with the nave, but arose chiefly from the widths to be spanned without raising the gallery floor excessively above the tops of the colonnades.

The solution entailed accepting arched interconnections between the main piers and those facing them across the aisles that were no more than one-and-a-half or two bricks thick at the crown (Fig 11)—far weaker than the deep continuous construction above the corresponding arches across the aisles of the Basilica Nova (Fig 14).

Whether or not on account of fears about the adequacy of transverse buttressing prompted by this break with precedent, iron tie rods were installed from pier to pier above the colonnades, though it is not known how continuous these ties effectively were where they disappear from sight over the capitals.

Though no further problems should have arisen in construction up to gallery level, we are faced, at this level, with a more surprising departure from precedent in what was actually built. The cross-sections of the piers around the nave were considerably reduced (Fig 13). And the columns between them, instead of being set directly over those below, as in the church

Fig 14. The surviving north aisle of the Basilica Nova (of Maxentius and Constantine), Rome (three high concrete groined vaults originally spanned the central space)

of Sts. Sergius and Bacchus (Fig 12), are smaller and more closely spaced (Figs 1 and 4).

There is some evidence that this departure was not the original intention but a change in plan, perhaps to speed construction and reduce costs[2]. Be this as it may, the reductions in cross-section of the piers certainly reduced their capacity to withstand future thrusts.

The change to brickwork in the upper parts of the piers, coupled with the high percentage of mortar and the rapid rate of construction, would have reduced this capacity further. So it is hardly surprising, in retrospect, that difficulties did arise when the construction of the arches that carry the dome was under way.

Two such difficulties, clearly caused by the outward thrusting actions of these arches and the ensuing tilting of the piers, are referred to by the historian Procopius, writing probably some 20 years later[6]. He tells us that the eastern arch threatened to collapse and that columns of the original infills below the north and south arches were spalling under the load they were having to bear as props to the arches. It must have been to halt the transverse tilting of the piers that the underpinning interconnections and other additions shown in Fig 11 were undertaken. The second difficulty was met by simply removing the columns until the arches had settled.

These expedients allowed construction to be completed without further known untoward incident. But the dome now had to be set on a non-circular base. And its shallow profile—probably of the same curvature as the pendentives, though raised slightly above them on a cornice—would have caused it to exert further thrusts far beyond the capacity of any circumferential tie provided by cramping together the individual marble blocks of the cornice.

Under the action of these thrusts, as well as those already acting, and twice shaken by earthquakes, the piers continued to incline outwards and finally allowed the crown of the eastern main arch (which had probably settled more than the western one at the outset) to fall. This led in 558-62, as we have already seen, to the rebuilding of the dome to its present raised profile. With the rebuilding, the church acquired the final Justinianic form shown in Figs 4 and 7.

Assessment and present relevance

To visualise fully the interior, it is necessary first to turn a blind eye to all the Turkish additions seen in Fig 1—the badly discoloured, crudely stencilled paint that now covers most of the higher wall and vault surfaces, the large inscribed discs set against the piers and the muslim furnishings distractingly aligned on Mecca. In their place, one must then imagine continuous expanses of gold-ground mosaic and, complementing the spatial setting and providing the appropriate centres of visual interest (Fig 4), a gold-and silver-encrusted sanctuary screen, altar and altar canopy, and a great elevated ambo standing nearer the centre of the nave for readers and singers. There would also have been much more light from larger windows beneath the north and south arches and from other windows in the aisles and galleries that are now blocked.

The abundance of light and its multiple reflection from the polished surfaces were stressed by Procopius, who also commented particularly on the way in which the high vaults seemed to 'float off from one another' and the dome 'not to rest on solid masonry but to cover the space below as if suspended from heaven by a golden chain' (a literary conceit borrowed from Homer)[7].

Even today, the overall impression is of enfolding surface rather than structural mass. It stems primarily from the supremely elegant plane and part-spherical surface geometry, in which visible faces of the piers are smoothly continuous with those of adjacent infills and those of the arches with those of infill or vault. It is reinforced, or was originally reinforced, by the revetments of marble and mosaic, by the cornices which run unbroken across the fronts of the piers, by the rings of windows in dome and semidomes, and by the manner of use of the only elements suggestive of support—the monolithic marble columns (Figs 1 and 3).

In this, Hagia Sophia differs markedly from most other great buildings of the past, including those later Ottoman mosques which most closely resemble it structurally. In these, piers, arches, vaults and other elements are clearly differentiated, and their surface detailing clearly expresses their loadbearing roles[8]. (Fig 15.)

In relation to contemporary design, however, Hagia Sophia's similarities with these other buildings are more important than the differences.

For all of them, structural stability must have been as much a major concern as it is today. The crucial difference is that, with all loads being carried by direct compression or tension, it could be, and was, achieved

Fig 15. The Suleymaniye Mosque, Istanbul

essentially in terms of geometry—structure was form and form was structure. Thus Hagia Sophia's high vaults were seen as an assemblage of part-spherical forms which met one another on plane circular arcs and were incapable of collapse unless spreading of the supports permitted major cracking and deformation. And from this approach stemmed the unifying geometric discipline referred to earlier.

The difficulties experienced in the later stages of construction showed its limitations when going beyond what previous practice had shown to be safe. Associated intuitive apprehensions of the forces that would be brought into play and the conditions of balance served as a guide. But, even when these were sharpened by professional skills in geometry and the limited statics of the time, they stopped short of making it possible to calculate thrusts and match them with opposing resistances before construction began. They did not do away with the need to test by actually building.

Today we are largely free of that need, even when exercising a far wider choice of form exploiting bending as well as direct compression and direct tension. The price we pay is that it becomes too easy to put choice of geometry in second place or even leave it to others.

Conclusions and final comment

We must be wary of drawing too facile conclusions. Justinian's commission has had few, if any, subsequent parallels and he was a very rare client in the ways in which he inspired the vision and provided the means and the drive to see it realised, despite all counsels of caution and difficulties of execution. And our present exploitation of bending inevitably leads to less legible structural forms and alters perceptions and reactions. If the paramount role of geometry in Hagia Sophia's design has a lesson for us today, it probably lies in its demonstration of the value of cultivating a good 3-dimensional imagination, statical as well as purely geometric, and seeking that simplicity that comes only from repeated critical appraisal and refinement of initial ideas. Comparison with other buildings shows, moreover, the great differences in visual character that can result from different detailing of similar forms.

A final comment must be made on structural conservation. Notwithstanding periodic earthquakes, the chief cause of deterioration and partial collapses over the past 1400 years has been neglect of normal maintenance. The basic adequacy of the structure is amply proven and its structural action is clear. Past remedial measures have not interfered with it, and any future interventions should similarly respect it.

Acknowledgements

Numerous debts are inevitably incurred in a study as extensive as this. In the initial phase, particular acknowledgement should be made to the then Director of the Ayasofya Museum, Bey Feridun Dirimtekin, the then Director of Dumbarton Oaks (Harvard University), Dr. John Thacher, and Robert Van Nice. More recently, I have been similarly indebted to Bey Erdem Yucel and Bey Sinasi Basegmez. Thanks are also due to all the individuals mentioned in the preface to ref. 2.

References

1. Downey, G.: 'Byzantine architects, their training and methods', *Byzantion*, **18**, 1946-8, p99
2. Mainstone, R. J.: *Hagia Sophia: Architecture, structure and liturgy of Justinian's Great Church*, London, Thames & Hudson, 1988 (includes fuller discussions of this and other topics of the present paper, together with full references)
3. for superb large-scale drawings of the church as it now is, see Van Nice, R. L., *St. Sophia in Istanbul; an architectural survey*,

Dumbarton Oaks, Washington, two instalments, 1965 and 1986

4. for the investigative procedure, see also Mainstone, R. J., 'The reconstruction of the tympana of St. Sophia at Istanbul', *Dumbarton Oaks Papers*, **23/24**, 1969-70, p355; id. 'Questioning Hagia Sophia', Colloquium on Hagia Sophia, Princeton University, May 1990 [now published in Mark, R. and Çakmak, A. S., eds, *Hagia Sophia from the Age of Justinian to the Present*, Cambridge University Press1992, pp158-176]
5. Mainstone, R. J.: 'The springs of structural invention', *Journal of the Royal Institute of British Architects*, **70**, No. 2, February 1963, p57. (Revised version 'Intuition and the springs of structural invention', *VIA*, 2, 1973, p42.); id. 'Structural theory and design before 1742', *Architectural Review*, **143**, No. 854, April 1968, p303
6. Procopius: *Buildings*, book I, i, 67-78, Heinemann: Loeb Classical Library, 1940
7. ibid., book I, i, 46-47
8. Mainstone, R. J.: 'The Suleymaniye Mosque and Hagia Sophia', *Int. symp. on Sinan the architect*, Ankara, October 1988 [see now chapter XIX, this volume, for a later paper on the same subject]

XI

Hagia Sophia's first dome

In his paper on the first dome of Justinian's Hagia Sophia,[1] Raybun Taylor states that the central dome of the eleventh-century church of S. Marco in Venice has a drum (note 43). It does not. It is correctly represented without a drum in the early thirteenth-century mosaics of the interior on the west wall of the south transept, and if any doubt remains this can be dispelled by climbing up to it or by consulting the recent photogrammetric survey.[2]

It is much less easy to establish the form of Hagia Sophia's first dome. We have only the description by Procopius and the accounts of the rebuilding in 559-562 given by Agathias and Malalas. Taylor rejects Kenneth Conant's and my readings of these texts and our proposed reconstructions.[3] He asserts that 'a close reading of Procopius suggests that the original dome had no windows, and instead rested upon a fenestrated drum' (p. 66). After summarizing my account of the structure and its early history and reproducing my analytical isometric, he offers an alternative reconstruction with a 10-foot drum. He bases this partly on a highly speculative reinterpretation of Malalas and sees support for it in his readings of other texts and his understanding of how domes behave structurally.

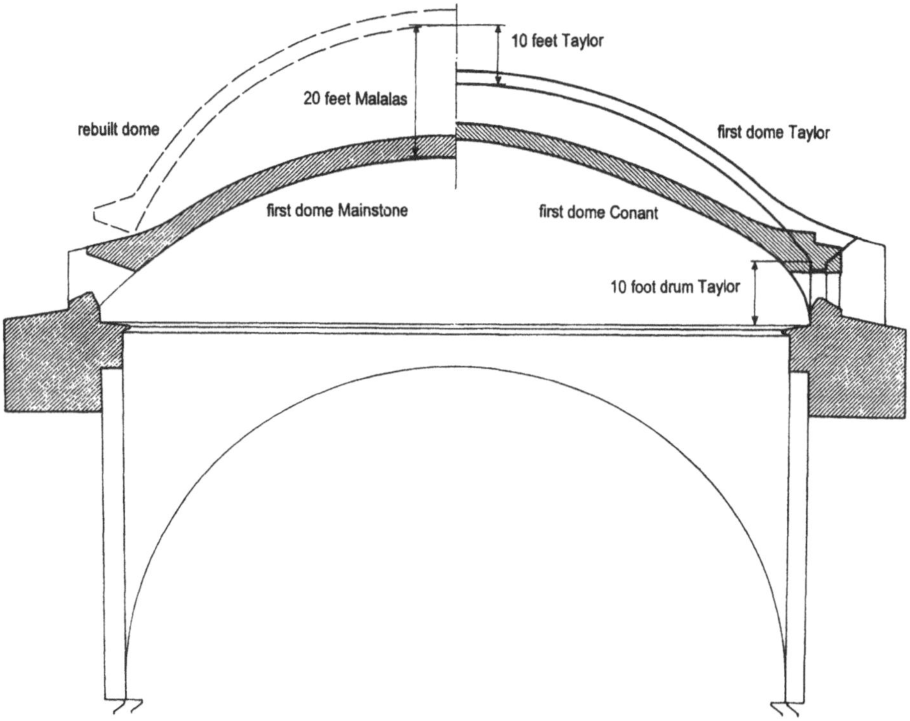

Profile of the present dome and suggested profiles for the first dome.
An earlier proposal by Antoniades was very similar to that by Mainstone.

Though I cannot speak for Conant, I can accept neither this alternative reconstruction nor the readings on which it is based and the arguments for its structural feasibility.

Procopius, though he never mentions the main and exedra semidomes, refers at length to the dome and its immediate supports.[4] After describing the main piers and the arches they carry – open to east and west but filled by tympana on north and south – he looks up to see a circular structure *(kykloteres oikodomia)* interrupted at short intervals by windows. He then goes back to the arches to interpolate the lengthy but perfectly clear description of the four pendentives and the circle *(kykloteres)* in which they terminate before returning to the dome *(sphairoeides tholos)* that rises above the circle *(kykloteres* again) and concluding with the oft-quoted passage about its appearance of not resting on solid masonry but being suspended by a golden chain from heaven.

The repeated use in these passages of the general term *kykloteres* denoting merely something rounded or circular does create problems. The *kykloteres oikodomia* in the first could be solely the structure containing the windows or it could be the entire dome containing them at its foot. But, whichever was intended, the addition here of the epithet *strongylon* to emphasise roundness rather than of the geometric term *kylindros*, which Procopius had used earlier for the semi-cylindrical wall of the apse, surely precludes reading it as a cylindrical drum – all the more so because such a drum would have been not only highly unusual but so unprecedented, as far as can be seen from surviving evidence, that it would have called as much for special attention as the pendentives.

In the later passage the intention can hardly have been to refer again twice over to the *kykloteres oikodomia* containing the windows because if this structure had been continuous with the pendentives these would no more have existed as separate elements than they do in the large semidomes. Here *kykloteres* must therefore be read as either the geometric circle that terminates the pendentives or the circular dome cornice that is its physical embodiment. But the presence of the cornice is implicit, even if only the former was intended, because it would have been contrary to all contemporary practice, both in Hagia Sophia and elsewhere, not to have had a cornice at such a juncture, and it may be noted that there is also no specific mention of the cornices elsewhere in the building.

Contrary to Taylor's claim (p. 72) that I drew attention to mention of a drum in one of these passages, I see no such reference. I interpret the *kykloteres oikodomia* of the first passage as the structure at the foot of the dome containing the windows, this structure being visually distinct from it but geometrically continuous with it in the manner described a little later that is so characteristic of the whole design. And I read *kykloteres* in the second passage as denoting both the geometric circle at the tops of the pendentives and the circular cornice. Taylor would perhaps have been less baffled if all this had been explained in my book and if the translation of Procopius on page 127 had not been shortened by omitting the interpolated reference to the pendentives, but space was everywhere at a premium in a book about such a complex building. Much else had likewise to be omitted or abbreviated.[5]

The later account of the church of the Holy Apostles tells us nothing relevant in itself since it merely describes the central dome in the same terms as those used for Hagia Sophia's dome.[6] Its emphasis on the similarity is relevant only if the form of this other dome can be independently established. Even setting aside the probability of some rebuilding of the domes in the centuries preceding the *ekphrasis* of Constantine of Rhodes, the illustrations in the *Menologion* and the Kokkinobaphos manuscripts, and the construction of the one extant likely direct filiation – the present S. Marco[7] – I do not think that any of these clearly indicates the presence of a fenestrated drum. S. Marco points in exactly the opposite direction, as already noted. In the *ekphrasis* the repeated play on the word *kylindros* does not suggest the precision seen in its use by Procopius. The earlier references must be to the vaults that carry the domes, and the last reference I take to be a poetic conceit for the over-arching of the domes themselves. The illustrations show only stylized exteriors and, as the domes of S. Marco and the semidomes of Hagia Sophia demonstrate, an exterior drum-like form does not necessarily signify a drum within.

As for the evidence of heightening at the first rebuilding, I must admit to being slightly sceptical about the accuracy of Malalas's figure. His 20 feet, though not impossible, could easily have been rounded off. But to ignore what the more reliable texts actually say in favour of a different and inherently less likely reading is pure speculation which I regard as inadmissible.

On all these grounds I maintain that the original dome had no separate drum, and I have little to add to what I have already written about my reconstruction of its likely form based on the only precise evidence we have about the change in profile at the rebuilding. I wish only to correct some of Taylor's misstatements before turning to matters of structural feasibility. He wrongly assumes (pp. 70-71) that I interpreted the *kykloteres oikodomia* as my low 'drum-like' section above the cornice and dismissed what he sees as the full implications of this term 'on physical grounds', and he places far too much emphasis on this drum-like section. It is of little real significance. I introduced it merely as a likely infilling of an otherwise fairly sharp re-entrant. He is equally wrong in stating (p. 71) that there would be inadequate headroom on the cornice. Even with no more projection of the lip than now, there would be headroom of 6 feet (compared with about 9 feet today) when standing almost a foot from the lip. And I take it for granted that the windows would be vertical as in the semidomes today and disagree with his assessment (note 30) that they would create more dazzle than similar windows set in a vertical drum. The deeper reveals would surely have the opposite effect by reducing direct light.

Turning now to the structural feasibility of Taylor's reconstruction, his commendable attempt to explore this unfortunately leads him into what is clearly unfamiliar territory and into further error.

Domes differ from arches in that they consist, in effect, of horizontal hoops as well as vertical arches. By virtue of stresses in these hoops and, *under certain conditions* but irrespective of the profile chosen, the thrusts in the vertical arches can be made to follow this profile as sketched in his fig. 6, and cracking can be avoided. These conditions are first that any necessary tensile stress in the hoops can be resisted without cracking and second that the support can both provide balancing counterforces for any horizontal thrusts exerted at the springings and move in or out just enough to match the contraction or extension of the lowest hoop. If either condition is not met, cracking will occur up to a certain height; and the structural behaviour up to this height will be that of a ring of arches separated from one another by the cracks, and the lines of thrust will change accordingly.[8]

Taylor recognizes the first condition but wrongly assumes that, with the use only of sixth-century technology, deficient tensile strength could have been made good by heavy cramping or tying around the base. Even a continuous circumferential iron tie would have to be about 100 millimetres square in cross section to serve this purpose. He simply ignores the second condition and fails to realize that the continued spreading of the dome base due to the thrusts generated by the weight bearing down on the main arches would soon lead to cracking whatever the profile. The addition of ribs could, moreover, do nothing to reduce the cracking, as he assumes it would (p. 74). They would indirectly have the opposite effect by adding to the weight on the main arches and hence to the thrusts on the piers.

For the profile that I suggested as being most likely and that Taylor is content to adopt above his cylindrical drum [see the figure above], the horizontal thrust would have been almost 0.1 meganewton per metre of circumference and would pass down undiminished to the base of the drum. Buttresses between the windows could resist it if they stood firm and, by adding their own downward thrusts, they could change the direction of the resultant thrust at the base. But they could do nothing to reduce its horizontal component either on the crown regions of the large semi-domes or on the main north and south arches. Their principal effect, like that of any ribs, would be to add to the vertical loads on all the main arches, again increasing the outward thrusts on the piers, the spreading at dome level, and the risk of collapse.[9]

The drum that Taylor envisages would therefore markedly diminish stability. This certainly does not argue in its favour, though I do not think that we can reject it on those grounds alone. Margins of stability in such structures are far more difficult to assess theoretically than many recent analyses assume. Still less do I think that we can argue either for or against such a drum on the basis of speculations about possible lost forerunners and the reasoning of designers, as in his final section, so I leave comment on these questionable speculations to another occasion.[10] My main reason for rejection is what I believe to be the correct reading of the textual evidence. I would add to this chiefly the departure that it would entail from the remarkably consistent character of the design elsewhere in the building, remarked upon even by Procopius.

Notes

1. Taylor, R., "A literary and structural analysis of the first dome on Justinian's Hagia Sophia, Constantinople", *Journal of the Society of Architectural Historians*, 55, 1996, 66-78.

2. The survey is not yet complete, but the principal cross sections were displayed at the conference on restoration of the basilica held under the auspices of the Istituto Veneto di Scienze, Lettere ed Arti and the Procuratoria in May 1995.

3. Conant, K. J, "The First Dome of St. Sophia and Its Rebuilding," *Bulletin of the Byzantine Institute of America*, vol. 1, 1946, 71-78; Mainstone, R. J., *Hagia Sophia*, London and New York, 1988, 209-212.

[4. Procopius, *De aedificiis*, revised edition with English translation, Dewing, H. B. and Downey, G., Loeb Classical Library, 1940, 20-21.]

5. Among the other omissions were details of several series of tests on the mortars undertaken for me at the UK Building Research Establishment, the results of which were reproduced in part by R. A. Livingston from a copy sent to Dumbarton Oaks as referred to in Taylor's note 60.

[6. Procopius, op.cit. 50-51.]

7. For recent discussion of S. Marco's construction history and the relationship between its design and that of Justinian's Holy Apostles see J. Warren, 'The First Church of San Marco in Venice', *Antiquaries Journal* 70 (1990), 327-359, and my reply, 'The First and Second Churches of San Marco Reconsidered', ibid. 71 (1991),123-137.

8. See also my *Developments in Structural Form*, London and Cambridge, Mass., 1975, [rev. 2nd edn, 1998], 42-44.

9. Estimates of thrusts for my reconstruction of the original dome are given in my *Hagia Sophia*, 212. Unlike those estimated by Swan and Mark and their associates, they take cracking into account.

[10. See now Mainstone, R. J., 'Structural analysis, structural insights and historical interpretation', *Journal of the Society of Architectural Historians*, vol. 56, 1997, 316-339 – chapter VI this volume]

XII

The structural conservation of Hagia Sophia: past history and present needs†

Introduction: the structural conservation of historic buildings today

The structural conservation of a historic masonry building presents very different problems today from the design of a typical new one. Such a building differs structurally in many ways from the buildings for which our present analytical tools were largely developed. It carries its loads largely by compression, the masonry usually being extensively cracked where primary tensions would otherwise arise or where high local primary compressions lead to high orthogonal tensions. It already has a particular form put together in a particular manner, often in several campaigns of construction and partial reconstruction, and it has usually been subjected to various other interventions during its life. Yet we can rarely know all we should like to know about this history and its present state and constituent materials. And although it has demonstrated its capacities and shortcomings over many years, the demonstrations are not in the measures that we use when designing a new building and may be difficult to interpret.

The Great Church of Hagia Sophia, in what is now Istanbul, furnishes an excellent example.

Hagia Sophia

It is shown as it exists today, crowned by a dome more than 30 m in span, in Fig. 1.[1] It was built to the orders of the Emperor Justinian between 532 and 537, so that it has now stood for almost 15 centuries. Yet its central space, vaulted in one continuous sweep from west to east, remains unequalled to this day. For the first nine centuries (apart from a relatively brief interlude during the Latin occupation of what was then Constantinople) it remained the principal church of the Byzantine Empire, and then for another five it served as the principal Ottoman mosque.

Justinian was no conservationist. Paying little respect to its century-old predecessor on the site, he had this razed to the ground after a fire and in its place he commissioned something as novel as anything built before or since. There were, of course, smaller-scale precedents for most elements of the design. They would have demonstrated (albeit within unknown limits) both the feasibility of these elements and some of their possible uses. They would also have served as stimuli to a creative imagination. But what is historically more significant is that they were brought together in a wholly new manner in what was really a superlative exercise in spatial geometry by men who were masters of that discipline.[2]

Innovation as far-reaching as this does, however, run great risks. These were all the greater on account of the scale of the enterprise and Justinian's impatience to see it finished. We know from a slightly later account that construction was not uneventful: difficulties arose soon after construction of the main arches began. An early partial collapse and two further ones more than four centuries and eight centuries later are also recorded, all them being triggered off partly by earthquakes. In addition there have been numerous campaigns of consolidation or repair and another major partial reconstruction to which the records make no explicit reference. After the first collapse the dome was raised and some later campaigns have considerably affected the outward appearance. But the reconstructions and other works have preserved most of the original structure intact to this day – a unique record for a building of its size and age.

What are the future prospects? Are the present strengths adequate to ensure survival for a further century or so? Or are there weaknesses that now call for intervention? If so, what should be the criteria, and how should we proceed?

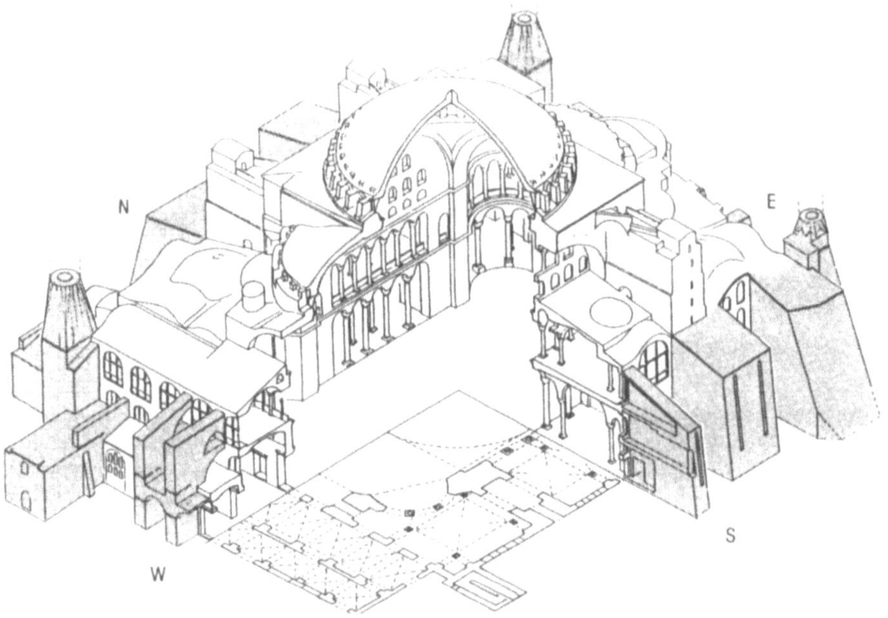

1. The structure today showing, except at the south west, the principal external additions.

Design, construction and subsequent history

To answer these questions we must first take a closer look at the structure, at its construction history and the changes subsequently undergone, and its behaviour hitherto – those demonstrations of capacities and shortcomings referred to at the outset. Its scale and complexity and the inaccessibility of most of the working masonry behind facings of marble, mosaic or rendering make this difficult. But nothing can hide the very marked departures from what could be presumed to have been the original setting out. Once it was demonstrated that they were indeed true deformations they provided invaluable clues to set alongside the documentary record.

Working from both existing survey data and fresh observation and measurement and concentrating first on the main arches and their supporting piers, it was possible to trace the growth of their deformations and by doing so to learn more about the early difficulties, the measures taken to overcome them, the reasons for the recorded collapses and the limits of the recorded partial reconstructions. In a similar way, but testing the conclusions now by limited probing behind the present renderings, it was also possible to identify and roughly date the unrecorded further partial reconstruction.[3]

However ideas evolved, the final design stemmed from the decision to adopt a novel main vaulting system consisting of a central dome abutted by semidomes of the same diameter at east and west. This called for different provisions to carry the dome and resist its thrusts at east and west and at north and south and led to the primary support system seen in Fig. 2. The support at east and west was the more

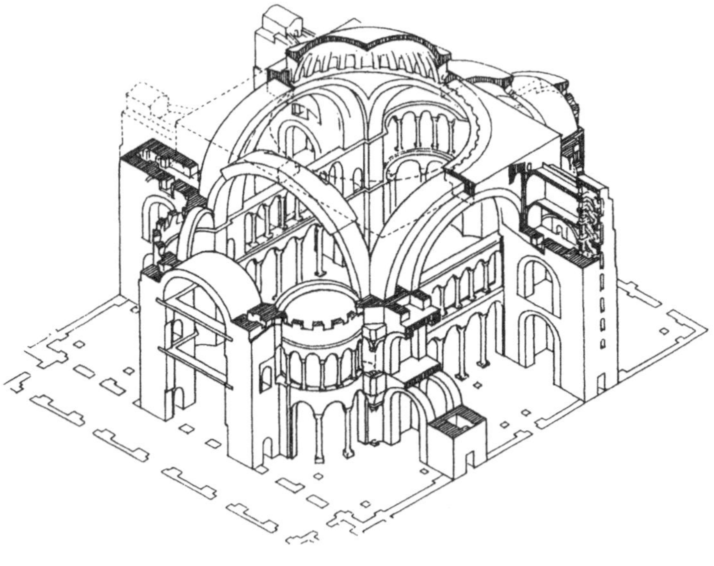

2. The original primary structural system with the vaults partly cut-away and the reconstructed section of the main west arch shown stippled.

novel, with the thrusts absorbed initially by the semidomes. But it was the support at north and south that proved troublesome, partly on account of a requirement for galleries over the side aisles as in the previous church. The closest precedent for it was to be found in the Basilica Nova in Rome. There, with no galleries, the main piers could be very substantially connected to others alongside them across the aisles. In Hagia Sophia the main piers were likewise supplemented by buttress piers across the aisles and they were similarly proportioned at ground level. But only very shallow interconnections were possible over the aisles without unduly interrupting their continuity or unduly raising the gallery floor and, above the gallery roof, there were initially only parallel pairs of arches carrying low walls spanned by a barrel vault. These interconnections were all the weaker because they were constructed of brickwork with wide mortar joints and were subjected to heavy load before the mortar had reached full strength. A further weakness was introduced by a considerable reduction in the cross section of the main piers at gallery level from that at ground level.

As a result of these weaknesses the supports began to yield alarmingly even before the main east and west arches had been completed. To slow down their yielding, bracing arches of cut stone carried on projections from the piers were hurriedly constructed across both aisles and galleries and the walls above were extended upwards with further barrel vaults spanning between them (Fig. 3). Work then proceeded to completion, though the dome had to be set on a base that was no longer circular as intended and on arches that were already deformed, particularly the eastern one. As far as can be seen, it was generally similar to the present dome except that it had a lower profile, so that its outward thrusts would have been about twice as great.

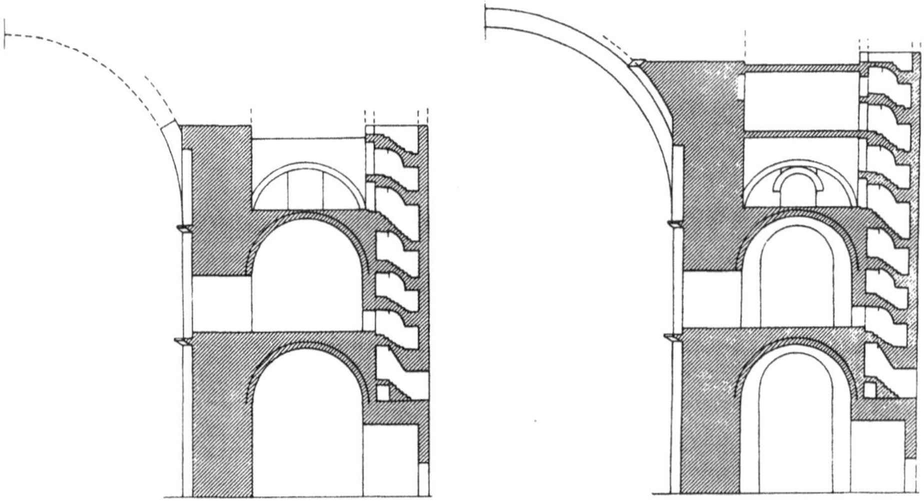

3. Transverse sections through a main pier and the linked buttress pier: on the left as originally built; on the right with additions made to their interconnections during construction of the main east and west arches.

The first partial collapse was primarily a collapse of the central part of the main eastern arch. That part of the dome which rested on the collapsed section also fell, but the rest remained standing – 'a wonder to behold' in the words of the contemporary account of Silentarius. It is notable that it followed one of several earthquakes in the preceding years but that it did not occur until 5 months after the last of them while unspecified remedial work was in progress. The state after this earthquake was probably similar to that observed recently in the Holy Trinity in Budva and the Metropolis in Mistra (Fig. 4).

Over the next 4 years the entire dome was rebuilt to a higher profile after reconstruction of the fallen part of the arch and some addition to the faces of the main north and south arches to reduce the excess of the north-south over the east-west diameter. There were also some additions to the dome base.

The second and third collapses likewise involved primarily the arches – first

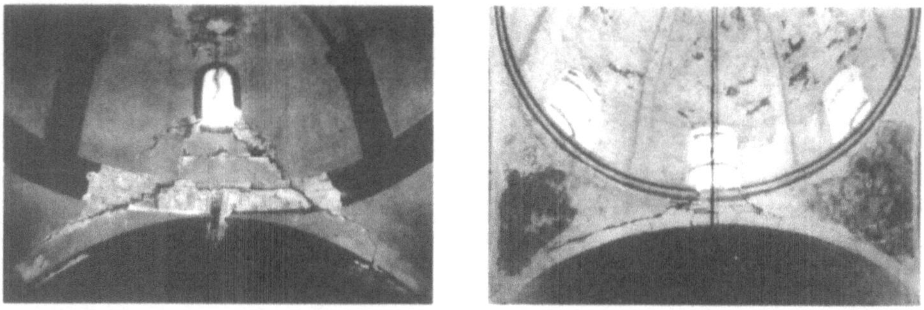

4. Earthquake damage: the Holy Trinity, Budva (left); the Metropolis, Mistra (right).

the main western arch then for the second time the main eastern one. Again they led to the fall of that part of the dome thereby deprived of support. The records are less informative than about the first collapse. But it is recorded that more than a century before the western arch fell it was already seriously cracked and some repair was undertaken, and that the third collapse, like the first, occurred not during an earthquake but some time later – on this occasion a year and a half later. Both reconstructions were limited to the replacement of what had fallen, leaving largely untouched to this day the 6th-century sections of the dome that had remained standing. Reconstruction of the eastern arch, the upper part of the eastern semidome, and the adjacent parts of the eastern pendentives seems to have been intended simply to reinstate what had fallen, though the work was less precise and there is a particularly clumsy join between the 6th-century and 14th-century portions of the northern pendentive. Reconstruction at the west was considerably more cautious. There was a significant change in longitudinal profile of the semidome above the level of the windows (Fig. 1), a considerable increase in depth and width of the reconstructed part of the arch (Fig. 2), and several additions including the blocking of two windows at each end of the dome reconstruction (Fig. 5).

The further major reconstruction that is not explicitly recorded was of the great tympana – the window-filled walls set within the main north and south arches. The original large upper window in each was replaced by a number of smaller ones and, thanks to the reconstruction, the tympana were relieved of load from the earlier settlement of the arch crowns and are now much less extensively cracked than they would otherwise be.

More significant than this reconstruction from the point of view of the present safety of the whole structure are some of the numerous additions, repairs and consolidations undertaken from a fairly early date. They include virtually all the buttressing masses (other than the corner access ramps) that now surround the exterior up to gallery roof level and numerous iron ties and cramps and infillings of both stone and brick (Figs 1 and 5).

The earlier buttressing masses are not solid but contain stairs, chapels and other rooms. The later ones range from the Gothic-looking (but probably much earlier) flying buttresses at the west to the more substantial ramped masses and fliers against the north, south and east sides. There are also four Ottoman minarets standing on substantial bases at the corners.

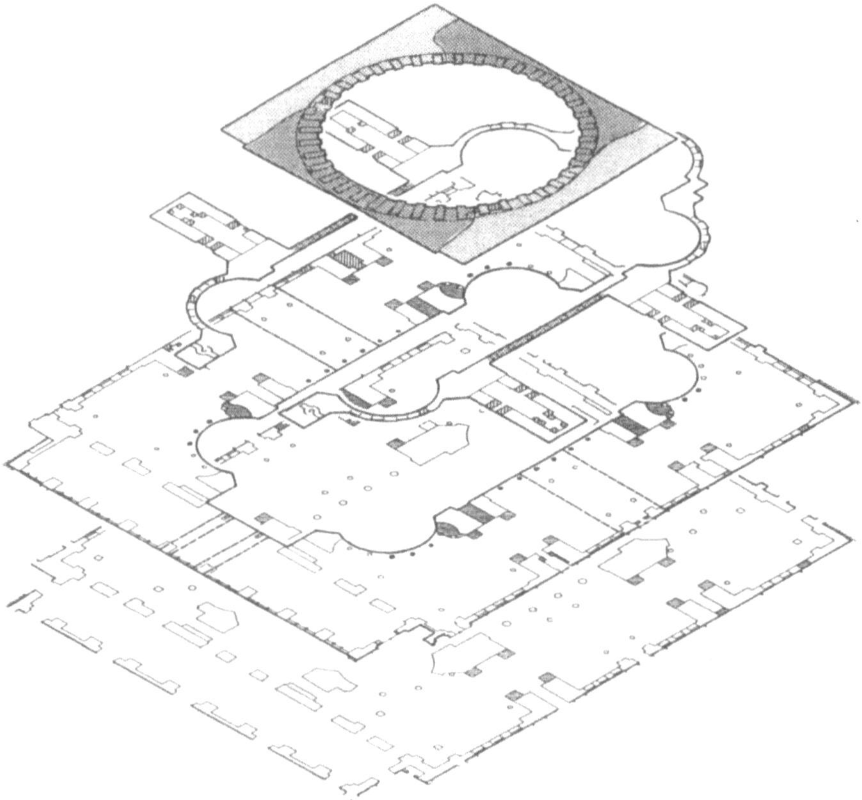

5. Exploded isometric plans at ground and gallery levels and at the bases of the semidomes and dome, showing infills and new ties inset in the gallery floor. Brick infills are shown by vertical hatching, stone infills by horizontal hatching, and ties by heavy broken line. Light and heavy stippling denote respectively 6th century and later reconstructions.

Most of the infillings are to be found in the buttress piers across the aisles from the main piers. These were constructed for most of their height in brick and initially had large voids serving as entrances or to accommodate stairs. More important probably are the infillings to the main piers at gallery level, made in at least two stages to bring their cross sections closer to those above and below. They are not identical in all piers: the north-south tunnel vault in the main north west pier is still unfilled whereas all its counterparts were filled, probably in the 16th century. There are also numerous fillings of windows and of previous voids in the dome base.

The total number of added ties and cramps is unknown. A few only, set into the gallery floor, are shown in Fig. 5. There are others conspicuously spanning between column heads or across the springings of vaults in both aisles and galleries. And it is known that circumferential ties were set around the base of the dome in 1848 and again in 1926. But no sign has been seen of ties also envisaged in 1848 between the springings of the main north and south arches.

The intention behind these works is rarely stated as being more than to make good the ravages of time and neglect or to forestall a supposed impending collapse.

Usually it was clearly to halt outward movements, either of the primary structure or of the secondary structure of the aisle and gallery vaults. Both ties and buttresses were used for this purpose. The infills to the buttress piers and other associated additions and reconstructions may well have had the same objective. But the infills to the main piers at gallery level were more likely intended as underpinning to reduce the vertical load on their original cross sections when splitting of the masonry suggested dangerous overloading. Restraint of a feared splitting away of one 6th-century projection from the main body of the south-west pier was clearly the intended purpose of long horizontal cramps that were exposed in the 1930s on its east face.

Present state

Together these works have left the structure an even more complex one than it was when first completed in 537. But, since the investigations on which the above account is based were conducted primarily to learn more about the church in the 6th century, many details relevant to present safety remain unexplored.

Doubts remain about foundation conditions and about internal details of the main structure and some of the additions. Nothing is known, for instance, about the interior of many of the outer buttresses or about the precise locations and present strengths of the dome ties. And, while many unbonded joints and voids – some filled or partly filled – have been identified, not all have been explored and there are probably others as yet unknown.

As for the present state of the masonry, the broad pattern of cracking in the primary structure can be inferred from proven deformations that are far in excess of possible elastic and early plastic ones. Though most cracks depicted in Fig. 6 were

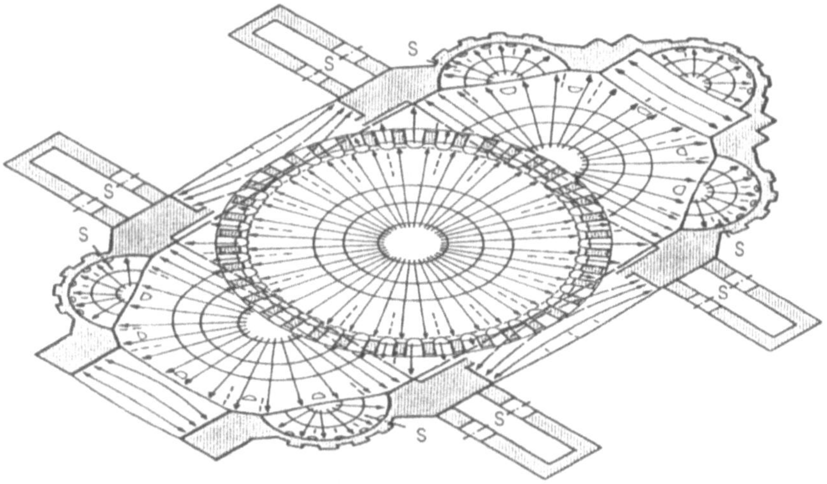

6. Thrust lines (arrowed) and circumferential compressions (full line) in the main vaults and the general pattern of associated cracking (broken line) at successive stages of construction. 'S' denotes a major shear failure above the level of the gallery floor.

hidden when investigation started and remain largely hidden today, this pattern has been checked by limited uncovering now recorded in the survey drawings, and it does indicate the most important features of the present transmission of outward thrusts. But while earlier exposures of the masonry of the southern main piers at gallery level revealed instances of the splitting referred to above under the high local compressions due to a reduced cross section and imperfect bearing between blocks, the full present state of these piers and their northern counterparts is not known. Nor is the present condition of those regions of the main transverse arches where high local compressions are also to be expected as a result of hinging deformations.

Present safety and conservation for the future

Because of these uncertainties and the impossibility of direct measurement of relevant material properties, a quantitative assessment of safety such as is now made when designing a new structure is not feasible even under static loading, and still less under dynamic loading in the absence of a relevant proven theory. Nor can useful guidance be obtained from inverse analysis of arbitrarily simplified models drawing on what has already been deduced about the growth of deformations and on such low-amplitude dynamic characteristics as can now be measured. Such analysis cannot discriminate between models that would be valid for predicting future behaviour and others that could be misleadingly invalid.[4]

It is therefore necessary to rely more directly on what the structure has disclosed by its behaviour hitherto.

The cracking of the vaults is not, in itself, a major cause for concern. The masonry dome, for instance, is an inherently stable form even when cracking extending upwards from windows around its base reduces it to a ring of arches with a common keystone in the uncracked crown region. With a thickness of 5% or more of the radius of curvature, collapse can occur only if there are large outward movements at the base or local failures under the radial compressions. There is a risk of the latter if water seepage over a long period seriously weakens the mortar, as does appear to have happened in 1824 in one of the minor domes over the galleries. But not otherwise. Much the same is true for the semidomes, abutted as they are at their boundaries.

Cracking in the structure as a whole even has one beneficial effect. As amplitudes increase during an earthquake, it leads to reductions in natural frequencies which make damaging resonance with a predominant ground motion highly unlikely. This may be why Hagia Sophia came through the 1766 earthquake unscathed while serious damage was done elsewhere.

The main cause for concern is continued growth of the cracks in the main support system. Slow growth results from repeated slight opening and closing of the cracks to accommodate thermal strains and from their progressive jamming further and further open by intrusive debris. Earthquakes hasten this process. Indeed this is probably the chief mechanism whereby they damage a masonry structure that is already extensively cracked in finding its own equilibrium state.

Further splitting of the masonry of the reduced sections of the main piers at gallery level could eventually result in sudden failure under purely vertical load.

The fillings and other additions shown in Fig. 3 should, however, have led to some net increase in safety over the past few centuries.

The chief risk seems, therefore, to stem from the increases in the transverse inclination of the piers that result from the crack growth. Previous collapses of the east and west arches occurred through excessive outward movements of their springings and excessive local compressions at the hinge regions at crown and haunches. Further movements could lead in the same way to further collapses.

A study of inclinations up to the present does suggest a continuing movement, though the recent history was merely interpolated and no distinction was made between slow long-term increases and the rapid increases during earthquakes.[5] A possible interpretation is a fairly uniform increase since the 6th century of about 1% of the present inclination every 80 years plus rapid growths of up to 3% during earthquakes. But the recent tying of the dome base and the fact that even the newer peripheral buttresses should now be providing fully effective support suggest that the present actual and potential rates of increase should be substantially lower.

The further increase possible without another collapse depends partly on the known geometry of the arches and partly on the largely unknown strengths in the critical crown and haunch regions already referred to. Taking into account the greatly increased cross section of the rebuilt part of the western arch and all other relevant factors, there should be no immediate danger. Indeed, since the much greater sections and lesser spans of the north and south arches make them much less vulnerable, it might be concluded that the whole structure, strengthened as it now is by added ties, buttresses and infills, will stand for many more years if there is no great deterioration of its materials.

To confirm this assessment, the chief need now seems to be some examination of the critical regions coupled with monitoring of critical spans and separations at cracks over a long enough period to allow cyclic movements to be discounted. Recording of strong-motion dynamic responses to indicate actual modes of response outside the elastic range would also be helpful.

If sufficient confirmation of present safety is not thereby obtained, making some precautionary intervention advisable, the aim should be to enhance safety without loss to the original structure or to the character of the present dynamic response. Measures meeting these criteria would be the consolidation (if inspection suggests that it is desirable) of some of the present infills and the renewal or supplementing of some of the present ties. Statical calculation, using models based on the observed modes of deformation, would be feasible and essential here to determine appropriate tie forces.

Whatever else is done, there is a continuing need for proper maintenance of the external weathershield, especially the lead covering of the vaults, and for the making good of any seriously weakened mortar. In the past this essential maintenance has been too much neglected. This neglect is referred to in the records even more frequently than earthquakes as the cause of damage. Even in recent years it has been too common to see figs and other vegetation growing through holes in the lead. A worse state is described in early 19th-century accounts and shown in photographs taken early in this century. It must not be allowed in the future. About this there can be no question.

† This is a slightly abridged version of the opening paper of the third conference on the Structural Repair and Maintenance of Historical Buildings, Bath, 1993 (first published in the conference proceedings, pp. 3-14)

Notes

1 See Van Nice, R. L., *St Sophia in Istanbul: an architectural survey*, Dumbarton Oaks, Washington, D.C., 1965 and 1986, and Mainstone, R. J., *Hagia Sophia*, Thames and Hudson, London, 1988, for further representations. Plates 149, 252 and A2 to A8 of the latter show the church in the 6th century.

2· The most relevant precedents are discussed in Mainstone, 1988, chapter 7. The church of St Polyeuctos has been claimed as a more complete forerunner by its excavator in Harrison, M., *A temple for Byzantium*, Harvey Miller, London, 1989. But no fallen dome masonry was found to support this claim, the suggested reconstruction poses far more problems than it solves, and other reconstructions seem more consistent with the evidence.

3 Mainstone, R. J., 'The reconstruction of the tympana of St Sophia at Istanbul' *Dumbarton Oaks Papers*, nos 23/24, 1969-1970, pp. 355-368; idem, 1988, chapters 4 and 8; idem, 'Questioning Hagia Sophia', *Hagia Sophia from the age of Justinian to the present*, ed. Mark, R. and Çakmak, A. S., pp. 158-176, Cambridge University Press, Cambridge, 1992.

4 Modelling possibilities are discussed further in Mainstone, 1992. [Now see also idem, 'Structural analysis, structural understanding, and historical interpretation', *Journal of the Society of Architectural Historians*, vol. 56, 1997, pp. 316-339, reprinted in chapter VI, this volume]

5 For a plot of the growth of the most significant inclinations on a logarithmic time scale, see Mainstone, 1988, pl. 105, Mainstone, 1992, fig. 95, and fig. 10 of chapter X, this volume.

XIII

THE FIRST AND SECOND CHURCHES OF SAN MARCO RECONSIDERED

THAT THE PRESENT SAN MARCO (fig. 1) either incorporates substantial remains of earlier structures or has had its design constrained by structures no longer extant can hardly be doubted. This much is obvious even from the ground plan, especially at the ends of the north and south transepts. Further anomalies apparent in the building itself also suggest that it is not all of a piece. Setting aside obvious later changes that have not affected the basic form, they include the existence of pointed arcades alongside semicircular ones and the curious way in which the barrel vaults to the north and south of the eastern dome oversail the side chapels.

It nevertheless guards its past well, whether or not the account of a complete rebuilding in the eleventh century is mere politically motived fabrication. Virtually all the working masonry, which might otherwise yield vital clues, is hidden behind marble or mosaic facings and sometimes rendered even more inaccessible by liturgical furnishings. Many columns, capitals and other details are *spolia* or purpose-made copies of earlier forms. And the evidence has been further confused by a long history of repairs and partial rebuilding.[1]

Undeterred, John Warren has presented an attractive new thesis which accounts for many of the anomalies.[2] The lead he has taken from the San Alipio depiction of non-figural mosaics over all the side portals of the west front is open to question. This depiction is hardly unmistakable evidence of an iconoclast date for these portals nor, by extension, for a similar date for the whole western narthex.[3] There are also instances later in the paper where alternative explanations cannot be so easily excluded as he would have us believe. But the thesis itself is an important one and deserves careful consideration.

Faced with some similar difficulties in an investigation of one of San Marco's forerunners, Justinian's Hagia Sophia, I found that the precise geometry (or stereometry) of the principal structure offered invaluable further clues. Partly by pin-pointing areas where direct evidence of past changes could be sought beneath the surface revetments, it finally allowed the sequence of construction and the subsequent changes to be established with a high degree of confidence.[4]

Lack of a comprehensive accurate survey of San Marco is a considerable obstacle to doing anything similar there. At John Warren's suggestion, I have nevertheless attempted it after going some way towards making good the lack of data by what might be called *ad hoc* photogrammetry.[5] Analysis of limited early data suggested that his thesis might be confirmed, at least partially. Rather different conclusions now seem inescapable.

In presenting them I shall, as far as possible, deal first with the evidence usually considered—architectural form, sculptural details and construction—and only then with the less direct inferences from the precise present geometry, though in the investigation it was essential to keep both types of evidence in mind alongside one another in formulating and testing hypotheses.[6]

Fig. 1. The interior today from the west

Initial assumptions: north, south and west ends of the main cross

To focus on those aspects that seemed most open to debate, it was assumed from the outset that the structure incorporating the main entrance at the west of the nave and those at the extremities of the two transepts were, as is generally agreed, either survivals of the first, Participacio, church or (at least in their original forms) of similarly early date.

Relative dates of the side and central apses at the east end

The situation at the east end is more problematical and unfortunately it is here that there are the greatest gaps in the available data. Even the facings of the working masonry are not fully accessible and there are numerous obstacles to direct observation and measurement. Thus it is necessary to rely largely on the usual archaeological evidence, including that furnished by the present crypt which extends over the whole of the relevant area.

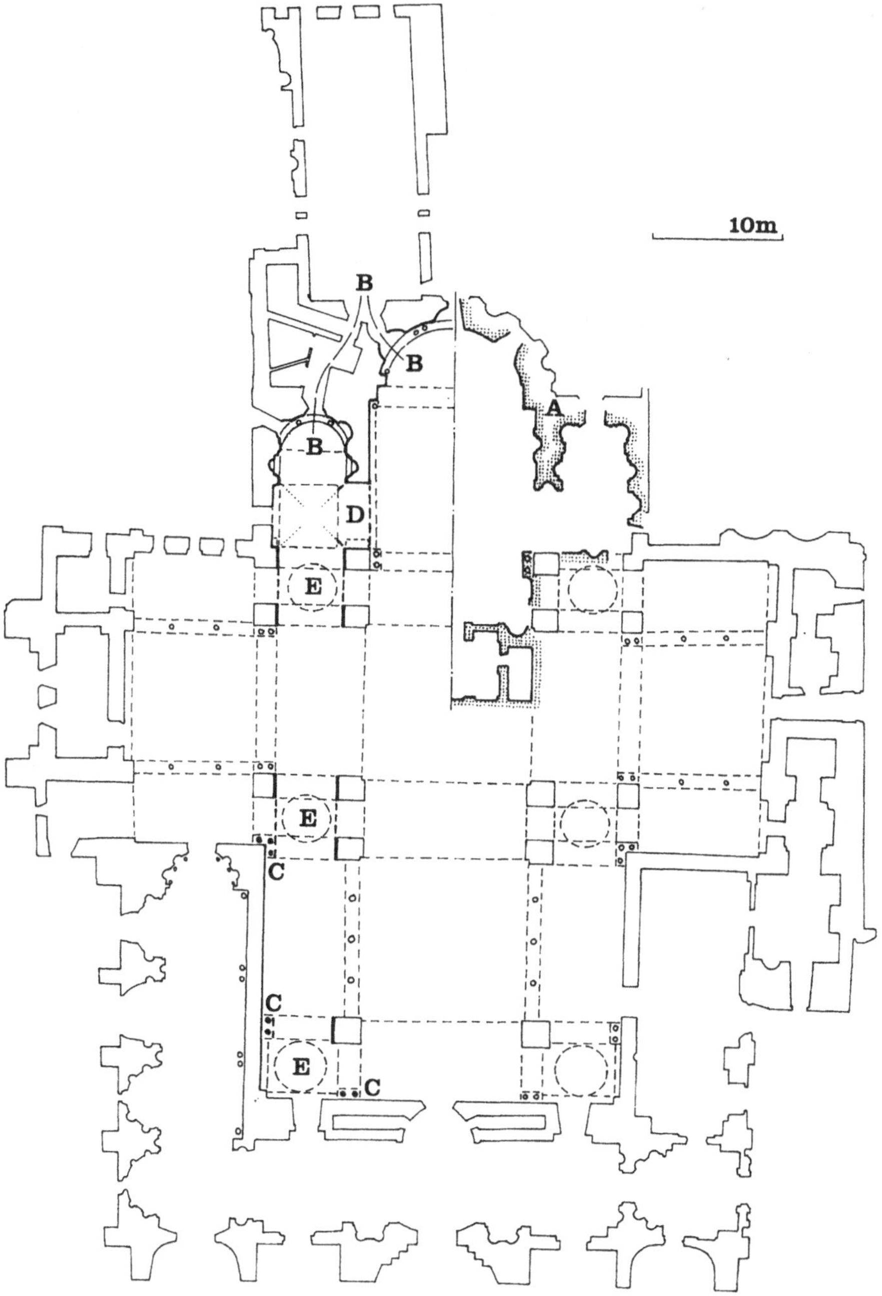

FIG. 2. Part plans of the church showing features referred to in the text. The ground plan is based on the plan in Demus *et al.* 1990, reproduced as fig. 1 in Warren 1990, and the plan of the crypt shown inset at the upper right is based on fig. 4 of Vio 1988

The walls of the main crypt (A, fig. 2) are very similar in plan to those of the sanctuary and chapels above, though doubts remain about the accuracy of available plans of the sanctuary and chapels and about the precise relationship of the two levels. It has been claimed that they are slightly differently orientated, with a slight northward displacement of the upper level.[7] But this has not yet been confirmed or refuted. If true, the upper level would rest, in part, on the crypt vaults rather than directly on its walls, and the altar and ciborium would also be displaced northwards. Both seem to me to be less likely than a more direct relationship.

Most significantly, the inspection that John Warren and I made in May 1991 convinced us both that the foundations and lower walls of this crypt are of one build throughout, and we saw no evidence suggesting an earlier central-apse foundation further west. The foundations are of the same character as those disclosed in earlier excavations, notably those of the quadripartite piers on the south side of the crossing.[8] With minor exceptions, the upper parts of the walls (where stone gives way to brick) are also of similar character throughout, and their small apsidal recesses have herringbone vaults identical with those of the recesses on the exterior of the main apse and the original San Alipio portal.[9] Likewise no significant difference could be detected between the vaults that carry the sanctuary and side-chapel floors. On this basis, it is difficult not to conclude that the present sanctuary apse and side chapels are also contemporary with one another, unless one or other has been rebuilt at some time on the same plan.

A later rebuilding of the chapels could explain the present difference in arch profiles. But neither possibility explains the 'collision' between the central apse and the present sacristy. Unless the current plans exaggerate the collision, I am inclined to regard it as the small price paid (perhaps around 1500 when the sacristy acquired its present form) for the easiest possible direct access to the church—a very reasonable objective (B, fig. 2).

The most relevant details within the sanctuary and chapels are probably the capitals in the apses. Careful study of these has shown that those of the south chapel are *spolia*, probably of early Byzantine origin, whereas those of the central and northern apses must have been carved for their present positions, taking the others as models.[10] The similarity of the carving of these purpose-made capitals strongly suggests that the central and northern apses are contemporary and hence that all three are.

The anomaly of the differences in arch profiles would then remain. There can be little doubt that the arch of the central apse was set out as a semicircle and that its present depressed profile is the result of deformation, mostly during and soon after construction. A similar deformation is almost invariably found in similar situations. Here it is merely more pronounced than is usual, suggesting that construction was too rapid and allowed load to act too soon on still plastic mortar. The smaller, lightly loaded arches of the side apses show less evidence of deformation, as is to be expected. They must intentionally have been given slightly pointed profiles, with separations of centres of about 0.3m on spans of about 5m (fig. 3b).

Profiles of the arches over the nave and transepts and over the aisles and galleries within the quadripartite piers

All the arches that span the nave and transepts prove, like that of the central apse, to have been set out as semicircles with centres a little above cornice level. They have all been similarly deformed in a manner consistent with the outward tilting of the supporting piers and the consequent separations of their springings.

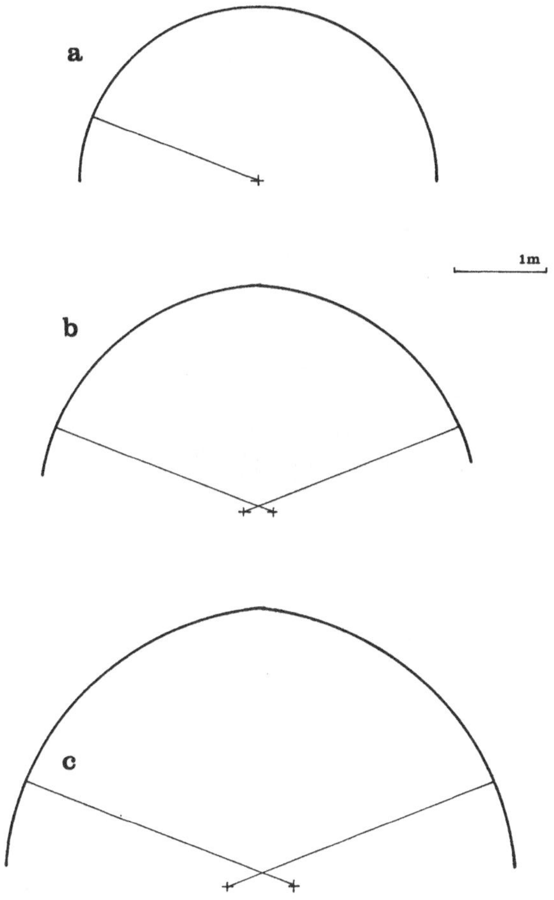

FIG. 3. Setting-out profiles of *a*, the north–south arches at gallery level in the quadripartite piers, *b*, the side-apse and chapel arches and north–south aisle arches in the quadripartite piers (the apparent springing levels being as marked on the left for the apse and chapel arches and on the right for the aisle arches), and *c*, the north–south arches of the north narthex. The scale should be regarded as approximate only since there are small variations in span from arch to arch

Fig. 4. The north-western leg of the quadripartite pier at the north-west of the crossing

If allowance is made for shearing deformations due to the same tilting of the piers—deformations clearly echoed in the string courses and cornices at gallery level—the north–south arches over the galleries also prove to have been constructed with semicircular profiles (fig. 3a).

The first analysis of the north–south arches over the aisles (made only for the arches facing one another across the north transept) suggested that their profiles did indeed result from a cutting back of earlier narrower arches, whether or not at the time when the original galleries were reduced to the present narrow passageways. But it was difficult to see how this cutting back could have been accomplished without seriously weakening the piers and still more difficult to see how there could have been further cutting in other places to replace up to three-quarters of the remaining masonry by isolated columns (C, fig. 2 and fig. 4). It no longer seems likely.

From the present fuller data, all these arches appear to have been set out to the same pointed profiles as the chapel arches (fig. 3b). Like the gallery arches they are now deformed, and their profiles are further disguised by the fact that their apparent springings are slightly higher above the centres of curvature. Variations from arch to arch suggest less precise or hastier construction than in the chapels unless they are partly attributable to varying thicknesses of plaster and mosaic and local detachments of these from the brickwork.[11]

Relative dates of the eastern side chapels and the central and western quadripartite piers

The adoption, from one end of the church to the other, of the same slightly pointed profile for all the arches spanning the side chapels and side aisles suggests a similar date throughout at this level. The galleries similarly seem to be of one date throughout.

This implies that all the quadripartite piers (including the isolated columns that make up part of the cross-sections of most of the legs of the piers set against the nave walls (C, fig. 2 and fig. 4)) belong to one building campaign, and probably the side chapels also.

There are several further pointers in the same direction:

1. The aisles are, throughout, of approximately the same width as the chapels. They are simply continuations of the chapels.
2. The fact that this is not a consequence of a later widening of the aisles is confirmed by the arch that spans along the north side of the sanctuary from the western termination of the south wall of the chapel to the nearest leg of the adjacent quadripartite pier (D, fig. 2, lower left fig. 5, and left side of Warren pl. XXXIII*b*). This arch carries on its own north side the apparent remains of an integral groined vault that originally spanned to the opposite wall as shown in dotted line in fig. 2; and when vault and arch were built it is unlikely that the leg of the pier was any wider than now from south to north. Moreover this must have been before the construction of the inner arch of the wider-spanning barrel vault that supports the dome at the north because this inner arch is carried partly by it at the west (upper left of fig. 5). And whether or not it is a separate skin of masonry, it seems to be roughly contemporary with the outer arch of the vault.
3. The arches and vaults linking the individual legs of the quadripartite piers are similar throughout at each level. This similarity extends even to the use of a similar small barely-visible dome as the final closure at the highest level (E, fig. 2).

Evidence of past structural movements indicative of relative dating and the sequence of construction of the domes

John Warren has described very well the ingenious means whereby the five domes are carried, without obvious incongruity, over the differently sized and shaped bays of the quincunx. I am much less happy with his arguments from past structural movements in favour of an earlier date for the central dome and its supports.

In drawing conclusions from the structural geometry it is necessary to consider all relevant aspects of behaviour both during construction and subsequently. For the supporting elements—chiefly the quadripartite piers—this means looking at both bodily settlements due to vertical load, and bodily rotations, shear and bending due to horizontal thrusts from the vaulting system. For the arches, pendentives and domes of the vaulting system it means looking chiefly at the deformations, both horizontal and vertical, resulting from the inevitable yielding of the supports under the thrusts just mentioned. All these movements or deformations will have taken place progressively, but at a diminishing rate in relation to the loading as the ground

FIG. 5. Looking north across the sanctuary

consolidated or the mortar hardened and stiffened. And through this they reveal sequences of loading and hence sequences of construction.

Relative vertical settlements are the easiest to see and interpret. Though undulations of the floor have too many possible causes to be helpful, undulations in the cornices over the arcades that run between the piers can safely be taken as indications of relative settlements of the piers and the much more lightly loaded columns of the arcades.

Most settlement would have taken place within a few decades of the full loading being reached. Therefore, if the central piers and dome had been built more than two centuries before the adjacent arcades, the cornices would run either horizontally to the nearest arcade columns or fall slightly if these columns had themselves sunk appreciably. Since they actually rise to at least the same extent as alongside the other piers (right hand side of figure 6), prior construction of the central piers seems highly unlikely.

Bodily rotations and shear and bending of the piers are more relevant to the behaviour of the vaulting system. They will have begun as soon as the arches began to exert their horizontal

FIG. 6. The eastern face of the pier at the north-west of the crossing

thrusts and will have increased when the domes also were built. Together, they will have led to the present observable outward horizontal displacements at the feet of the arches and the associated deformations of the arch profiles already referred to. Construction of an arch on one side of a pier before that on the opposite side will have led to variations in the net forces and resultant displacements as construction proceeded.

Reasonably accurate estimates for the displacements at the springings of the arches carrying the central dome have been made by correlating estimates from photographs taken from different directions, and more approximate estimates have been made for other displacements at this level and for some of the displacements at lower levels. A marked asymmetry of the movements of the central piers in relation to the axes of the crossing could indicate different histories of construction and subsequent loading but could equally result from different foundation conditions. Further measurements would be needed to choose between these possibilities and to distinguish confidently between the contributions to the displacements of rotations at foundation level and combined bending and shear above this level.

The final deformations to be considered are the transverse horizontal deformations of the main arches and the associated radial horizontal deformations of the dome cornices. These will have arisen from thrusts exerted by the apse semidome and the domes, especially from thrusts that were not balanced initially by opposing thrusts from an adjacent dome or that have never been balanced in this way. There will also have been some straightening of the cornices over the crowns of the arches when the domes were built, on account of the local spreading associated with the further depressions of these crowns.

To assess fully the significance of these deformations it would be necessary to have complete reflected plans of all domes and their supporting arches. Sufficient data on the piers to provide a basis for establishing the probable setting-out curves of the dome cornices would also be required. It is, however, reasonable to assume that the arches were constructed with faces that were straight in plan, so that their present reflected plans are more directly revealing than those of the domes.

Figure 7 reproduces those reflected plans that it has been possible to deduce from the arch profiles already referred to and photographs looking vertically up the domes. Unfortunately the plans of the arches are less reliable than those of the domes.[12]

Superimposed on the cornice plans are best-fit ellipses (for the eastern dome a circle). These are not the original setting-out curves (whose diameters would have been slightly smaller), but they provide the best available basis for estimating likely changes in shape.

In spite of possible errors in the plans, it seems fairly clear that:

1. The cornice of the eastern dome is deformed much more than any other, and the forward arch of the apse is likewise bowed much more than any other. In relation to the hypothetical circular base, the east side of the cornice is pushed inwards by the bowing of the arch. The western side is pushed outwards towards the crossing, suggesting that there is probably a westwards bow in the western arch also.[13]
2. To a much smaller extent, the cornices of the western, northern and southern domes are also bowed towards the crossing in relation to hypothetical ellipses, this bowing being accompanied by similar bowing of the adjacent arch faces. Initially it will have been greater because of the slight straightening referred to above.
3. The central dome shows only slight movements in relation to a hypothetical ellipse and slight inward bowing of the east and west arches. There is certainly not the major outward bowing that would be expected if this dome had been constructed long before the others.

These observations imply the following sequence of construction: first, the eastern apse and eastern dome base (arches, pendentives and cornice). These seem to have been constructed fairly rapidly, leaving little time for the mortar to stiffen as construction proceeded. Second, but very shortly afterwards, the eastern dome. Third, the remaining dome bases and the northern, southern and western domes. Finally, the central dome.

Narthexes

I do not think that a ninth-century date for the structure containing the main entrance to the nave necessarily entails accepting the present narthexes as substantially the remains of an earlier atrium.

Several problems arise with such an atrium. A defensive gateway would surely have been integral with its outer walls. Within, it is difficult to envisage the transformation (whether in one stage or two) of the open columnar boundaries of which John Warren detects a partial survival on the south wall of the present north narthex into the present nave walls and outer legs of four of the quadripartite piers. And in surviving early atriums the bays are much more uniform than are those of the San Marco narthexes.

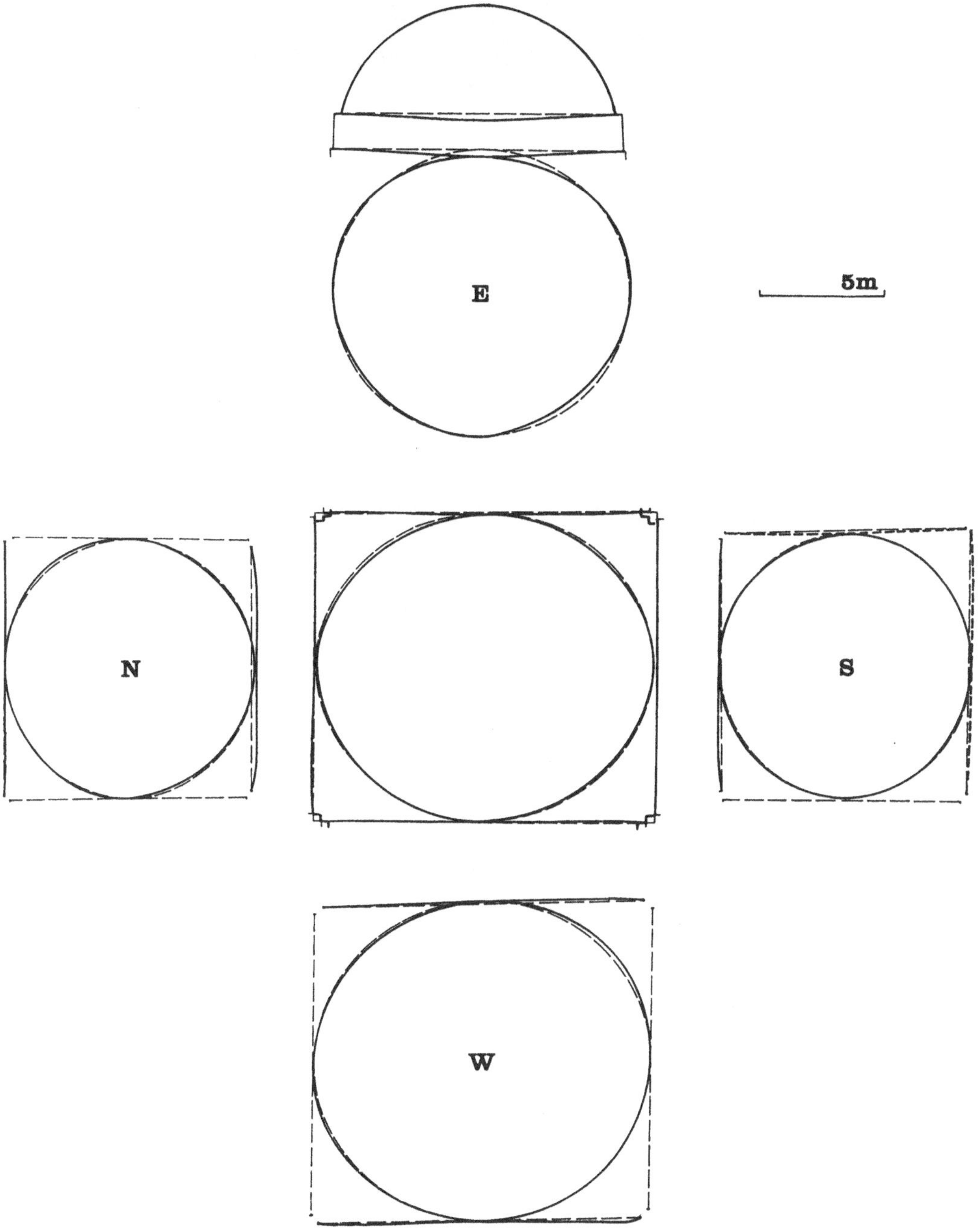

FIG. 7. Reflected plans of the dome cornices and principal arches compared with best-fit ellipses and straight lines shown in lighter broken line. The positions at floor level of the corners of the quadripartite piers corresponding to the springing points of the arches carrying the central dome are also shown in heavy line on the basis of estimated horizontal displacements at cornice level

FIG. 8. The northern narthex looking east

The lack of uniformity at the west and south suggests several phases of construction here. At the north (where the only detailed examination was made) I prefer a date after that of the eastern side chapels for the vaulting. The arches are equally consistently slightly pointed, but they now have centres nearer the springing level and twice the separation (fig. 8 and fig. 2c). The capitals of the columns along the present south wall suggest to me that they were both set in place against the wall rather than paired with others now built into it or displaced by it.

The second (Contarini) church

References to dates have hitherto been largely to relative dates. On the basis of the conclusions reached above, I have no difficulty in accepting the record of a virtual rebuilding in the Contarini period and assigning most of the present church to it—excepting chiefly the narthexes, those structures at north, south and west assumed at the outset to be earlier, and possibly the lower parts or foundations of the remaining walls of the nave and parts of the east and west walls of the transepts.

The adoption at this date of a pointed arch profile in the eastern chapels and side aisles is puzzling. If it was a deliberate importation from the east, this could as well have occurred in the eleventh century as the ninth, or it could have been carried over from earlier use in the Participacio church. A further possibility, bearing in mind the unusually small separation of centres, is that it had its origin in an attempt to forestall or disguise the type of deformation that actually occurred in the central apse and its arch.

The sequence of construction at the highest level seems to have been as noted above. I would merely add here that some priority of emphasis on the eastern bay at the planning stage also seems to be indicated by the fact that it alone seems to have been intended to be square. The central bay has a north–south measurement at floor level 1·04 times its east–west measurement and a present north–south diameter at dome level 1·07 times its east–west diameter, not much different from the ratios in the transepts, though in the opposite sense.

The strange way in which the eastern chapels are enclosed within the structure carrying the eastern dome could be explained as part of the original design if these chapels were envisaged, together with the galleries above, as two-storeyed on the pattern of some roughly contemporary palace 'Doppelkapellen'.[14] Otherwise I see it as the result of a change in plans at gallery level. This would be all the more likely if the Participacio church did have a substantially different form. For its successor would then have been highly novel for its time and place and its design could well have had to be developed and modified as construction proceeded.[15]

I find it more difficult to express an opinion on the date when the crypt assumed essentially its present form. The similarity of the masonry of its foundations to that exposed elsewhere is possibly of limited value for dating purposes, because earlier material could have been reused and the techniques may have changed little from the ninth to the eleventh centuries, likewise with the herringbone vaulting. I therefore leave it open for further investigation, merely repeating that it does seem all of one build at each level if one excludes the small, almost separate, portion under the crossing.

The first (Participacio) church

Though the investigation outlined here is necessarily incomplete, I am forced to reject John Warren's central thesis that this first church has survived largely intact in the present one, despite its recorded destruction. Moreover I see no need to strain the basis of comparison to establish a likeness to the Apostoleion.[16]

The most likely candidates for survival seem to be the west wall and some of the foundations—notably those uncovered by Forlati in the south transept and possibly those of the present main crypt. Though they have not been exposed, I suspect that Participacio foundations may also exist below the side walls of the nave and perhaps below their continuations in the west walls of the transepts and the corresponding walls further eastwards.

Taking at its face value Forlati's demonstration that no foundations exist in the south transept on the east–west axes of the quadripartite piers, a simple basilican plan is precluded. The most likely form then becomes that best exemplified by the now largely ruined cathedral of Jesolo—a timber roofed basilica with staggered triple apses over a crypt like that of the present San Marco and transepts of relatively shallow projection.[17] The church at Jesolo also has similarities in detail (notably similar niches with the same herringbone brickwork vaults) that could indicate a fairly direct copying of the first San Marco.[18]

NOTES

[1] Summarized in Forlati 1960. The present ties in the galleries, for instance, could well be those added in about 1744.

[2] Warren 1990. What follows is an invited comment written to appear with this paper but subsequently held back to the present volume at the editor's request.

[3] Non-figural work does not necessarily denote work of the period of iconoclasm. The San Alipio mosaic also depicts non-figural designs in the indubitably later lunettes at gallery level; others of later date are found within the church; and it would have been impossible to reproduce any but the simplest figural subjects (as over the central portal) on such a reduced scale.

[4] See chiefly Mainstone 1969–70; 1988, chapters 4 and 8; and 1992.

It has to be recognized, however, that interpretation of the evidence furnished by precise stereometric data is less straightforward than might at first sight appear and calls also for a clear understanding of the structural behaviour giving rise to any deformations.

In the simplest possible case of sections of wall or pier built against one another and similarly constructed, supported and loaded, but inclined from the vertical to different extents, it might reasonably be deduced without further enquiry that the more inclined one is the older, the other having been built vertically against it after it had already inclined somewhat.

In principle, sequences of construction can be similarly deduced in more complex situations. But, in a structure as complex as Hagia Sophia or San Marco, one is confronted with interrelated deformations in three dimensions, with variations from element to element in the manner of construction and in loading and support, and not only with upright forms but also with curved ones whose initial geometry cannot so confidently be assumed *a priori*.

[5] Three main series of photographs formed the basis; photographs from gallery level looking horizontally north, south, east and west along the main axes of the church; photographs from both gallery and floor levels looking horizontally towards the opposite pier from all openings in the quadripartite piers; and photographs looking vertically up from beneath the arch of the apse and the centres of all but the eastern dome. In working from the first and second series, verticals were established from the chains of hanging lights and corrections were made for distortions of the projected images due to different focal lengths of camera and projector lenses. The last series was supplemented by photographs reproduced in Demus *et al.* 1990.

In allowing and facilitating the necessary photography and presenting copies of the publication just referred to and of other publications referring, *inter alia*, to a more comprehensive photogrammetric survey now in progress, the present Proto, Arch. Ettore Vio., was most helpful and I am deeply indebted to him, as I am also to John Warren for the stimulus he has provided to undertake this investigation and persevere with some very tedious data reduction.

[6] If, in terms solely of structural behaviour and the resultant deformations, it is possible to think of one explanation for what is observed, it will usually be possible to think of others that are equally valid for other assumptions about the inevitable unknowns. At best one can seek an interpretation that is self-consistent and consistent with all other relevant evidence. Only by taking this other evidence fully into account can the risk of wrong interpretations be minimized.

[7] Dorigo 1983, vol. 2, fig. 346.

[8] Forlati, 1951; and *idem* 1975, 48–50. Now see also Rhône-Poulenc 1992, which appeared only after this paper was written.

[9] See Marangoni 1933 for the last.

[10] Buchwald 1964.

[11] In the course of recent restoration work in the dome of Hagia Sophia, Thessaloniki, the mosaic was found to be detached in some areas (though still safely *in situ*) by up to 40mm. Detachment becomes more likely with single curvatures as in these arches.

[12] This is because the only correction needed to the photographs of the domes was for dips in the cornices associated with depressions of the arch crowns after the cornices were completed. (The existence of these dips is confirmed in the west and south domes by drawings made during nineteenth-century restorations; see Dalla Costa 1983, figs. 6, 7).

Plotting of the true reflected plans of the arches had to be based on estimated heights above floor level and estimated springing points of the arches where (in the vertical photographs) these were hidden by the cornices. Some errors must also have arisen in correlating the vertical photographs and arch profiles on account of the rounded edges of the arches.

[13] An axial vertical photograph of this dome together with its supporting arches was not possible on account of obstruction by the baldacchino.

[14] See also Demus 1960, 85–6 and note 112.

[15] Justinian's Hagia Sophia presents a comparable instance of a change in plan at gallery level. See Mainstone 1988, 191–4 and fig. 222.

[16] The early documents liken the church only to the Anastasis. Demus seems to have first suggested a similarity to the Apostoleion on the basis of Forlati's reconstruction of this first church as a domed cross. The 'conclusive proof' that he cites for an acknowledgement of this similarity already in the ninth century is no more than a reference to Venice having become '*quodammodo sedes apostolica*' by harbouring the relics of St Mark. See Demus 1960, 67, 96.

[17] This is essentially the reconstruction suggested recently in Herzner 1985.

[18] For Jesolo see Artico 1977.*

BIBLIOGRAPHY

Artico, L. 1977. 'Novita su Jesolo', *Arte Veneta*, 31, 16–26
Buchwald, H. 1964. 'The carved stone ornament of the High Middle Ages in San Marco, Venice', pt. 2, *Jahrbuch der Österreichischen Byzantischen Gesellschaft*, 13, 137–70
Dalla Costa, M. 1983. *La Basilica di San Marco e i restauri dell'Ottocento*, Venice
Demus, O. 1960. *The Church of San Marco in Venice*, Dumbarton Oaks, Washington
Demus, O. *et al.* 1990. *San Marco: Mosaics, History, Lighting*, Milan
Dorigo, W. 1983 *Venezia origini*, Milan
Forlati, F. 1951. 'Il primo San Marco', *Arte Veneta*, 5, 73–6
——, 1960. 'The work of restoration in San Marco' in Demus 1960, 193–201
——, 1975. *La Basilica di San Marco attraversso i suoi restauri*, Trieste
Herzner, V. 1985. 'Die Baugeschichte von San Marco und der Aufstieg Venedigs zur Grossmacht', *Wiener Jahrbuch für Kunstgeschichte*, 38, 1–58
Mainstone, R. J. 1969–70. 'The reconstruction of the tympana of St Sophia at Istanbul', *Dumbarton Oaks Paps.*, 23/24, 355–68
——, 1988. *Hagia Sophia: Architecture, Structure and Liturgy of Justinian's Great Church*, London
——, 1992, 'Questioning Hagia Sophia' in *Hagia Sophia from the Age of Justinian to the Present* (edited by R. Mark and A. Ş. Çakmak), 158–76, Cambridge
Marangoni, L. 1933. *L'architetto ignoto di San Marco*, Venice
Rhône-Poulenc, 1992. *San Marco: La crypta la storia la conservazione*, Milan
Vio, E. 1988. 'Le levate fotogrammetriche della Basilica di S. Marco', *Venetia Arti*, 2, 157–62
Warren, J. 1990. 'The First Church of San Marco in Venice', *Antiq. J.*, 70, 327–59

* [now see also Dorigo, W., *Venezie sepolte nella Terra del Piave*, Viella, Rome, 1994, pp. 259–298, 'La cattedrale di S. Maria Maggiore di Equilo'.]

XIV

Squinches and pendentives: comments on problems of definition

The two definitions quoted from the *Penguin Dictionary of Architecture* at the beginning of a paper on squinches and pendentives by Daria Jones and George Michell [1] describe with exemplary brevity and clarity the simple basic forms of pendentive and squinch as seen in Byzantine, Romanesque, Renaissance, and later Western architecture. But, as the authors of that paper pointed out, the transitional elements adopted in Islamic architecture often departed widely from these simple norms. So great was the variety of the Islamic elements that the inconsistencies referred to between the terminologies used by different authors to classify and describe them are hardly surprising. Perhaps the most that could be expected was that each author should be consistent in his use of whatever terminology he chose and that it should be a meaningful and useful one in relation to his particular choice of subject matter. The possibilities of confusion introduced by differing and conflicting terminologies are, nevertheless, greatly to be regretted now that we are in a position to take a more overall view of the development of this type of structural element. The attempt by Daria Jones and George Michell to clarify the underlying problems of definition is therefore to be welcomed, and their preliminary proposals for a more broadly based and more definitive terminology deserve careful consideration. These comments are offered, at their invitation, in that spirit.

What is new about the proposed terminology is its explicit primary basis in the recognition of different modes of load transfer from the dome or drum to the supporting walls or arches below. Since these modes are, inter alia, dependent on the geometry of the inner surface of the element and on the manner of construction, these characteristics are also relevant and cannot be ignored. But neither is taken as a primary basis, as has been usual hitherto.

It we accept that choice of primary basis as being the most fundamental and widely applicable one available, how valid is the proposed new functional classification? And what improvements, if any, can be suggested in the proposed terminology and definitions of the terms?

Before attempting to answer these questions, it may be helpful to make a few general observations whose validity is not, in the light of modern understanding of structural behaviour, really open to doubt:

1. Since we are only concerned with masonry construction, the mode of load transfer will always be almost entirely compressive. In other words it will be almost entirely a matter of balanced thrust and counter-thrust. There may be very limited tensile actions within individual blocks of stone where these are incorporated in such a way as to enable them to act as beams or cantilevers. But there will

be very little possibility of tensile actions between block and block or brick and brick: such interactions will be largely limited to direct pressure and tangential friction.

2. In these circumstances, horizontal coursing of the masonry can transform the overall mode of load transfer into that of a beam or cantilever only in certain special circumstances. Usually it will be necessary for the angles between adjacent supporting walls or arches to be spanned diagonally by single blocks of stone, or for single blocks of stone not only to project outwards for the full distance at each level but also to project back far enough behind the faces of the supporting walls or arches to gain independent anchorage against overturning. Where these conditions are not met, the only possible modes of load transfer will usually be those available to elements of the same geometric form but with the individual blocks or bricks set more as voussoirs; though the stability will be more precarious because the thrusts and counter-thrusts will act at appreciable angles to the bed joints and thereby increase the risks of collapse initiated by relative slips and rotations of the blocks or bricks. By way of illustration, it may be noted that none of the systems seen here in fig. 1 (sketched for Jones and Michell by Garry Martin as examples of the corbelling principle) could remain in equilibrium solely under the action of vertical compressive loads between the elements: all would immediately collapse as a result of relative slips and rotations. They could be made stable without extensive tensile reinforcement only by means of heavy additional loading of the haunches, to permit the development of large horizontal frictional forces and corresponding horizontal forces within the blocks.[2]
3. The very limited tensile strength of its masonry, and of any circumferential ties built into its springings, will mean that the dome will inevitably thrust outwards on its supports. Under the thrust, these supports will almost certainly move outwards to some extent, and the outward movement will be accompanied by some more-or-less radial cracking of the lower part of the dome that will extend into the transitional elements. Both for this reason and for other related ones, the mode of load transfer through the transitional elements will be determined, to some degree, by characteristics of the dome and of the supporting walls or arches as well as by those of the transitional elements themselves.
4. So far as the mode of load transfer is determined by characteristics of the transitional element itself, both the total geometry of the form (external as well as internal) and the choice of materials and manner and sequence of construction must be taken into account. The total geometric form may be regarded as the primary determinant, but the thicknesses adopted have usually been sufficient to leave open a wide range of possibilities. The choice of materials and manner of construction may result in some parts of the element being stiffer – or less easily deformed – than others. If so, these parts will take more load than they would otherwise do in terms of their dimensions. The sequence of construction may be important in that, if infills between ribs were added only at a late stage in construction, or if masonry piled up externally was a similarly late addition, these will carry less load than if they had been built *pari passu* with the rest of the element. The coursing and bonding of the masonry are, from this point of

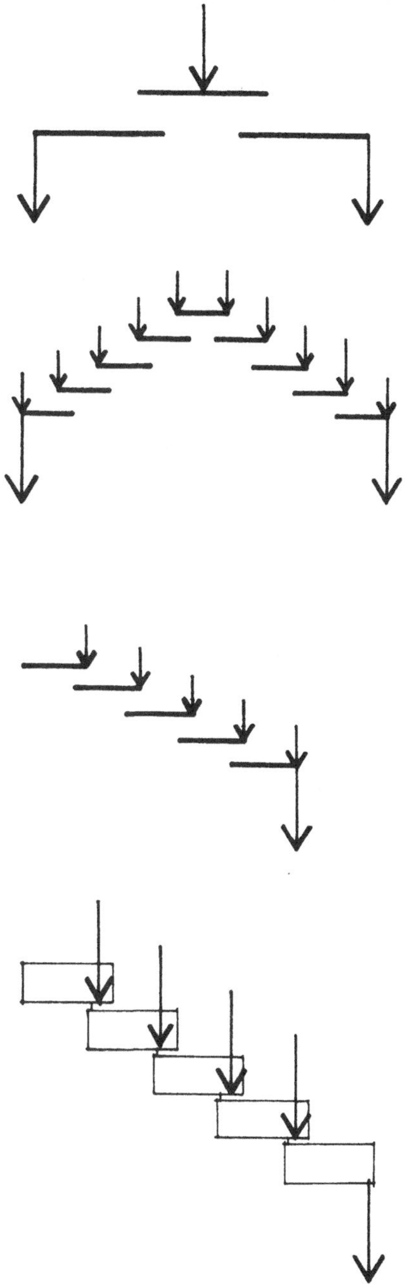

A series of elements forming a stepped beam or arch which transmits loads perpendicular to the elements to one or two points of support.

Detail of part of a corbelling system illustrating the transmission of vertical load from element to element. That part of each which is projected without support is cantilevered.

1. Schematic diagrams by Garry Martin of the corbelling principle (from reference 1).

view, merely one aspect of the manner of construction, and, for the reasons given above, their influence will usually be a relatively minor one.

In the light of these observations, it will be obvious that a classification based on modes of load transfer is, unfortunately, far less straightforward than might be hoped. It is certainly less straightforward than would be a similar classification of modern counterparts of the same elements, since these modern counterparts – constructed in reinforced concrete – would be relatively so much thinner that a given choice of surface geometry would leave open only a much narrower range of possible modes. Equally certainly, particularly when the mode must be inferred from purely surface indications, it must leave a considerable number of examples about whose rightful classification there is an inescapable margin of doubt. But that is an inherent difficulty with many types of classification. It should not, in itself, be accepted as a reason for rejection unless there is an obviously preferable alternative. We should rather try, as far as possible, to minimize it.

My own suggestions for doing this are to make the following, admittedly somewhat arbitrary, assumptions:

1. That the mode of load transfer to be considered is that which, as far as we can judge, was most likely to have developed immediately on completion of the dome by virtue of the geometric form, choice of materials, and manner of construction of the transition element itself. In other words it should ignore subsequent changes in the mode attributable primarily to the kinds of behaviour of the dome and the supports referred to in observation 3 above.
2. That, when we make the necessary judgement, we should take full account of clear indications of simultaneous construction of and effective load sharing by, for instance, ribs and infill panels or the inner skin of the element and the backing masonry. But, in the absence of such clear indications, we should regard all infills as playing, at most, a very secondary role, and all backing material as doing no more than applying an inert vertical load to the inner skin.

These assumptions will remove some of the potential sources of ambiguity and confusion in the indeterminacies, purely from indications visible on the surface, of the structural actions or modes of load transfer and in the ever-changing character of these actions. By emphasising the actions of which the inner skin – as distinct from the backing masonry – would have been capable, they will moreover tend to result in more sharply defined functional categories that are also closer to those that might be arrived at in terms of the geometry of the surface alone and that will probably reflect, as well as any modern functional categories can be expected to do, the broad concepts of the structural actions that the designers might have had. Similar assumptions seem to be implicit in the proposals of Daria Jones and George Michell. But I think that it is important to state them explicitly.

Turning now to the proposals by these authors for a new functional classification and its associated terminology and definitions, I can accept without reservation their category of *corner arch.* I accept also the need for the categories of *pendentive and squinch.* But I am not entirely happy with the ways in which these have been distinguished from one another in the definitions and diagrams, or with the intermediate category of *squinch arch.*

As defined, the *squinch* is distinguished from the *pendentive* as being 'three dimensional' and as transferring loads 'to the two intersecting walls or arches which it bridges diagonally and loads continuously' rather than 'to one point situated at the intersection of two supporting arches or walls'. Three-dimensionality should, I submit, be allowed to be equally characteristic of both the geometry of the inner surface and the mode of load transfer of the *pendentive.* If it is not, we exclude the prototypical spherical triangular form. And both forms surely load the supporting walls or arches continuously – as shown in Garry Martin's diagrams of these forms reproduced in fig. 2 here. In fact, in the simple forms sketched, the loading would be more nearly uniform along the lower boundary in the case of the *pendentive.*

Without, I hope, departing radically from the concepts behind these definitions and their diagrammatic illustrations, but seeking rather to clarify them, I should like to consider first what I see as the most fundamental geometric and functional distinctions. Geometrically I see the *pendentive* as an element whose inner surface runs continuously, in all directions radiating outwards from the vertical axis of the dome, from a horizontal upper termination to its lower terminations where it meets the supporting walls or arches, or whose effective thickness (discounting usually any backing masonry in accordance with assumption 2 above) is sufficient to contain such a surface. It may 'run continuously' either in straight lines or in curves that are concave downwards throughout their length. I see the *squinch* as an element whose inner surface may, at most, run continuously in this sense as far as an upper termination that is not horizontal but prominently arched upwards when viewed from the front.

Functionally – in terms of the transfer of load from the dome to its supports – both forms (if we exclude the special case of the merging pendentive under a very shallow dome) call for some change in direction of the downward and outward thrusts of the dome as these thrusts pass into a more nearly horizontal part of the element. This change in direction is illustrated in the cross sections at the right hand side of my diagrams A and B in fig. 4. The significant difference is that the change takes place in the *pendentive* at a horizontal upper termination, but, in the *squinch,* along the line that is prominently arched upwards. In both cases additional forces must be brought into play to maintain equilibrium, but this last difference means that they are of fundamentally different character. In the *pendentive* they consist simply of a continuous arch-like thrust in the complete horizontal ring of which the upper term-ination of the element is one part. They play no part in sharing the vertically downward thrusts of the dome, though they do add something to the outward thrusts. In the *squinch* they are similar in one respect: they are again arch-like thrusts acting in a horizontal rather than a vertical sense. But these thrusts must now act within the vertically arched form of the forward edge of the squinch, which must be strong enough over a sufficient depth to permit this, and which will inevitably then find itself acting also to some extent as a vertical arch and directly sharing some of the vertically downward thrusts of the dome. In other words the mode of load transfer for the *squinch* must include the *corner arch* mode sketched

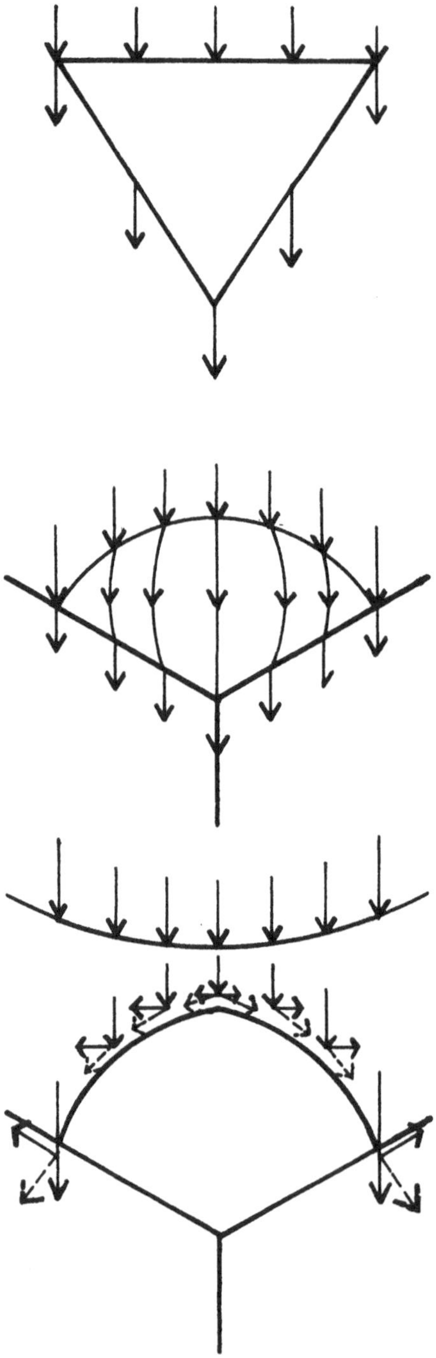

(a) The Pendentive.

Loads transmitted from base of dome or drum.

Continuous loads transmitted to two intersecting walls or arches.

(b) The Squinch.

Loads transmitted from dome or drum.

Continuous loads transmitted to two intersecting walls or arches.

(c) The Corner Arch.

Loads transmitted from base of dome.

Loads transmitted to a point on each of the intersecting walls.

2 Schematic diagrams by Garry Martin of the pendentive, squinch and corner arch (from reference 1).

(d) The Squinch Arch.

The squinch arch combines the principles of (b) and (c) in fig. 2.

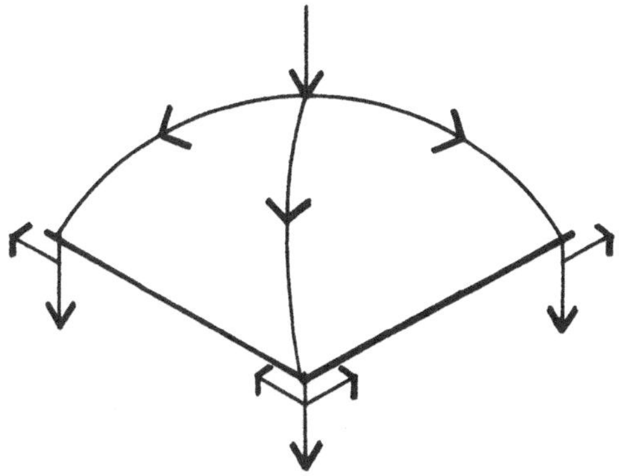

α A three-dimensional extension of the arch system.

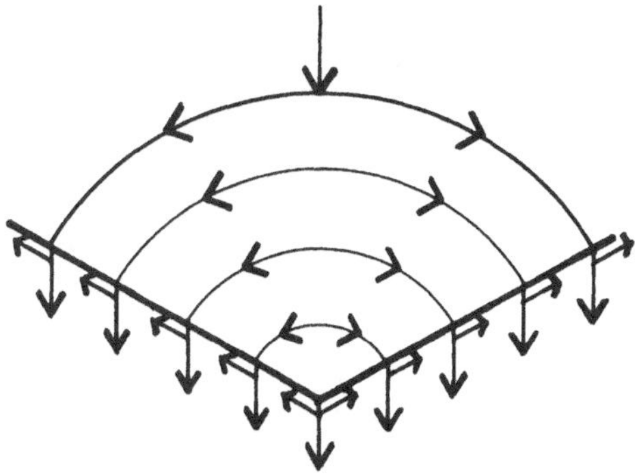

β A series of concentric arches based on the vaulting principle.

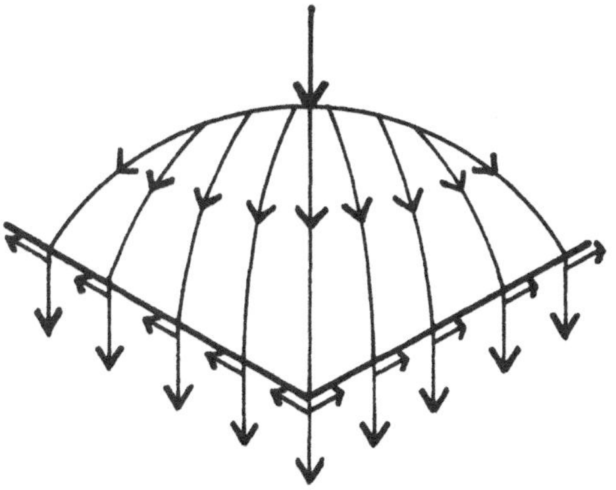

χ A combination of the arch and corbelling principles.

3. Schematic diagrams by Garry Martin of the squinch arch (from reference 1).

by Garry Martin at (c) fig. 2, as well as the mode sketched there at (b), as shown at B in my fig. 4. The *corner arch* merely lacks the further transmission of load through the rear part of the *squinch.*

It is because the *squinch* must always, in effect, incorporate a *corner arch* that I am reluctant to accept the *squinch arch* as a valid intermediate category. I can see no way in which it could be clearly differentiated functionally from the other two – even if our estimates of the structural actions immediately on completion of the dome were much more precisely accurate than they are ever likely to be. I should prefer therefore to retain the term *squinch* for all elements in which the corner arch is not completely or very largely open; and to qualify it, where qualification seems necessary, by some such words as 'perforated' (denoting a relatively small window opening), 'with heavy front arch', 'with receding and diminishing concentric corner arches' (whose structural action will be much closer to that of any other squinch than Garry Martin's diagrams (d) in the present fig. 3 suggest), or 'with heavy front arch and rear half arch'.

The one intermediate category that does seem to me to be desirable is that of the *multiple-squinch pendentive or pendentive with intersecting diagonal ribs.* In the former, several tiers of squinches are superimposed, with the number of squinches becoming greater in each superimposed row so that the overall geometry approximates to that of a pendentive. In the latter there is a similar approximation to the overall geometry of a pendentive, but the arched front edges of the separate squinches of the former become continuous ribs running diagonally over the surface. By virtue of the overall geometric similarity, the modes of load transfer will also be generally similar to that of the simple pendentive. For this reason it seems correct to refer functionally to both as special categories of pendentive. But there will be differences in that the simple downward and outward transmission of the load from above will be modified to some extent by the arching action of the front edges of the squinches or of the diagonal ribs.

In addition there is possibly room in a truly comprehensive functional classification for two further primary categories: the *corner beam* and the *corner corbel.* Irrespective of the shapes into which their soffits may be cut, these will be elements which truly span diagonally across a corner as a beam, or which truly cantilever out (in the structural sense) from a corner. In the light of observation 2 above, they are likely to be encountered only as quite small elements in places like Syria where good stone is used in large finely dressed blocks.

A possible set of purely functional definitions would now seem to be as follows:

Pendentive (A in figure 4)

A transitional element between part of the base of a dome, domical vault, spire, or drum and two supporting walls or arches meeting at an angle outside the base, in which the outwardly inclined downward thrusts from the dome, domical vault, spire, or drum are carried down to the supporting walls or arches without any marked change in direction except, in most cases, where their outward inclination is initially increased (i.e. they become more horizontal) at the horizontal upper termination of the transition element.

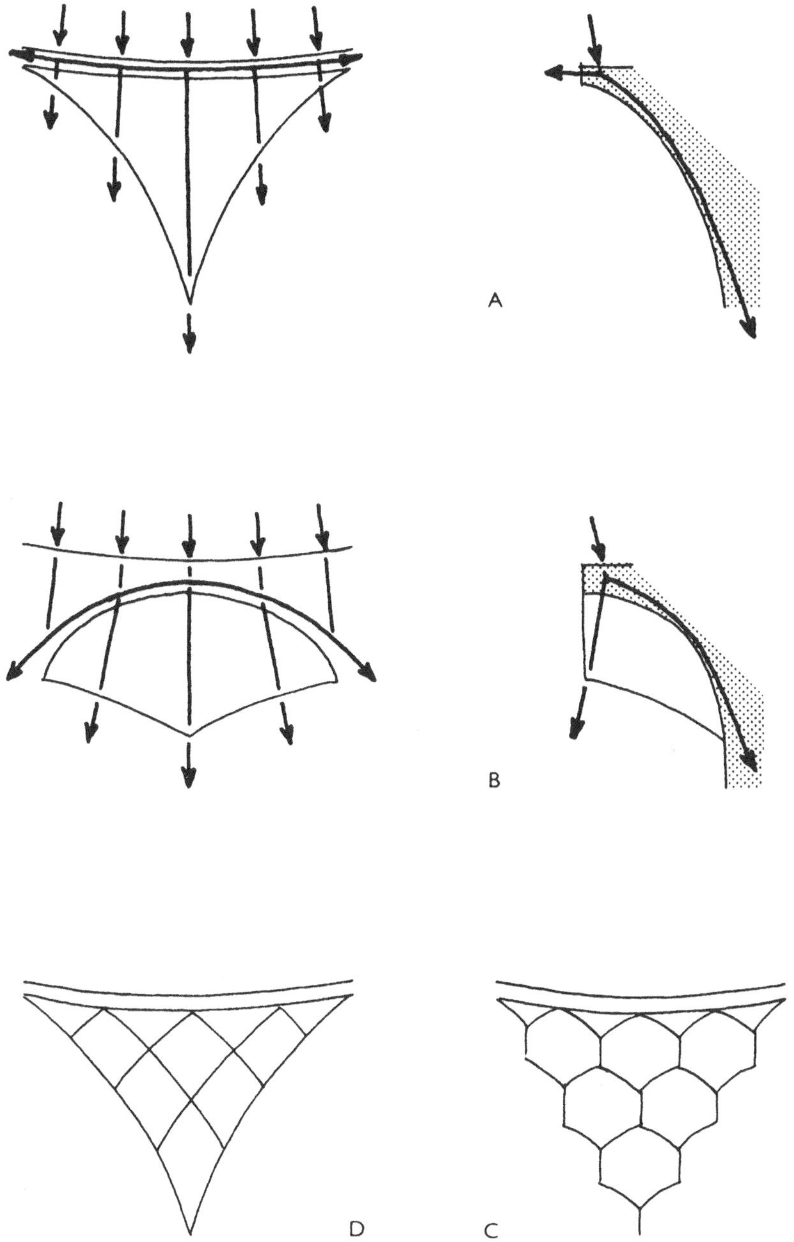

4. Alternative functional classification of transition elements in terms of the mode of transfer of load from the dome: A – pendentive; B – squinch; C – multiple squinch pendentive; D – pendentive with intersecting diagonal ribs.

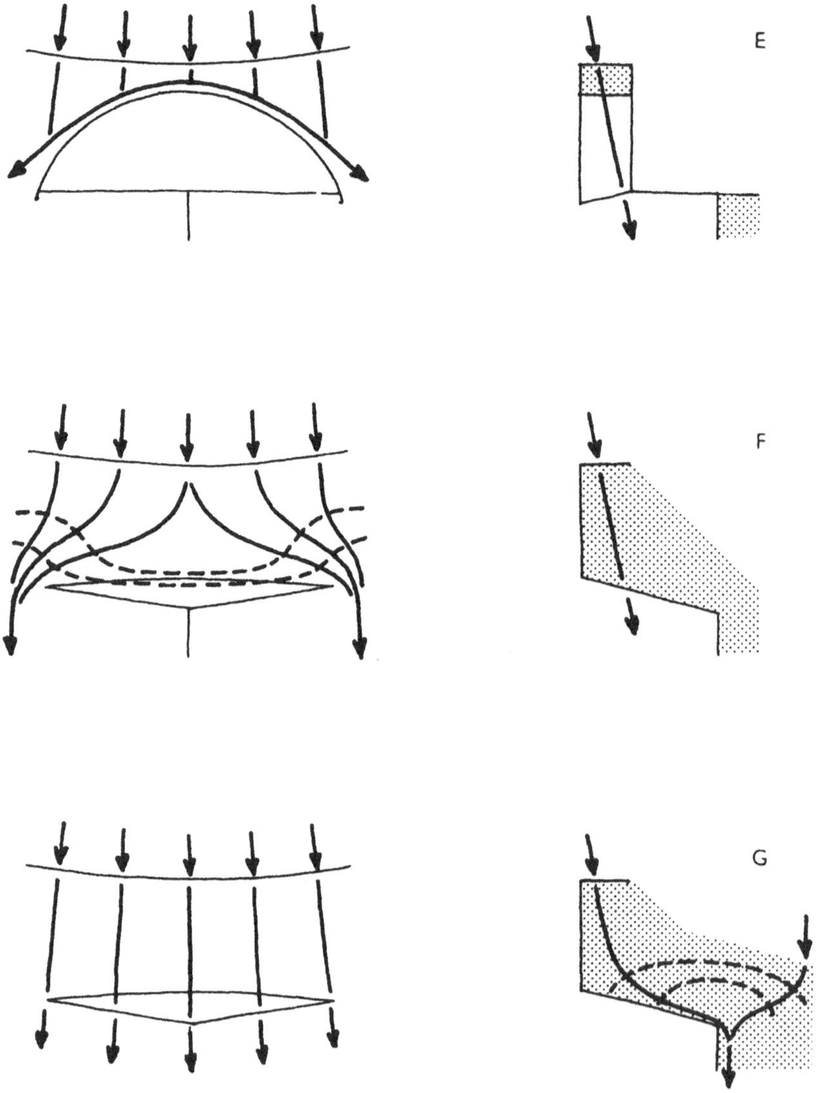

5. Alternative functional classification of transition elements in terms of the mode of transfer of load from the dome: E – corner arch; F – corner beam; G – corner corbel. (Broken lines in diagrams F and G denote internal tensions.)

Squinch (B in figure 4)

A transitional element between part of the base of a dome, domical vault, spire, or drum and two supporting walls or arches meeting at an angle outside the base, in which the outwardly inclined downward thrusts from the dome, domical vault, spire, or drum suffer marked local changes in direction as they are carried down to the supporting walls or arches but nevertheless apply a continuous load to these supporting walls or arches. The local changes in direction occur principally at a vertically arched front termination of the element running diagonally across the corner, and they involve the downward transmission of part of the thrusts from above by a vertical arching action within the arched front accompanied by outward horizontal thrusts at its feet.

Multiple-squinch pendentive (C in figure 4)

A special class of *pendentive* consisting of superimposed tiers of small *squinches* whose number increases in each superimposed tier so that the overall mode of transfer of the thrusts from above to the supporting walls and arches generally resembles that in a simple *pendentive* except that it has superimposed on it a network of local arching actions corresponding to the front terminations of the constituent *squinches.*

Pendentive with intersecting diagonal ribs (D in figure 4)

A special class of *pendentive* in which the inner surface is traversed by a network of ribs running diagonally downwards in directions roughly parallel to both lower terminations of the element so that the overall mode of transfer of the thrusts from above to the supporting walls and arches generally resembles that in a simple *pendentive* except that it has superimposed on it a network of diagonally inclined local thrusts corresponding to the ribs.

Corner arch (E in figure 5)

A transitional element between part of the base of a dome, domical vault, spire, or drum and two supporting walls or arches meeting at an angle outside the base, in which the outwardly inclined downward thrusts from the dome, domical vault, spire, or drum are carried down to the supporting walls or arches solely by a slightly inclined arching action between two diagonally opposite points on these walls or arches. Any filling of the arch plays a negligible part in transmitting the thrusts.

Corner beam (F in figure 5)

A transitional element between part of the base of a dome, domical vault, spire, or drum and two supporting walls or arches meeting at an angle outside the base, in which the outwardly inclined downward thrusts from the dome, domical vault, spire, or drum are carried diagonally across to the supporting walls or arches solely by a horizontal beam action between these walls or arches.

Corner corbel (G in figure 5)

A transitional element between part of the base of a dome, domical vault, spire, or drum and two supporting walls or arches meeting at an angle outside the base, in which the outwardly inclined downward thrusts from the dome, domical vault, spire, or drum are carried back to the supporting walls or arches solely by cantilever actions, there being sufficient weight on those parts of the element projecting outside the supporting walls or arches to provide all the necessary anchorage against overturning of the element.

1. Jones, D. and Michell, G. with drawings by Garry Martin, 'Squinches and pendentives: problems and definitions', *art and archaeology research papers,* no.1, June 1972, pp. 9-25. Some of the drawings are reproduced again here, in figs 1-3, for greater intelligibility of the present text.
2. This is discussed further, with illustrations, in chapter 6 of my book *Developments in Structural Form,* Allen Lane, 1975, [rev. 2nd edn, 1998].

XV

Westminster Hall Roof

1. Westminster Hall

[The timber roof of Westminster Hall, as reconstructed around 1400 (fig. 1), has a single clear span of over 20 m. This span is achieved by combining two different structural systems in each transverse frame. An arched rib springing from corbels set well down on the walls is combined with a more complex hammer-beam truss linked by wall posts to the same corbels but also connected to wall plates at the top of the walls. The hammer beams project inwards from the wall plates and wall posts and, at their inner extremities, support hammer posts that rise to the main purlins. These posts support both the principal upper rafters and a collar that links their feet. In the centre, this collar meets the tops of the arch ribs and a crown post rises from there to the ridge beam. The lower rafters sit on the outer ends of the hammer beams and lean against the hammer posts near their tops. In addition there are curved braces rising to the hammer beams from the same seatings on the wall posts as the main arched ribs and other shorter braces between the hammer posts and intermediate points on the main arched ribs and there is a secondary queen-post system above the collar. Light vertical tracery fills the spaces that remain.

In any such system, geometry alone allows many potential paths for the loads to follow. Just which paths they will choose depends on such things as how the members are joined together, how (in relation to one another) they give way to internal stress, and where they receive their main support. This is true for all structures with more members than the bare minimum. For timber structures like these roof frames there is the further complication that behaviour will certainly change with time more than for most others because of the susceptibility of the material to both short and long term dimensional changes with changes in moisture content and, most seriously, to weakening by local decay or insect attack. For the Hall roof, the behaviour had already changed so much in the last century that it had been extensively reinforced to relieve badly weakened members. Further reinforcement was added early in this century, changing the behaviour again. At best, therefore, one can hope only for a general understanding of the principal aspects of behaviour at some particular time.

In 1914, at the time of the latter restoration, Baines attempted to reach such an understanding of the roof as he found it, using graphic statics.[1] But he abandoned it as impracticable. Two further attempts, concentrating more on the original state, were then made by Pippard in 1926 [2] and by Heyman in 1967 [3], both using strain energy methods which are more suitable where there are alternative paths for the loads if a valid calculation model can be devised. Their objectives were, however, different.

Pippard had recently developed a method for estimating the effective stiffnesses of curved compression members like the arched ribs and curved braces of the roof and wished chiefly to illustrate its use. For this purpose he identified what he took to be the principal members of the roof frames, ignoring queen posts, upper collar and all tracery, and introducing other simplifications, and then idealised the resulting frames as pin-jointed trusses, all of whose members carried only direct tension or compression. Recognising the importance of the nature of the support provided at the walls, he considered three possibilities: all support at corbels at the feet of the wall posts and arch ribs, all support at the wall plates, and rigid support at both levels. Since the first assumption led to the best agreement between the calculated forces in the members and what Baines had earlier observed, he took it as being the more reasonable. But, on account of his limited objective, he did not discuss the validity of his idealisations or discuss further his results.

Heyman adopted Pippard's idealisations and used his method of calculating the stiffnesses of the curved members. But he started by assuming that all support, at least horizontally, must be provided at wall-plate level except under wind loading, doing so chiefly on the basis of considerable observed decay at the feet of the wall posts at the time of the last restoration and certain features of roofs elsewhere. Although his main objective appears to have been to demonstrate an application of the safe theorem of current plastic theory to timber frames, he nevertheless asserted that the forces he calculated in the members for this idealisation and this condition of support were not only reasonable but could 'with some certainty' be ascribed to them.

Fig. 3 illustrates the idealised pin-jointed frame and the assumed loads, and figs 4 and 5 illustrate, by the thicknesses of the lines, the calculated forces in the members for these analyses, while fig. 2 illustrates similarly a preliminary analysis by Heyman for a much simplified model. Solid lines denote compression and broken lines tension. Pippard's calculations for all support at wall-plate level differed only slightly from Heyman's seen in fig. 5, the difference being due to the slightly different figures that he arrived at for some of the stiffnesses.

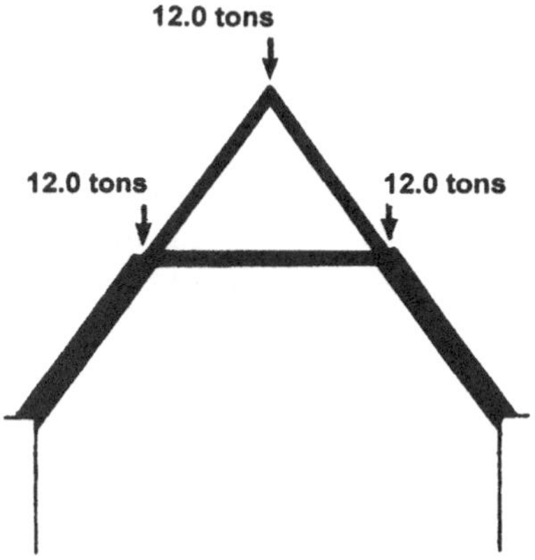

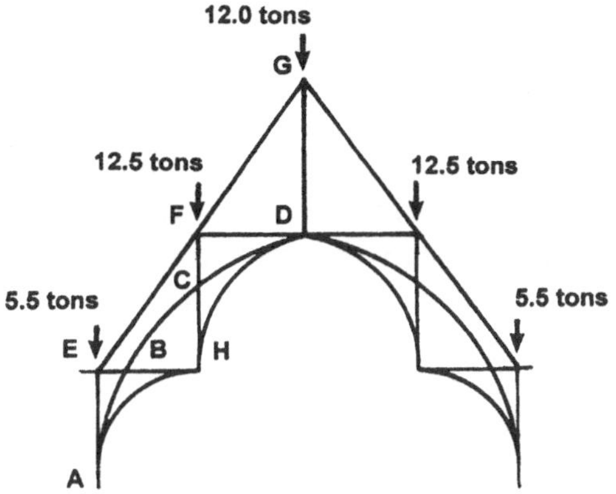

2. (left) Heyman's solution *(a)* for a basic truss (for purposes of comparison). 3. (right) Heyman's pin-jointed analogue and the loading assumed in deriving his full solution *(b)*

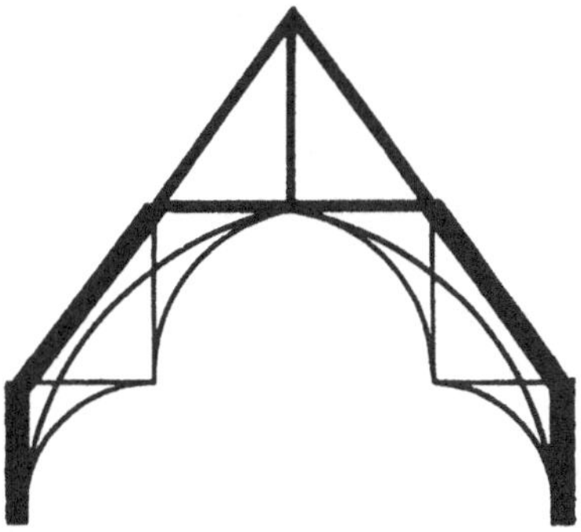

4. (left) Pippard's first solution *(g)* for a similar analogue, recalculated for the same loading. 5. (right) Heyman's full solution *(b)*

Not being satisfied with the basic idealisation of fig. 3 and the ruling out of any horizontal support to the roof at the corbel level before decay weakened the feet of the wall posts, I undertook fresh analyses. They were first published as a discussion of Heyman's paper and are reprinted here in that form. Hence the inclusion of some general comments on the relevance of Heyman's 'plastic' approach. All figures have, however, been re-numbered in a single sequence.]

4 WESTMINSTER HALL ROOF

For more than 500 years the great hammerbeam trusses of Westminster Hall have been a *fait accompli,* demonstrating by their support of the roof (until gravely weakened by decay) the adequacy of the structural principle embodied. Dr Heyman has nevertheless responded, like Harvey and Pippard, to the challenge they present to our understanding and he has added to the interest of his paper with some provocative general comments in the conclusions.

I hope that, in his reply, he will expand some of these comments and, in particular, both explain the distinction drawn in § 58 between 'elastic' solutions and those which he has presented and comment further on the relevance, in the present instance, of plastic theory. It is difficult to follow the argument here and to reconcile the confident attribution of functions to the members in §§ 61-67 with the earlier suggestion that the plausible solutions 'are merely possible equilibrium states for the truss'. And was the situation observed by the Steel Structures Research Committee (and later in my own fuller tests at Whitehall Gardens[4]) indeed the same as that which, in the analyses reported, led to different loads in the members for different assumed boundary conditions or even as that which would, in practice, introduce a measure of indeterminancy into the behaviour of the roof ? The differences between calculated and observed stresses in steel-framed buildings are either purely local or the result of self-straining or of interactions between the frame and its cladding. Both local effects and self-straining have, however, been ignored in the analyses and are unlikely to be important in practice because of the type of connexions, and there is no cladding to interact with the truss.

I have wondered whether sufficient distinction has been drawn, both here and elsewhere in the paper, between the theoretical model of the truss which was analysed and the real truss, between the real truss as built and the much weakened and patched one which can now be seen, and between this Westminster truss and others which bear some resemblance to it. To 'construct reasonable equilibrium distributions' the engineer has to simplify and idealize but, if it is desired to understand how a structure carries its loads and not merely to demonstrate that it can do so somehow, it is essential to ensure that the theoretical model is sufficiently representative of the particular structure in the state considered. It is unfortunate, therefore, that the details of the Westminster truss were illustrated only in Viollet-le-Duc's partly conjectural drawings and that considerable weight has been placed both on details of other roofs which differ in important respects (as correctly noted by Baines in the case of the Eltham Palace roof) and on the present decay at the foot of the wall posts which is no more relevant than the more extensive decay in the neighbourhood of the wall plates to the behaviour of the truss as originally built.

Very complete data on the construction are available in Baines's report and an attempt has been made to illustrate it more adequately in Fig. 6. The following features are notable in relation to the pin-jointed truss analogue used by Pippard and Heyman:

(a) the hammerbeam, collar beam, hammerpost and wall post are all continuous members and are stiff enough to develop considerable moments if subject to flexure at intermediate nodes;

(b) the lower arch rib and curved brace terminate at about mid height of the wall post; they can carry load only if the wall post provides a vertical reaction (by compression at the foot) and a horizontal one (by flexure associated with tension in the hammerbeam), and they are both stiffer, by an order of magnitude, than the notional equivalents of Pippard's analogue;

(e) only between E and H' and in the collar can appreciable tension be developed and there can be little horizontal reaction between the hammerposts and the collar where the section of the post is greatly reduced and that of the tenon is only 2% of that of the post. Since, also, the upper and lower rafters are connected only through the post there can be little transmission of thrust from one to the other.

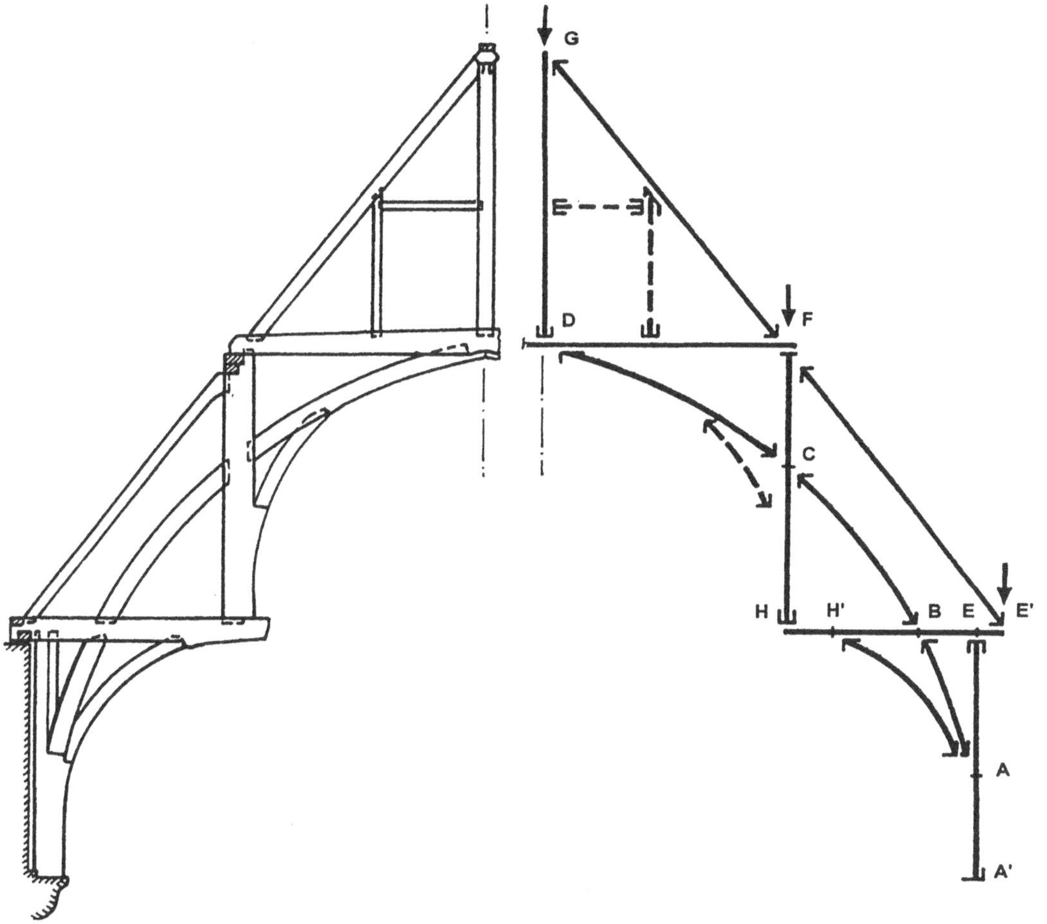

6. (left) details of the actual roof construction based on Baines's survey.
7. (right) a linear analogue in which members shown there in broken line are obviously of secondary importance and have been ignored in the discussion.

Clearly the pin-jointed truss analogue misrepresents the range of behaviour of which the real truss was capable in several important respects and it is difficult to regard any of the solutions presented as a 'possible equilibrium state'. Consideration of the deflexions of the linear analogue in Fig. 7 suggests, however, that it can be considered as two largely independent parts – an upper triangle including the collar, and a lower part comprising chiefly the arch ribs, hammerbeams and hammerposts, the lower braces and the wall posts – and that the behaviour of the lower part will approximate closely to that of the statically determinate system shown in Fig. 8 under vertical load. The lower rafters can be ignored in relation to the primary action of the truss because the relative stiffnesses are such that they will go slack and, though it is less certain that there will be no external reactions at E or E' the relative stiffnesses again ensure that these reactions will be small and hardly significant. At the tops of the hammer posts and at D, the ridge load will be distributed by the collar in whatever ratio is required for equilibrium of the lower part, there being little associated deflexion of the collar unless there is considerable lack of fit between the principal members of the upper triangle.

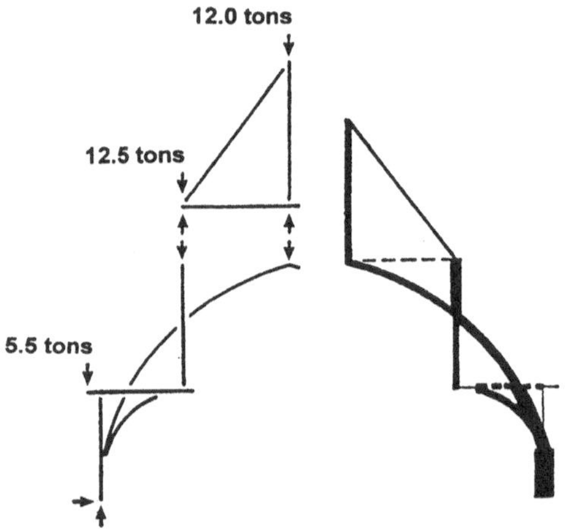

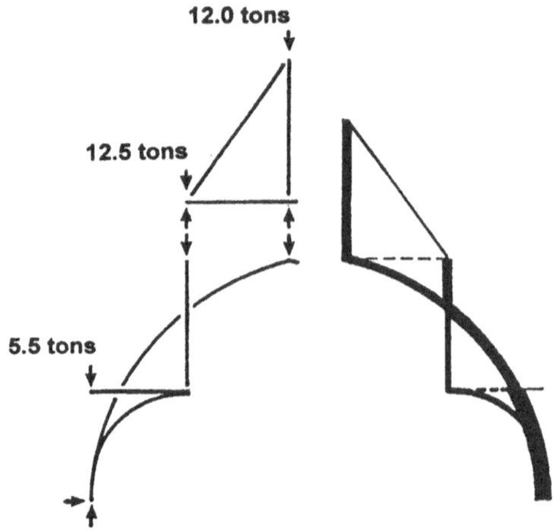

8. (left). Statically determinate analogue assumed in deriving solution *(m)* for the same loading as in Fig. 3, and, on the right, the estimated axial forces.
9. (right). Alternative analogue, with same geometry as Pippard's, assumed in deriving solution (*n*) and, on the right, the estimated axial forces.

The axial forces in the members calculated for the system of Fig. 8 are represented on the right half of the figure in the same manner as in Figs. 4 and 5, and both they and the moments at the intermediate nodes of the hammerbeam and wall post are given in the Table as solution (*m*). The calculations were made with lengths and sections of members scaled *ab initio* from Baines's drawings and the full validity of the solution is sensitive to some of these. Accepting them as correct and ignoring initial lack of fit and relative shrinkage, the hammerbeam would be just clear of the wall post at E and just in contact with the wall plate at E', but the reaction at E'

Member	Node	Flexibility *in/ton* × 10^3	Heyman basic (*a*)	Heyman full (*b*)	Pippard 1 (*g*)	Flexibility *in./ton* × 10^3 or EI* *ton in.*^{2} × 10^6	Mainstone Fig. 19 (*m*)	Mainstone Fig. 20 (*n*)
Crown post	DG	3·12	0	−4·6	−14·0		−8·6†	−11·7†
Upper rafter	FG	3·08	−7·29	−4·5	1·0		−2·1†	−0·15†
Collar	FD	1·26	−8·56	−6·7	4·0		1·25†	0·08†
	D						*0*†	*0*†
Lower rafter	EF	2·59	−22·46	−16·4	8·5		0	0
Hammerpost	FC	0·31	0	−2·7	−18·5	0·31	−14·2	−12·6
	C						*0*	*0*
	CH	0·43	0	−0·1	−7·1	0·41	−8·25	−6·0
Hammerbeam	EB	0·19	0	−2·7	−4·8	0·17	5·7	0
	B					*5·44*	*429*	*330*
	BH′‡	0·42	0	0·1	6·4	0·26	6·7	2·9
	H′					*5·44*	*297*	
	H′H						0	
Arch rib	AB	7·10	0	−5·0	−20·3	0·69	−18·5	−21·05
	BC	2·68	0	−6·3	−25·4	2·74	−12·9	−16·1
	CD	13·2	0	−4·6	−18·6	11·3	−9·0	−11·75
Lower brace	AH′‡	696	0	−0·13	−7·5	58·0	−9·5	−4·65
Upper brace	HD′‡	960	0	0·00	2·8		0	0
Wall post	EA	3·20	(−24·00)	−19·0	1·5		0	0
	A					*5·11*	*786*	
	AA′						−24·0	
Thrust at wall plate			12·7	12·0	0		0	0
Thrust at wall post corbel			0	1·3	9·7		7·7	10·15
Deflexion of hammerbeam, *in*			0·08	0·11	6·7		0·28 + 0·09*x*§	

Loads in tons (tension positive); moments in tons inches in italics; * values of EI in italics; † for no deflexion of collar beam; § *x* = spread of walls at corbel level, *in*; ‡ BH, AH, HD respectively for solutions (*a*), (*b*), (*g*) and (*n*).

would be negligible and the effect on other loads correspondingly small. The horizontal movement of F relative to D is sensitive to vertical deflexion of the collar but, for the near-zero deflexion to be expected, this movement is zero for a spread at the wall corbels of about 1 in. E and F separate by about 0.4 in.

The solution is compared in the Table both with Heyman's basic and full solutions (*a* and *b*), with Pippard's No. 1 (*g*) and with a further solution (*n*), illustrated in Fig. 9, in which the geometry of Pippard's frame is retained but a statically determinate mode of behaviour similar to that of solution (*m*) is assumed. Solution (*n*) is, of course, open to many of the objections made to the original solutions but, being intermediate between these and solution (*m*), it helps to clarify one of the most important features of the truss – the manner in which the horizontal thrust is both carried well below the top of the wall and much reduced in magnitude by the combined adoption of the hammerbeam principle and of an arch rib and brace meeting the wall post at about mid height.

I agree that the 'actual' state of the truss at any time is essentially indeterminate. Baines's survey showed, moreover, that there were differences between the trusses. Solution (*m*) is not, therefore, an 'absolute' one. I consider, nevertheless, that it represents better the functions of the principal members than any of the solutions presented in the Paper and, if I am correct in this, it is worth noting that it shows these functions to be much closer to those previously ascribed by men like Viollet-le-Duc and Baines. They are also closer to what could be inferred about Herland's intentions from the relative proportions of the members and details of the connections. In defence of Viollet-le-Duc, it should also be stated that there is no evidence whatever that he was misled by the common mediaeval (and earlier) conceit of the angels.

Finally, if it is desired to estimate the collapse load of a structure like the Westminster roof, is it necessary to invoke the safe theorem of plastic theory? Since appreciable self-straining is not possible it seems (intuitively) that the collapse load cannot be less than that estimated for any plausible statically determinate analogue of the structure arrived at by setting redundant loads at zero, and, if so, there is no need to assume ductility where none existed. Would Dr Heyman care to comment on this ?

References

1. Baines, F. *Westminster Hall: report to the first commissioner of H. M. Works, &c., on the condition of the roof timbers of Westminster Hall, with suggestions for maintaining the stability of the roof*, (Cd.7436), London, H.M.S.O., 1914.

2. Pippard, A. J. S. and Glanville, W. H., 'Primary stresses in timber roofs with special reference to curved bracing members', *Building Research Technical Paper No. 2*, London, H.M.S.O., 1926.

3. Heyman, J., 'Westminster Hall Roof', *Proceedings of the Institution of Civil Engineers*, vol. 37, 1967, pp. 137-162.

4. Mainstone, R. J. Studies in composite construction, Part III: tests on the new Government offices, Whitehall Gardens. *National Building Studies Research Paper No. 28.* London, H.M.S.O., 1960.

XVI

Brunelleschi's Dome of S. Maria del Fiore and some Related Structures

The completion, between A.D. 125 and 128, of the dome of the Roman Pantheon[1] marked the main culmination of a period of intensive development of wide-spanning concrete vaults that had begun some 60 years previously. This dome was emulated but never challenged in the subsequent years of the Western Empire. In the East, in Justinian's church of St. Sophia in Constantinople,[2] something that was in several ways more daring was attempted between 532 and 537, but the central dome which there crowned a complex system of billowing semidomes had a diameter little more than two thirds of that of the Pantheon. Only with the completion, in 1434, of the dome of Santa Maria del Fiore in Florence—the new cathedral that replaced the earlier church of Santa Reparata—was this pre-eminence of the Pantheon's dome lost. Brunelleschi's Florentine dome is both slightly greater in mean span than that of the Pantheon and, being carried at a considerable height on a relatively slender drum, called for a greater refinement in its structural design. Today it still dominates the city skyline in a manner unequalled anywhere else and it remains the greatest masonry dome in existence (Plate XII).

Unlike the domes of both the Pantheon and St. Sophia, it was built on piers commenced long previously by other men. Its span was fixed, its basic form largely predetermined and even its drum had been completed before detailed design began in 1417. Brunelleschi's achievement was, therefore, primarily a technical one with little scope for spatial or formal invention and it was thus that it was recognised and acclaimed by his contemporaries. What chiefly caught their imagination was the fact that the small (but increasingly important) city of Florence had at last demonstrated to the whole world that, in vaulting the vast space of the crossing of the new cathedral, it could equal and even outdo the greatest achievements of the ancient Roman Empire. This feat had been all the more remarkable because none of the massive temporary supports normally considered necessary at the time had been used and because both the unsightly external buttresses of northern gothic and the visible internal ties generally preferred in Italy had been dispensed with.

In the records which have come down to us, the acclamation is best summed up in the dedication to Brunelleschi of the 1436 manuscript of the Vernacular version of Alberti's *De Pictura*. "Who is so stubborn or so jealous as not to praise Pippo the architect when he sees here such a great structure, rising above the heavens, broad enough to cover with its shade all the people of Tuscany, made without any aid of falsework or other quantity of timber (*facta sanza alcuno ajuto di travamenti o di copia di legnami*), an achievement which, if I am not mistaken, was believed to be impossible in our time even as, among the ancients, it may well have been unknown and unheard of."[3] It was given

[1] The best account of the Pantheon is now Kjeld de Fine Licht, *The Rotunda in Rome,* Jutland Archeological Society, Copenhagen, 1968.

[2] See R. J. Mainstone, "The structure of the church of St. Sophia, Istanbul," *Transactions of the Newcomen Society,* XXXVIII (1965–66), 23–49, and idem, "Justinian's church of St. Sophia, Istanbul: Recent studies of its design, construction and first partial reconstruction," *Architectural History,* 12 (1969), 39–49.

[3] Quoted here from the edition of Luigi Malle, *Leon Battista Alberti: Della Pittura,* Florence, 1950, 54.

more formal expression in the cathedral records and in the epitaph to "*Philippo Brunellesco antiquae architecturae instauratori*" set up in the cathedral after his death in 1446 and was echoed finally, at much greater length by his biographers Manetti[1] and Vasari.[2] Both refer to the more purely architectural achievements of the other building commissions, culminating in the design for S. Spirito, as well as to the cathedral dome, but treat the latter as unquestionably the central and major achievement. Both also emphasise again the feeling, which must have been almost overwhelming at the time, of having matched up to the Romans; of having, in Manetti's words, brought back to life "*questo modo de'muramenti, che si dicono alla Romana ed all'antica.*"[3]

I shall concentrate in this paper on some structural aspects of the design and construction of the dome, largely passing over other almost equally important aspects of the total technical achievement such as the organisation of the work (including the supply of materials) and the design of new equipment for handling and lifting.[4] Nothing will be said of the vexed question of the roles of Ghiberti and other collaborators. I do not think that anything significant would be added to our understanding of the nature of the technical achievement if it could be definitely established that certain contributions were made by others, so it seems sufficient, for the present purpose, to note that, as early as 1423, Brunelleschi was officially recognised (in a prize award of 100 gold florins) as "*inventor et ghubernator maiori Cupule*"[5] and that he almost certainly took the major initiative from the start. Subsequent references to Brunelleschi reflect this judgement but should not be read as necessarily excluding collaborators.

Today the dome still stands virtually untouched save for the renewal of the circumferential timber tie (which had rotted badly by the seventeenth century) and extensive reconstruction more than once of the crowning lantern. Thus it is still possible to inspect at close quarters nearly all details of the original construction that are visible on the surface. This I have done on several occasions spread over the past 17 years, most recently and most fully in the summer of 1967 at the invitation and in the company of Professor Howard Saalman of Carnegie-Mellon University, to whose stimulus and interest this paper largely owes its origin. Much, however, remains hidden beneath the surface, so that considerable reliance must also be placed on documentary evidence. Fortunately this is far more comprehensive than for almost any other comparable structure, though, for various reasons, it contains serious ambiguities that have given rise to much conflicting speculation. To the extent that these ambiguities arise from the difficulties of giving clear expression in the legal Latin or Vernacular of the time to new structural concepts, any translation into the more precise vocabulary of Modern English necessarily involves a particular narrowing of the meaning and makes it essential to return to the original if a new understanding is sought. This I have endeavoured always to do,

[1] Antonio Manetti, *Vita di Filippo di Ser Brunellesco,* probably written soon after Brunelleschi's death but unpublished until 1812. I have used the editions of Carl Frey, *Le Vite di Filippo Brunelleschi Scultore e Architetto Fiorentino scritta da Giorgio Vasari e da Anonimo Autore,* Berlin (1887), 59–118, and Elena Toesca, *Antonio Manetti: Vita. . .,* Rome, 1927. These are cited below as Frey and Toesca. Since the paper was written a new edition has appeared edited by Howard Saalman, *The life of Brunelleschi by Antonio di Tuccio Manetti,* Pennsylvania State University Press, 1970. This has an English translation of the Italian text by Catherine Enggass.

[2] Giorgio Vasari, *Le Vite de piu eccellenti Architetti, Pittori et Scultori Italiani. . .,* Florence, 1550. English translation by William Gaunt, Everyman's Library, London, 1927.

[3] Frey, p. 61; Toesca, p. 2.

[4] For these aspects see especially Piero Sanpaolesi, *La cupola di Santa Maria del Fiore,* Florence, 1941 (cited below as Sanpaolesi); F. D. Prager, "Brunelleschi's inventions and the 'Renewal of Roman masonry work'," *Osiris,* IX (1950), 457–554 (cited below as Prager) and Gustina Scaglia, "Drawings of Brunelleschi's mechanical inventions for the construction of the cupola," *Marsyas,* X (1961), 45–68. Revised versions of the latter two papers with some additional related material have now appeared in a single volume: Frank D. Prager and Gustina Scaglia, *Brunelleschi: Studies of his technology and inventions,* MIT Press, 1970.

[5] Cesare Guasti, *La cupola di Santa Maria del Fiore illustrata con i documenti dell'archivio dell'Opera Secolare,* Florence, 1857, doc. 177, p. 71. This basic source is cited below as Guasti, *Cupola.*

though it will be apparent to those who have made similar attempts that my debt to previous scholars is considerable.[1]

The Construction Below the Dome

Construction of the new cathedral began at the west end as far back as 1294.[2] The overall width of the nave was then established, but the bays were initially much narrower than at present, suggesting that a timber roof may have been intended as at S. Croce and the cathedral of Pisa. Beyond this the intentions are obscure, though it is probable that some sort of dome over the crossing was envisaged from the start, such as already existed at both Pisa and Siena. Progress was slow and almost came to a halt when energies were turned to the campanile in the second quarter of the next century.

The building, as it now exists (Plate XIII) dates effectively from 1355–57, when the present almost square nave bays were decided upon and the first explicit mention of a dome over the crossing appears in the official records. Even this plan was not definitive though. Renewed intensive discussion and debate in 1365–67 led to fresh proposals and a significant enlargement of the whole project. A fourth bay was added to the nave, the central octagon over which the dome was to rise was increased in width to the full width of nave plus aisles (72 braccia = 42 m.), and it was decided to raise the entire dome above the rest of the church by interposing an octagonal drum between it and the supporting arches. This design was embodied in a substantial brick model erected alongside the church and, after it had been formally approved in 1367, all those responsible for work on the church were required, up to 1421, to swear to adhere to it.

The model no longer exists, though there are two oblique reflections of the discussions that led up to it in a fresco in the Spanish chapel of S. Maria Novella and in a large tabernacle in the church of Or San Michele, both of which show a pointed octagonal dome, the first with a lantern but no drum and the second with a drum pierced by round arches. Other evidence suggests that, without indicating any details of construction the model did at least establish all the major dimensions including the basic curvature of the dome.[3] The problem eventually to be faced in constructing the dome was thus closely defined in its main essentials without being solved, indeed without any real assurance that it was soluble. A hope was expressed, nevertheless, that there would be no need for the visible tie bars just decided upon in the nave as a result of ominous cracks in the recently completed vaults there. More generally, the character of the structure now envisaged was no longer that of the purely gothic west end of 70 years earlier, nor even that of the somewhat later gothic of Milan cathedral. In the centralised planning of the octagonal crossing and three identical radiating arms to south, east and north, and in the insistence on a structure of maximum formal clarity, unencumbered either externally or internally by obtrusive props or ties, a character later to be recognised as Renaissance was already emerging. This cannot have been without influence on Brunelleschi, whose first

[1] Only the most directly relevant recent studies are cited in these notes. Fuller references are given in Howard Saalman, "Giovanni di Gherardo da Prato's designs concerning the cupola of Santa Maria del Fiore in Florence," *Journal of the Society of Architectural Historians,* XVIII (1959), 11–20; idem, "Santa Maria del Fiore: 1294–1418," *Art Bulletin,* XLVI (1964), 471–500 (respectively cited below as Saalman, *Cupola* and Saalman, *S.M.F.*), and in Sanpaolesi and Prager. There is no recent survey as comprehensive as that by G. B. Nelli in S. B. Sgrilli, *Descrizione e studi dell' insigne fabbrica di S. Maria del Fiore,* Florence 1733. For the dome itself, however, some more accurate data may be found in *Rilievi e studi sulla cupola del Brunelleschi eseguiti dalla Commissione nominata il* 12 *Gennaio* 1934, Florence, 1939 (cited below as *Rilievi e studi*) and further measurements are now in progress. Piero Sanpaolesi, *La cupola del Brunelleschi* (in the series *Forma e colore*), Florence, 1965, is an excellent album of recent photographs.

[2] The basic source for the period prior to the construction of the dome is Cesare Guasti, *Santa Maria del Fiore : La costruzione della chiesa e del campanile secondo i documenti tratti dall'archivio dell'Opera Secolare,* Florence 1887 (cited below as Guasti, *S.M.F.*). The present brief summary is based largely on the analysis in Saalman, *S.M.F.*

[3] Saalman, *S.M.F.*, pp. 491; idem, *Cupola,* p. 14.

direct involvement with the construction was to advise, with other artists, on a departure from the 1367 model that seemed to jeopardise this character.[1]

The drum was complete by 1413 and, as work on the side arms continued, the problem of the detailed design of the dome must have come increasingly to the fore. Brunelleschi was certainly involved by 1417.[2] In 1418 an open competition was announced[3] and a period of protracted discussions and debates ensued. It ended in 1420 with the appointment of Brunelleschi, together with Ghiberti and Battista d'Antonio, to superintend the construction, and the approval of the design to be followed.[4]

Specification for the Dome

The design had already been embodied in a substantial masonry model, but this time the model was supplemented by a written specification.[5] It is a document of outstanding importance since, with modifications agreed in 1422 and 1426, it is the primary source for following the development of the design and for interpreting those features of it whose nature or intention is not clear from inspection of the structure today. Its twelve clauses are too long to quote here in full but the main provisions may be summarised as follows:

(1) to (3) define the basic form of the dome. It is to have two octagonal shells, the inner being initially $3\frac{3}{4}$ braccia thick and the outer $1\frac{1}{4}$ braccia, separated by a void initially 2 braccia wide. (1 braccia = 0·584 m.) Both shells are to taper uniformly[6] to about two-thirds of their initial thicknesses at a central eye (later to be covered by the lantern). The curve of the inner shell at the internal corners is to be "*quinto acuto*"—i.e. it is to be a circular arc of radius one fifth less than the diameter across the corners. The outer shell is to keep the inner one dry and, at the same time, to make the dome more imposing (*piu magnifica e gonfiante*) and to provide covered access within the intermediate void.

(4) calls for 24 internal ribs to bind the two shells together (*e legano insieme le decte due volte*). Eight of these, in the angles, are to be 7 braccia wide at the foot and the remaining 16, two to each side, 4 braccia wide at the foot. These ribs also are to taper uniformly[6] to the eye.

(5) and (6) indicate that the separate shells and ribs are to commence at a height of $5\frac{1}{4}$ braccia, up to which the construction is to be solid, but they are mostly concerned with the specification of 6 rings of large sandstone blocks well clamped together and having, in addition, iron chains set above them. Successive pairs are to be 2, $1\frac{1}{3}$ and 1 braccia high and at the foot there are also to be long transverse blocks of sandstone to provide a base for both shells.

(7) calls for further interconnections between the shells in the form of small barrel vaults between adjacent ribs at 12 braccia intervals to carry ambulatories and, below these barrel vaults, of oak and iron chains to tie the ribs together (*che legano i decti sproni*) and encircle the inner shell

(8) specifies the materials to be used—briefly stone up to a height of 24 braccia and, thereafter, a lighter material such as brick or tufa.

(9) to (11) are concerned with matters of limited structural importance.

[1] Guasti, S.M.F., doc. 425, pp. 299, 300. It should be pointed out, however, that there is no direct reference to the architectural character in the document. This refers only to a departure from the model in the construction of one of the triangular buttresses at the lowest level visible in Plate 1. The buttress had been built too high and was to be lowered.
[2] Guasti, *Cupola,* doc. 16, p. 17.
[3] Guasti, *Cupola,* doc. 11, p. 16.
[4] Guasti, *Cupola,* doc. 71, pp. 35–7.
[5] Two copies of this exist, one published by Guasti, *Cupola,* doc. 51, pp. 28–30 and the other by Alfred Doren, "Zum bau der Florentiner Domkuppel," *Repertorium fur Kunstwissenschaft,* XXI (1898), 249–62. I am quoting from Doren's definitive version.
[6] More accurately "like a pyramid (*piramidalmente* or *piramidalmente . . . per iguale proporzione*)."

(12) deals with the method of erection. "The shells are to be built as specified without any supporting centering (*sanza alcuna armadura*) up to at least 30 braccia, but with such working platforms as may be decided by those in charge of construction; and thereafter as considered advisable, since in building practice teaches what should be done (*perche nel murare la praticha insegnera quello che ss'ara a seguire*)."

The amendment of 1422[1] introduced two modifications, both for the purpose of reducing weight. The intermediate ribs were reduced in thickness to 3 braccia and the height of the change from stone was reduced to about 12 braccia, above which brick was to be used.

That of 1426[2] was wider in scope. It called first for small openings in the faces to permit access to the interior, etc. Then, in place of the small barrel vaults between the ribs, arch-like bands of brickwork were to be constructed above the second ambulatory, 1 braccia high, "for the perfection of the circle of the outer shell so that the live arch shall be complete and not broken (*per perfectione del cerchio che gira intorno la cupola di fuori, accio che detto archo vivo sia intero e non rotto*)." If judged to be unsightly or an obstruction, these additions could be removed upon completion of construction, their primary purpose being to allow the construction to proceed with greater safety. Revised details of the final pair of "stone chains" (*catene di macigno*) and superimposed iron chains follow. Finally comes a call for the manufacture of special large bricks to be laid in a herringbone (*spinapescie*) bond as determined by the master in charge, and a confirmation of the earlier decision to dispense with a supporting centering. In the absence of this, a device with three cords (*gualandrino con tre corde*) is to be used to control the curvatures.

As usual, some discretion over details is left to those directly in charge of construction. The formal latin authority (to Brunelleschi and his associates) that accompanies this last amendment specifically confers a wider discretion though,[3] and this has naturally led to speculation about the extent to which the specification, as amended, was subsequently followed. There are three principal matters of interest: the ring of horizontal arches; the "stone chains" and the oak chains and barrel vaults previously called for but passed over in silence in 1426; and the manner of construction.

Construction of the Dome

It is immediately apparent from an inspection of the structure today (Fig. 1) that not merely one but no less than nine rings of horizontal arches of about the size specified were built in the annular void between the two shells. They were never removed and, since they seem to be bonded into the outer shell, it was possibly decided at an early stage that they would be left.

The omission of the small barrel vaults previously called for followed inevitably from the decision to construct these other additions and all the oak chains were likewise omitted except the first—which had already been constructed in 1423 of chestnut. The ostensible purpose of the barrel vaults had been to carry ambulatories. These ambulatories are, in fact, carried instead by regularly spaced heavy sandstone blocks spanning between the two shells and strongly reminiscent of the interconnecting blocks between the first pair of stone chains called for in 1420. It is difficult to escape the conclusion that they are indeed performing a similar function in relation to the second and third pairs of stone chains. If so, it seems certain that both pairs were constructed, as further documents suggest. Professor Saalman and I found final confirmation of this in 1967 in a detail which has not previously been noticed in this connection. We believe that, though it has hitherto been assumed that all six chains lie wholly within the thickness of the masonry, a length of almost 4 m. of the second outer chain is exposed where the outer shell is cut back to allow the internal

[1] Guasti, *Cupola*, doc. 52, pp. 30–1.
[2] Guasti, *Cupola*, doc. 75, pp. 38–41.
[3] Guasti, *Cupola*, doc. 75, p. 40. Authority is given to vary the design "*in addendo, minuendo ac disponendo, plus et minus . . . et prout et sicut dictis operariis . . . videbitur et placebit, non obstante dicto raporto.*"

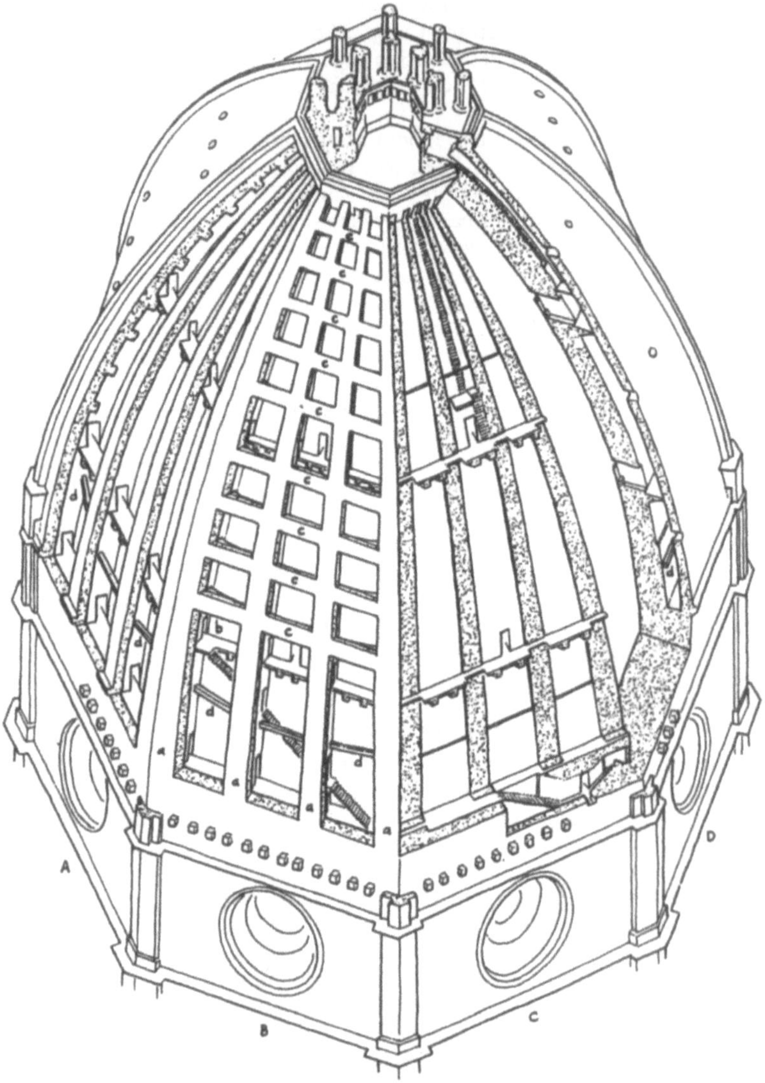

Fig. 1. S. Maria del Fiore, Florence. View of the dome partly cut away. The timber chain is marked *d*.

Shown above A and B: the outer shell cut away between radial ribs, *a*.
A: the circumferential arches *c* also cut away.
C: only the inner shell and the ambulatories.
D: a complete cross-section through both shells.

Drawn by the Author. Partial sources—an unpublished drawing by Siebenhuener, a drawing by Sanpaolesi, the survey published by Sgrilli.

stairway to ascend to the second ambulatory.[1] (Plate XIV (*c*)) It is reasonable to assume that the superimposed iron chains were also constructed in all cases, but there is no means of telling what form they took without exposing them.

The documentary evidence and a continuous literary tradition are unanimous in stating that the construction of the dome was completed without *armadura*. Alberti's contemporary testimony has already been quoted. Alongside it may be placed that of the official cathedral records on Brunelleschi's death 10 years later. There he is described as having completed the dome "*sine aliqua armatura . . . ut clare asseruerunt dicti operarii, viderunt et notum est in civitate predicta*" (as clearly stated by the operarii and seen and known in the city).[2] This apparently simple fact has, nevertheless, been repeatedly questioned in more recent times because, among other reasons, there are references in the documents to the making of centers (*cientine*).[3] I believe that the description of the method of construction given by Nelli,[4] a later architect to the cathedral, is correct in describing the herringbone bond of the brickwork as the feature which made it possible to dispense with supporting centering, and that the centers referred to were used only to assist in keeping to the specified profile.

The herringbone bond, which is today such a characteristic feature of the construction above the second ambulatory (Plate XIV (*a*)) involves essentially the setting of a brick vertically on edge at regular intervals in each otherwise horizontal course. The special large bricks that were called for in 1426 are only about 50 mm. thick, so that each vertical brick rises through four or five horizontal courses. Being set with one face against the most exposed face of the corresponding brick of the previous course, it forms part of a diagonally ascending band. In a circular dome this would be quite straightforward and the ascending bands, being set radially in plan, would automatically converge towards the top (Fig. 2).[5] Here the same convergence is seen but nothing was straightforward on account of the octagonal plan, the consequent elliptical profile of each side of the dome and the introduction of ribs built integrally with the shells. In part the problems were overcome by the manufacture of a number of special non-rectangular bricks and by the cutting in-situ of others and in part by the introduction in places of special levelling courses of tapered bricks. But they must have remained a troublesome preoccupation to the end.[6] There was also a further drawback which is unlikely to have been overlooked. This is that the inclined bands of vertical bricks would inevitably constitute planes of weakness in resisting tensile hoop stresses. A much greater tensile strength could have been achieved in the brickwork by laying all the bricks horizontally with all vertical joints lapped and the tensions transmitted from brick to brick by friction and bond shear at the horizontal joints.

It is inconceivable that these difficulties and drawbacks were accepted right to the end unless they were indeed directly linked with (and justified by) the avoidance of the expense, difficulties, and risks of constructing a heavy vault-spanning supporting centering. More positively, the herringbone bond would have made it possible to proceed without such support. Since every complete course

[1] Parts of two sandstone blocks, 470 mm. high, are visible towards the upper left of the photograph, notched over one of the transverse beams that has been cut to allow the stairway to pass. The other (inner) end of this transverse beam is seen projecting at the right to carry the ambulatory, here cantilevered out from the inner shell.

[2] Guasti, *Cupola*, doc. 120, pp. 56–7.

[3] Guasti, *Cupola*, docs. 170–2, p. 70.

[4] G. B. Nelli, *Ragionamento sopra la maniera di voltar le cupole senza adoperavi le centine* in *Discorsi di Architettura del Senatore Giovan Batista Nelli*, Florence, 1753, 53–74. This account is very discursive, the relevant passage appearing on pp. 65–7. The word *spinapescie* never appears, but Nelli gives what seems to be a fairly detailed description of the bond actually adopted in the upper part of the dome.

[5] Piero Sanpaolesi, "Le cupole e gli edifici a cupola del Brunelleschi e la loro derivazione da edifici Romani," *Atti del I Congresso nazionale di storia dell'architettura*, Florence, 1936, 37–41, first drew attention to this drawing which illustrates this simple application very well.

[6] Sanpaolesi (pp. 25–9) gives a detailed account of the practical problems with illustrations of some of the expedients adopted. His Plate Vb attempts to show the internal details of the bond as they would have been seen during construction, but I do not find it wholly convincing.

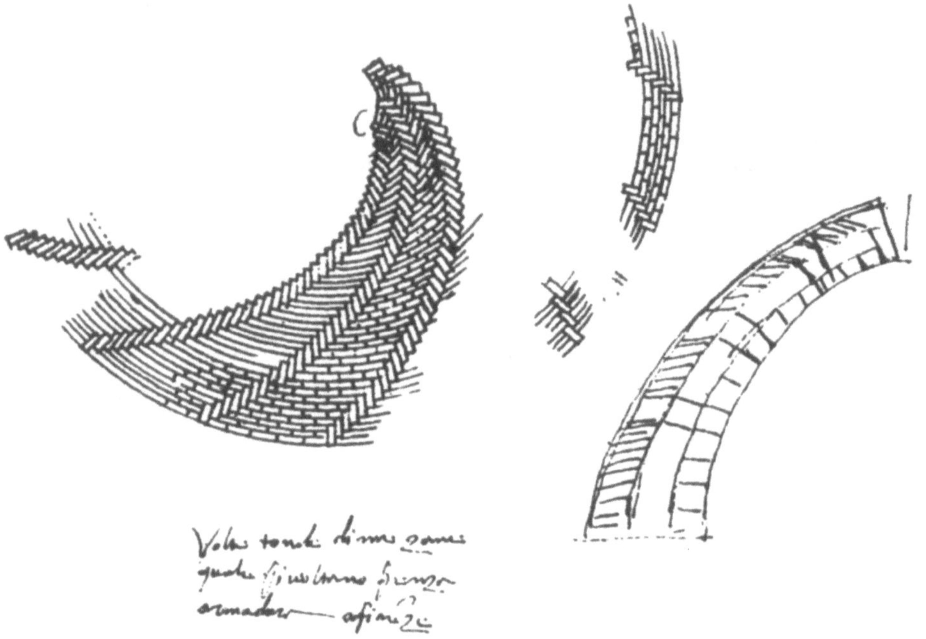

Fig. 2. Detail of a drawing by Antonio da Sangallo the Younger.
(*Uffizi* 900—*courtesy of the Soprintendenza alle Gallerie*)

of a dome is, in effect, a closed horizontal arch, it has no need of support other than that of the structure below. Only an incomplete course needs, temporarily, some further support. The inclined bands of vertical bricks do not eliminate this need, but they subdivide each course into a large number of short sections never more than about five bricks long and reducing to single horizontal brick at the eye. Each of these sections, when completed, would constitute a short flat arch which would be self-supporting as soon as the mortar had hardened sufficiently. It may be assumed that, working together, teams of two or three bricklayers spaced around the dome on light working platforms would each be responsible for one section and would complete this from the intrados outwards, the only difficulty being to keep the bricks on the intrados in place until the short flat arch was complete and self-supporting. There are several ways in which this might have been done, depending partly on the nature of the mortar used. The mortar beds, as well as the bricks, are thin and they are inclined to the horizontal at an angle never more than 60°. Friction alone would keep them in place up to an angle of, say, 30° and even at the eye the shear on the joint would be only 0·001 N/mm^2. This may be compared with a bond shear of at least two orders of magnitude greater for a modern cement-lime-sand mortar after 28 days. One possibility, therefore, was simply for assistants to hold the bricks in place for a time in much the same way as they still do when constructing vaults freehand in the Middle East today. Another simple possibility suggests itself however. This is the use of timber boards just long enough to span from one band of upright bricks to the next and either held against these bricks as each course was laid or tied back to them. Since the average rate of construction was less than one course a week, there would be ample time for each course to become self-supporting before the next was added and thus to make this procedure practicable. Possible confirmation of its actual use may be found in a passage in Alberti's "*De re aedificatoria*".[1] Its precise range of possible meaning turns on the words *amenta* and *ansas*, the second of which could be read as referring to something like the bricks that project vertically. Leoni's translation here seems as good as any: "Yet it will be necessary, when you have laid one or two rows of stone, to make little light stays, or catchers jutting out (*amenta et ansas*), on which, when those rows have settled, you may set just framework enough to support the courses next above . . . ."[2] The context is a general discussion of the construction of polygonal domes without centering and there is no explicit reference to Brunelleschi's dome, but the procedure described could have been suggested to Alberti by what has been proposed here.[3]

Control of the form without first replicating it in a complete system of self-supporting centering called for some system of measurement. The *gualandrino* with three cords specified in 1426 can be envisaged thus: two of the cords (or chains) have one end of each fixed at opposite ends of a horizontal line, drawn at right angles to a corner diagonal of the octagon and passing through the centre of the circular arc that traces, in the vertical plane of that diagonal, one corner of the intrados of the inner shell. The other two ends are joined together so that they would sweep out this arc and define it uniquely in space. The fixed ends would be attached to eyes on the ambulatory inside the dome at its springing. The third cord was presumably associated with the control of thicknesses and of the profile of the outer shell. This device would be cumbersome and was presumably used only at intervals in each corner in turn.[4] Day-to-day control would be the role of the *cientine*. There was apparently one for each corner and their low cost[5] indicated that they cannot have been more

---

[1] L. B. Alberti, *De re aedificatoria*, Florence, 1485, Book III, chapter XIV (but not so subdivided until late editions and with the pages unnumbered): "*Tamen conferet ductis iam una atque alteris lapideis et duratis coroni levia illic subillaqueare amenta et ansas; quibus tantum armamenti committas; quantum sat sit ferendis coronis hi quae aliquot inde in pedes superastruantur quoad siccescant. . . .*" Prager has since made the interesting suggestion to me that the passage should have read "*Tamen conferes . . . levia sub illaque* [*testudine*] *ar*[*m*]*amenta et ansas*"

[2] James Leoni, *Ten books on architecture by Leone Battista Alberti*, 1726, 59. (Facsimile edition edited by Joseph Rykwert, London 1955). Cosimo Bartoli in the Italian translation on which Leoni's English one is based, *L'architettura di Leonbatista Alberti*, Florence, 1550, 90–1, renders the words "*amenta et ansas*" as "*spranghe o perni non gravi.*" Portoghesi has more recently translated them as "*leggeri lacci e grappe.*"

[3] There is one difference between the procedure I have suggested and those described by Nelli and Alberti. I have assumed that one course at a time would be laid and left to harden. Nelli (p. 65) says that three courses are constructed at a time and Alberti (*loc. cit.*,) even speaks of a few feet. The former would be equally reasonable, but the latter would not unless the "*ansas*" were something other than the projecting vertical bricks of the herringbone. Another possibility is that Alberti had in mind the built-in iron rings set in horizontal rows about 2·2m, apart in the inner surface of the inner shell that are described and illustrated by Sanpaolesi (pp. 16–18). This spacing corresponds, roughly, to a year's work, so I think that Sanpaolesi is correct in interpreting them as supports for working platforms for the following year rather than for any kind of formwork. Since Alberti is only giving general advice, it is quite likely that he ran together two distinct aspects of Brunelleschi's method which he may not even have fully understood.

[4] This system of control of the curvature in the corners would virtually preclude the use of any large working platform such as those suggested to Sanpaolesi (Plate VI*b*) and Prager (Fig. 11, p. 508) by the drawing published by C. von Stegmann and H. von Geymuller in *Die Architektur der Renaissance in Toskana* (Munich, 1885, Fig. 12, p. 48), and, as an engraving, in Nelli (*op. cit*). This drawing was said to be by Brunelleschi, but is clearly much later and is inaccurate in important respects. I take it to be a later record of a structure associated with the construction of the lantern and perhaps with the final stage of work on the dome at the level of the eye. In relation to the suggested points of attachment on the ambulatory, it should also be pointed out that recent surveys show the centres of curvature of the circular arcs that best fit the present internal profiles in the corners to lie below the level of the ambulatory. This would seem, *prima facie*, to call for lower points of attachment. As yet, however, the surveys have not been properly correlated with one another. (The only one yet published has even claimed that the present profiles are elliptical arcs with small and widely varying eccentricities.) Also there is insufficient further data to permit the necessary allowances to be made for the movements (and consequent changes in profile) that have occurred since construction—let alone the divergences likely to have occurred during construction.

[5] Professor Saalman has pointed out to me that the recorded payment for the first eight amounts to less than the average daily wage of the lowest paid labourer for each and was only slightly higher for the remaining two (presumably replacements).

than light templates formed to the appropriate curve and used in exactly the same way as the similar templates employed in 1909 in constructing the large dome of the cathedral of St. John the Divine in New York.[1] The most effective control of the straightness of the sides would have been simply by sighting from one corner to the next along each course. If the procedure suggested above is correct, this would have been done when setting each vertical brick. The horizontal planks, or short lengths of string would have given the line between the vertical bricks.

The dome was completed up to the eye in 1434. Brunelleschi then made models and other preparations for the construction of the lantern, but work on this had barely started when he died in 1446.

Structural Assessment

Before turning to a discussion of the likely origins and development of the design, it is worth considering the probable structural actions, both during construction and afterwards, of the structure as built.

The proportions of the dome are very massive, even in relation to its span and rise. Wind loading is therefore relatively unimportant. Since earthquakes were not a serious threat either, it will be sufficient to consider only gravity loading. Under such loading the actual cross sections, even of the shells considered independently between a pair of ribs, are quite thick enough for the entire dome to stand as two rings of independent vertical arches, each ring with a shared keystone at the central eye—providing that the supporting structure does not yield too much under the thrusts generated. But this thickness does not, in itself, preclude the alternative possiblity of a more shell-like action in which the outward thrusts developed in the upper parts of the shells and ribs are contained lower down by internal horizontal tensions. Evidence of cracking referred to below shows that the present behaviour of the dome is closer to the first of these possibilities than to the second, but it still leaves open the question whether this was also the case right from the completion of construction.

The construction of possible thrust lines for action of the shells as rings of independent arches merely provides a theoretical demonstration of the first possibility, with reference to the dome only and not its supports—a demonstration that is hardly necessary in view of the fact that it does still stand despite some radial cracking.[2] A more revealing insight—and one that is more relevant to the design problem that was posed after completion of the drum in 1413—is given by considering the second in terms of a simple thin-shell analogue. In plan, even at the base, the thickness of the inner shell exceeds by about 25 per cent the difference between the internal radii measured at the corners and at the centres of the sides. It is therefore possible to inscribe within the thickness two circles spaced apart a distance equal to this excess, and to consider the actual octagonal shell as containing within itself another shell of the same profile but circular in plan and only one-fifth of the thickness. Such a shell is statically determinate, so that it may be analysed without calling for unknown effective moduli of elasticity of the materials. It is, of course, a pointed shell of revolution open at the top throughout construction, but eventually carrying a relatively light lantern—i.e. a distributed more-or-less vertical load—on the uppermost ring of a central eye.

The radial symmetry of the gravity loading, as of the shell's geometry, means that it will be subject only to direct tensions and compressions in the meridional and horizontal circumferential directions. Adopting Flügge's notation (Figures 3 and 4) the equations of equilibrium for an element of the shell are:[3]

[1] "Erecting a large dome without falsework," *Engineering Record,* 60 (1909), 508–10.

[2] W. B. Parsons, *Engineers and Engineering in the Renaissance* (Baltimore (1939), reprinted, M. I. T. Press (1968), 594–8) gives two such analyses and some related caclulations.

[3] Wilhelm Flügge, *Stresses in shells,* Berlin, 1960, chapter 2, "Direct stresses in shells of revolution," pp. 18–24. For an analysis of an alternative analogue of the actual dome, see Guido Guerra, *La statica delle cupole a padiglione in muratura,* Rome, 1951 (reprinted from *Giornale del Genio Civile,* Sept. 1951) and idem, *Statica e tecnica costruttiva delle cupole antiche e moderne,* Naples (1958) 48–9.

$$\frac{d}{d\phi}(rN_\phi) - r_1N_\theta\cos\phi = - p_\phi rr_1, \qquad (1)$$

and

$$\frac{N_\phi}{r_1} + \frac{N_\theta}{r_2} = p_r \qquad (2)$$

Solving for N_ϕ and integrating:

$$N_\phi = \frac{1}{r_2\sin^2\phi}\left[\int r_1r_2(p_r\cos\phi - p_\phi\sin\phi)\sin\phi\, d\phi + C\right] \qquad (3)$$

N_θ is then given by equation (2). The constant of integration is given by the known value of N_ϕ at the open top ring at any intermediate stage of construction or at the uppermost ring of the central eye on completion of construction. Thus, at any intermediate stage of construction, $N_\phi = 0$ for $\phi = \phi_c$, where ϕ_c is the value of ϕ at the then uppermost ring. Similarly on completion of the dome to the central eye but before construction of the lantern, $N_\phi = 0$ for $\phi = \phi_e$, where ϕ_e is the value of ϕ at the eye; and, on completion of the lantern, $N_\phi = -P/\sin\phi_e$, where P is the weight per unit length (measured around the eye) of the lantern.

Since the springings may be assumed to be vertical, the meridional stresses there will, of course, also be vertical, their outward components higher up being wholly contained by horizontal circumferential or hoop tensions. These hoop tensions are the really interesting stresses, since they are the only ones likely to approach the ultimate strengths of the materials used.

Fig. 5 shows a number of plots of the hoop stresses, $\sigma_\theta = N_\theta/t$, both tensile and compressive, expressed non-dimensionally as multiples of ρr_1 where ρ is the density, assumed uniform, and r_1 is the radius of curvature of the generating arc as in Fig. 3. Of the two curves passing through the origin, the upper one represents the progressive increase in tension at the base as construction proceeds,

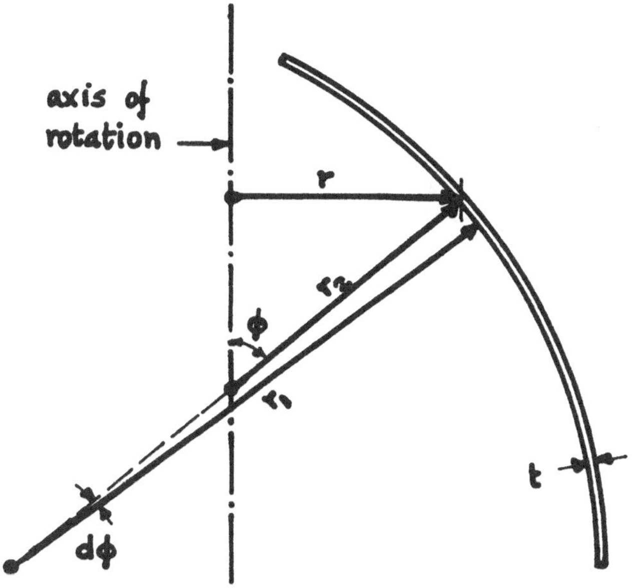

ϕ = angle between normal to middle surface of shell and axis of rotation

r = distance of point on middle surface of shell from axis of rotation measured normal to axis

r_1 = radius of curvature of middle surface of shell

r_2 = distance of point on middle surface of shell from axis of rotation measured normal to surface

t = thickness of shell.

Fig. 3.

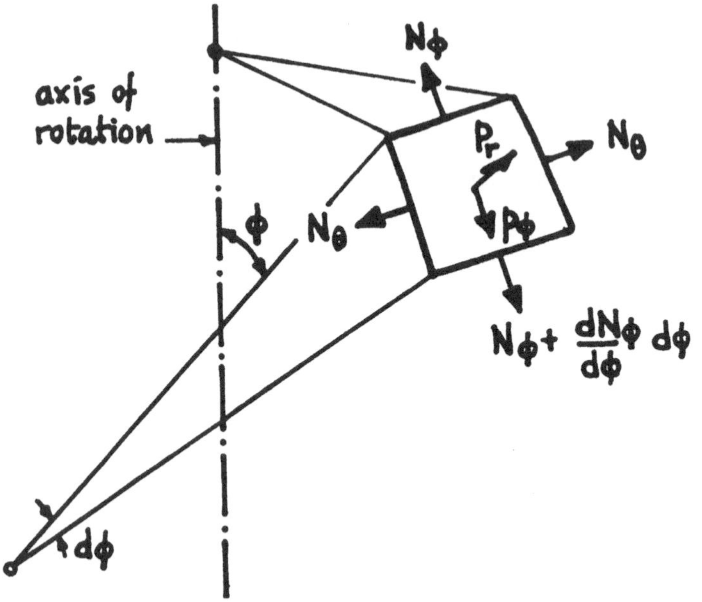

N_ϕ = meridional force in shell per unit length

N_θ = hoop force in shell per unit length

p_ϕ = meridional camponent of load per unit area

p_r = normal component of load per unit area

Fig. 4.

and the lower one the initial compression at the open top ring at any stage of construction. The family of curves roughly parallel to the first similarly represent the increases in tension (appearing for a time as reductions in the initial compressions) at various higher levels corresponding to the change from stone to brick, the timber tie, and the second ambulatory. These levels are denoted, in terms of ϕ, by the abscissae of the lowest points on the curves. The other family of curves, shown in broken line, represent the variations in stress with height throughout the dome when construction has just reached the level denoted again by the abscissa of the lowest point on the curve. The additional stresses due to construction of the lantern cannot be shown in the same way since they depend on the thickness of the shell, but, in the dome as built, they are spread over such a thickness near the base as to be negligible there in relation to the stresses plotted.

From the lower curve passing through the origin it will be seen that the hoop stress in the ring that at any stage terminates the construction will always be compressive, being virtually zero for the first ring. It will increase with the height of this ring to a maximum of 0·33 pr_1, when the structure is about three-quarters complete. Thereafter it will reduce again to about half this value. But at any particular level this initial compression will continually decrease as construction advances higher, the rate of decrease being roughly the same at all levels. Up to a height rather more than midway between the second and third ambulatories the stress will, at some stage, change to tension: indeed at the very foot it will be tension almost from the start. The final value of the tension at the foot will be 0·48 pr_1, which may be compared with 0·99pr_1 for the corresponding hemispherical shell. At the level of the change to brick it will be 0·25pr_1, at that of the timber tie 0·21pr_1 and at that of the second ambulatory, where the herringbone starts, 0·12pr_1. Taking pr_1 as roughly 0·7 N/mm^2, these maximum tensions are, respectively, about 0·34, 0·18, 0·15 and 0·08 N/mm^2 and, if construction proceeds at a uniform rate and takes 14 years, they will develop from the initial compressions by

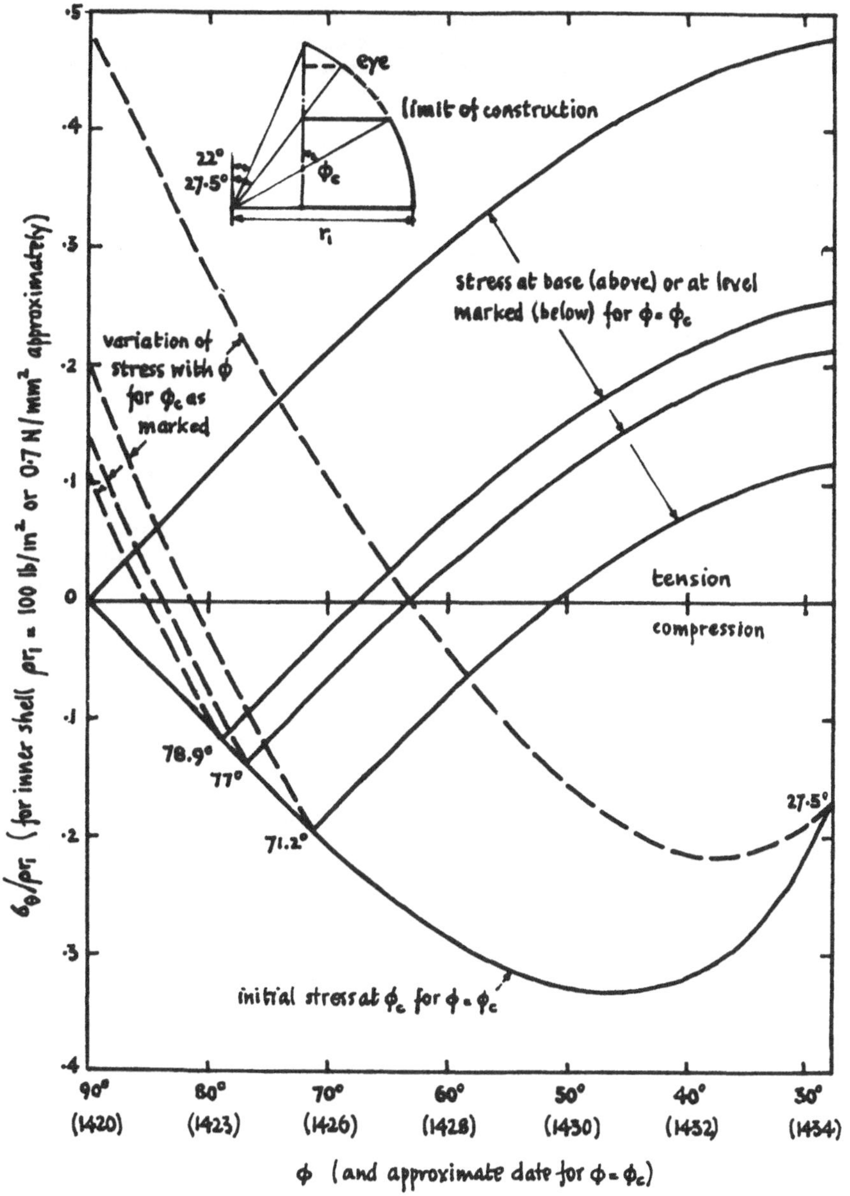

Fig. 5.

The development of hoop stresses $\sigma\theta$ (expressed as $\sigma\theta/\rho r_1$) in a simple statically determinate analogue of the inner shell of the dome.

(*Author*)

changes of about 0·025 N/mm² each year. (With a supporting centering the stress changes would be indeterminate and would include an unpredictable change at the time of decentering.)

The highly indeterminate actual structure could be considered, conservatively, as if it contained this hypothetical shell wholly inscribed within its own inner shell, as its only active element on which the rest was merely a dead weight. On this basis, and making some allowances for the reductions in thickness and density in the upper parts, and for the fact that much of the lower parts of the outer shell and the ribs could be properly considered as a direct vertical load on the drum, the maximum hoop tensions might be as high as five times those calculated. It is nearer the truth, though, to consider the full widths of both shells as polygonal domes interconnected by and subject to some external load through the ribs, the horizontal arches in the upper part being effectively reinforcements of the outer shell. From this point of view the structure is much more efficiently designed and, ignoring possible local concentrations, the maximum tensions will be perhaps only twice the calculated ones though greater, at any level, in the outer shell than in the inner one because of its greater radius of curvature.

These stresses may be compared with possible ultimate unit tensile strengths of the order of 5 and 0·5 N/mm² for the stone and brickwork respectively, the former taking no account of the weakening effect of the joints and the latter referring primarily to the horizontally coursed brickwork below the start of the herring bone. In the completed structure the frictional forces on the bed joints of the stone should have been sufficient, if all the joints are well lapped, to develop an overall strength about half that of the individual blocks. The final stresses seem, therefore, to be within the ultimate capacities of the materials, even ignoring the cramps and the continuous iron and timber ties. These will have assisted, however, in resisting the lesser hoop tensions at earlier stages of construction. Taking account also of the very slow rate of construction, which would allow the mortar of the brickwork to develop its own bond and tensile strengths before being subjected to tensile stress, it can be further concluded that the strengths should have been adequate at all intermediate stages of construction. These conclusions ignore the effects of shrinkage and thermal strains, but it is only when movement is restrained that such strains give rise to stress. Restraints should have been slight during construction and will always have been minimised by the interpolation of the drum between the dome and its more rigid substructure.

Turning to present observable fact, however, there are today conspicuous cracks on three of the diagonally aligned faces of the octagon.[1] Each extends from below the drum to near the eye, initially almost on the vertical centreline but subsequently tending to follow the inclined bands of vertical bricks as might be expected. (Plate XIV (*b*)). If these cracks developed with or before the completion of construction they would show the conclusions to be mistaken. Did they do so? The documentary evidence suggests not. Considerable concern was expressed about the cracks in the seventeenth century,[2] but prior to that there is no mention of them despite numerous references to damage to the lantern by lightning.[3] This silence may be contrasted also with the immediate concern and almost panic remedial measures provoked in 1366 by the discovery of cracks in some recently completed aisle vaults.[4] If the cracks do indeed date only from about 200 years after the completion of construction, the probable immediate cause was thermal expansion. Contributory factors could have been deterioration of the masonry of the outer shell, of the timber tie, and perhaps also the swelling of embedded iron cramps and chains, all as a result of ingress of water and general dampness.[5] Once a crack had formed through the thickness of either shell it would, of course, tend

[1] Illustrated in *Rilievi e studi,* Plates V–VII.

[2] Guasti, *Cupola,* docs. 382–391, pp. 169–83. See also Giovanni Poleni, *Memorie istoriche della gran cupola del Tempio Vaticano,* Padua, 1748, 101–10; *Rilievi e studi,* 18–24.

[3] Guasti, *Cupola,* docs. 337–339, pp. 117–21. See also *Rilievi e studi,* 16–20.

[4] Guasti, *S.M.F.,* docs. 143–149, pp. 168–74.

[5] Guasti, *Cupola,* docs. 382–383, pp. 169–71.

to spread fairly rapidly as a new pattern of equilibrium was established with some of the outward meridional forces generated in the upper part being resisted by inward horizontal forces at the base rather than by hoop tensions as previously.

Origins and Development of the Design

It may be assumed that the octagonal dome of the nearby baptistery of S. Giovanni, constructed in the late twelfth century in a remodelling of an earlier structure (Plate XV) was the immediate prototype for the cathedral dome.[1] Its basic structure is, likewise, a pointed octagonal shell with a central eye closed by a lantern. This shell springs from what is, in effect, the inner wall of a double-walled octagonal drum. To span the greater width of the outer wall and provide a unified external aspect, there is also an outer shell carried, very much in the manner proposed in the 1420 specification, by small barrel vaults spanning between more-or-less radial walls above the extrados of the inner shell (Plate XVI (*a*)). There is also, at a level corresponding roughly to that in the cathedral dome, a circumferential timber chain between the two shells.

Even apart from a 70 per cent increase in scale, the cathedral dome as built is nevertheless a fundamentally different structure. The baptistery structure is not a true double dome but rather a pyramidal roof of stone superimposed on, partly carried by, and near the top merging with, the inner shell, and the walls which carry this roof, though analogous to the ribs in the cathedral dome, do not appear to have been conceived with any intention of integrating the two shells into a single structure. It is possible to identify any other prototypes?

Manetti and Vasari both place considerable emphasis on the years which Brunelleschi is said to have spent in Rome between his failure in the competition for new bronze doors for the baptistery and his preparation of models for the dome.[2] Today it is difficult to assess precisely what he could have learnt there from studies of ancient remains, since much has since been lost or made less revealing by subsequent adaption or restoration. Something is known, however, about the extent of the losses and it seems unlikely that he could have seen any constructional techniques radically different from those exhibited by the remains we know or know of now. If the basic structural characteristics of the cathedral dome are taken to be the double-shell form with intermediate ribs, the horizontal arches that reinforce the upper part of the outer shell, the reinforcements of both shells by iron cramps and stone, timber and iron chains, and the construction without supporting centering using brickwork in a herringbone bond, the only close precedents to be found in Roman construction are for the use of stone chains and iron cramps.[3] Brunelleschi's special bricks are not unlike Roman ones but the use made of them in Rome was very different. All wide-spanning Roman vaults were constructed of concrete cast in horizontal layers and are of the very massive form best exemplified by the Pantheon dome—essentially hemispherical internally but externally much flatter, and composed only of a single shell, although its immensely thick base is lightened by leaving voids within it.[4] All must have been cast on supporting centering, and the nearest approach to dispensing with this is the adoption of various devices to lessen the load upon it while the concrete is still green. In later structures these devices included an increasing use of embedded ribs of brick and in one instance—the so-called

---

[1] The parallels drawn in this concluding section of the paper are based largely on personal observation, but references to the most relevant literature are given below. For the baptistery see E. Isabelle, *Les édifices circulaires et les dômes*, Paris, 1855, Plate 53; Josef Durm, *Die Domkuppel in Florenz und die Kuppel der Peterskirche in Rom*, Berlin, 1887, Plate II (illustrating also the rather similar dome in Cremona) and A. Nardini Despotti Mospignotti, *Il Duomo di San Giovanni*, Florence, 1902.

[2] Frey, pp. 73–5; Toesca, pp. 18–21; Vasari, *op cit.*, Part II, 300–1 (Gaunt, pp. 274–5).

[3] See, for instance, M. E. Blake, *Ancient Roman construction in Italy* and *Roman construction in Italy*, Carnegie Institution of Washington, 1947 and 1959, *passim;* G. Lugli, *La tecnica edilizia romana*, Rome, 1957, I, 235–42.

[4] Lugli, *op. cit.*, pp. 663–93; William Macdonald, *The architecture of the Roman empire*, Yale University Press, 1965, *passim;* Licht, *op. cit.*

temple of Minerva Medica (Plate XVI (*b*))—these ribs have a remarkable superficial resemblance to the ribs and horizontal arches of Brunelleschi's dome.[1] It is even possible that, in the fifteenth century, some had been left free standing by the fall of the concrete infill.[2] In general, though, these ribs became an integral part of the concrete mass and always appear so today. If, in terms of structural function, they have a counterpart in Florence, it is to be found in the inclined bands of vertical bricks of the herringbone bond.[3]

Looking further afield, however, it is possible to find fairly close precedents for most of the characteristics, some even during the Roman period. The empire was widespread and, in the Balkans and Asia Minor, brick rather than concrete, was commonly used for vaults.[4] Here, continuing probably a very ancient tradition seen for instance in the store rooms of the Ramesseum at Thebes (Plate XVII (*a*)),[5] there are examples of domes or vaults constructed without centering at Spalato, (Plate XVII(b)) Salonika, Aspendos and in the Theodosian walls of Constantinople.[6] None of these exhibits a herringbone bond, but each exhibits a variant of the principle of building up the vault by means of a succession of short arches pitched as flatly as possible. The tradition continued and is very widespread in both Byzantine and Islamic architecture, still being characteristic of the latter where brick has remained the normal material of construction and timber is scarce.[7] With the use of brick is associated also the adoption of a more uniform thickness for all vaults, and usually the need for a separate weathering skin. This skin took various forms but, as early as the eleventh century, the true double dome (with pointed profile) emerged in Persia.[8] By the early fourteenth century it had

---

[1] G. Giovannoni, *La sala termale della Villa Liciniana e le cupole romane,* Rome, 1904; Guido Caraffa, *La cupola della sala decagona degli Horti Liciniani,* Rome, 1944.

[2] This possibility is suggested by a number of drawings etchings, etc., dating from the early sixteenth to the mid-nineteenth centuries. I have found so far a total of eight such representations by (in order of date) G. B. del Porto, M. van Heemskerk, J. d. A. Frankaert, G. B. Piranesi (three), F. J. B. Kobell and E. Isabelle. The best is that by Kobell. This is reproduced in Richard Krautheimer, *Early Christian and Byzantine architecture,* Penguin Books, 1965, Plate 80A, and in my forthcoming book, *Developments in structural form.* For the last see Isabelle, *op. cit.,* Plate 23.

[3] I do not, however, consider that a drawing attributed to G. Guidetti (Uffizi, 1330) and showing a concrete semidome with two intersecting sets of inclined bands of vertical bricks embedded in it represents, as Sanpaolesi suggested, a Roman precedent for the Florentine bond. It seems much more likely, for various reasons, to be a rather ill-considered reinterpretation by the artist of the lozenge-shaped coffering of the semidomes of the Temple of Venus and Rome, in terms of Brunelleschi's system. Nor can I find anything relevant in the sketchbook of Guiliano da Sangallo (Vatican codice Barberiniano latino 4424) referred to in this connection by G. C. Mars, *Brickwork in Italy,* Chicago, 1925, 23.

[4] J. B. Ward-Perkins, "Notes on the structure and building methods of early Byzantine architecture" in *The Great Palace of the Byzantine Emperors. Second Report,* Walker Trust, Edinburgh, 1958, 52–104; William Macdonald, "Some implications of later Roman construction," *Journal of the Society of Architectural Historians,* XVII (1958), 2–8.

[5] The Egyptian process of construction is described in Somers Clarke and R. Englebach, *Ancient Egyptian masonry,* Oxford University Presss, 1930, 181–3.

[6] G. Niemann, *Der Palast Diokletians in Spalato,* Vienna, 1910, 75–6, and Plate XIII; E. Herbard and J. Zeiller, *Spalato, le palais de Dioclétien,* Paris, 1912, 89–93, and Plates X and XII; J. B. Ward-Perkins, *op. cit.,* 89–95; idem, "The aqueduct of Aspendos," *Papers of the British School at Rome,* **23** (1955) 115–23.

[7] See, for instance, André Godard, "Voutes Iraniennes: les cupoles," *Athàr-é Iran,* IV (1949) 259–325. To avoid possible confusion it seems desirable, however, to distinguish between different principles of centerless construction—breaking up the continuous horizontal rings into shorter elements that can be constructed independently in succession, pitching the bricks at a flatter angle than the radial one, obtaining adhesion to what has already been completed by the use of a gypsum of fast-setting cement mortar, and mechanically keying each brick or unit to the existing work. These principles are not mutually exclusive, but we are concerned here largely with the first. The method used at the cathedral of St John the Divine was based on the third. See also G. R. Collins, "The transfer of thin masonry vaulting from Spain to America," *Journal of the Society of Architectural Historians,* XXXVII (1968) 176–210. The well-known Italian system using terracotta tubes was based on the last. See, for instance, Paolo Verzone, "Le cupole di tubi fittili nel V e VI secolo in Italia," *Atti del I Congresso nazionale di storia dell'architecttura,* Florence (1936), 7-11; and G. Angelis d'Ossat, Nuovi dati sulle volte costruite con vasi fittili', *Palladio,* **5** (1941), 241–51.

[8] David Stronach and T. C. Young, "Three Seljuk tomb towers," *Iran,* IV (1966), 1–20.

reached a truly monumental form in structures like the mausoleum of Oljeitu at Sultaniya.[1] Here, as in Florence, the inner shell is stepped back at a number of points to reduce its thickness and weight and it is covered by a lighter outer shell of uniform thickness carried by the inner one by means of radial and circumferential ribs. Circumferential timber ties are usually built into the bases of these domes. Finally, there is at least one fairly close parallel to the circumferential iron chains set above "chains" of stone blocks cramped together in the western exedrae of the church of St. Sophia in Constantinople.[2] The chains, composed there of long iron bars, probably date from the mid-sixth century.

I cannot say how many of these latter precedents were known to Brunelleschi. At best, however, they could only have helped to point the way towards a possible solution of the problems posed by the decisions of 1367. Even a synthesis of known forms and techniques on an accustomed scale (as seen elsewhere in Brunelleschi's architecture) demands considerable imagination. If, for example, the dome of the mausoleum of Oljeitu was a second prototype alongside the baptistery dome, it would still have been a major achievment merely to combine the two forms in the manner seen in the cathedral without any change in scale (the mausoleum dome being, like all Persian domes, circular in plan) and to devise a method of construction that did not rely on centering for temporary support. To simultaneously almost double the scale would have introduced problems of an altogether new order.

That the final design and the full elaboration of the technique of erection were not arrived at without careful preparation and continued thought and experiment is shown both by what is known about preparatory studies and by the progressive evolution of the design as construction proceeded. Apart from the masonry models of the dome itself, Brunelleschi tried out the principle of vaulting without centering in a small cupola in a nearby church.[3] Later he made further models of the circumferential chains and other details and Gelli records that, for this purpose, he had a level site prepared large enough to work on the design at full scale.[4]

The way in which the design evolved shows that he must have acquired, perhaps partly through the study of Roman remains,[5] a keen awareness of the general pattern of stresses within a circular dome—of the tensile hoop stresses in the lower part, the corresponding compressive ones above, the way in which the latter made it possible to leave an open eye of any diameter at the top, and the outward thrusts which would be exerted at the base if the tensile hoop stresses led to radial cracking.[6]

---

[1] The best account of this development is in D. N. Wilbur, *The architecture of Islamic Iran*, Princeton, 1955, 61–7. For the mausoleum of Oljeitu see also André Godard, "The mausoleum of Oljeitu at Sultaniya" in A. U. Pope, *A survey of Persian art*, London, 1938–39, II, 1103–18. By the early fifteenth century development had proceeded to the point where the outer shell became bulbous, as in the Gur-i-Mir at Samarkand, inspired by perhaps earlier wooden double domes.

[2] R. L. Van Nice, *Saint Sophia in Istanbul, an architectural survey*, Washington (1966) Plate 21.

[3] Frey, pp. 87–8; Toesca, pp. 37; Vasari, *op. cit.*, Part II, p. 310.

[4] G. B. Gelli, *Vita d'artisti di Giovanni Battista Gelli* in *Archivio storico Italiano*, 5th ser., XVII (1896) 54. "*Pippo . . . fecie fare uno ispianata in su renaio d'Arno circha d'un mezo miglio per ogni verso et quivi disegniata in terra questa cupola quanto ella haveva a esser grande appunto, et fatto un punto nel mezo disegnio tutte le pietre . . . che tiravano . . . et colta la misura della grandeza et qualita loro che ve n'era di varie sorte che incastravono l'una nell' altra, ne facie alchuni modegli di rape et mettendovi la misura comincio a farla lavorare di quella maniera a scharppelini et con quelle comincio a voltare detta cupola . . . .*"

[5] Particularly from the characteristic patterns of radial cracks extending up and down from the springings of nearly all large domes or semidomes, but stopping short of any open eye at the top. Today these cracks are most apparent in some of the remains of the Baths of Trajan, having mostly been patched or covered with stucco elsewhere. They are equally pronounced in the Pantheon though, as can be seen in the hidden chambers just above the springing line of the dome. Alberto Terenzio, "La restauration du Pantheon de Rome," *La conservation des monuments d'art et d'histoire*, Paris, 1933, 280–5, gives a detailed drawing of these cracks as observed during a restoration (Plate XXVI).

[6] This insight may be contrasted with some of the ideas expressed much later when a proposal to add iron chains was being debated. See Alessandro Cecchini, "Due discorsi sopra la cupola di S. Maria del Fiore" in G. B. Nelli, *op. cit.*, pp. 77–103; and the summary of the debate in Poleni, *loc. cit.*

The use of stronger materials at the base and of various chains were obviously intended to prevent such cracking and outward thrusts. The thickness of the inner shell was, almost certainly, determined so that it would contain within it the circular one analysed above.[1] Initially the outer shell seems to have been regarded structurally as little more than an unavoidable extra load on the inner one, but the replacement of the small barrel vaults by the horizontal arches of the 1426 amendment seems to have been directed at making it more self supporting. At the same time an attempt was made to bind the two shells together for mutual support in much the same way as parallel arch ribs and even ribs of different radii were interconnected in early iron bridges and with equally little understanding of the strains that might result.[2]

Brunelleschi's greatest problem would have been to match the tensile strengths of the structure to the stresses generated. In tackling it, the models and small-scale tests would have been of little assistance without some intuitive recognition that the stresses would increase in proportion to the radius of curvature even if geometrical similarity was maintained.[3] A dual approach seems to have been adopted. On the one hand the stresses were minimised by reductions in thickness and density in the upper parts and by the adoption of a pointed profile. On the other care was taken to provide as much strength as possible where it was most needed and a possibility was left open of adding additional hoop reinforcement here if, as construction advanced, it appeared to be necessary. This last, in the light of events, I take to be the significance of the reference, in 1420, to unexecuted iron chains above the timber ones. As has been shown, tensile stress would develop only slowly at the level of the one executed timber chain and an iron chain here could serve a useful purpose, if at all, only at a fairly late stage of construction. The individual timbers would have had considerable tensile strength—probably in the range 2–4 MN depending on the extent of defects and on moisture content—but the joints, if the present ones can be taken as a guide, would have been much weaker. It seems quite likely that, since the chain remained visible and accessible and there would have been no means either of determining its effective strength or calculating the strength that it ought to have had, it was simply kept under observation as construction proceeded above it with the intention of adding the iron chain if and when signs of excessive strain appeared.[4] The tightness of the bolts would have been one rough index of the load transmitted. If this was the case, the chain might indeed be regarded as having served as "a tool of observation and test," as Prager suggested.[5]

Taking into account also those aspects of the total construction process which have not been considered here, the final impression is of a highly systematic and unusually comprehensive empirical

---

[1] Indirect confirmation of this is provided both by the stated reason for introducing the horizontal-arch reinforcements to the outer shell in 1426 and by the opening sentence of the passage in Alberti, *De re aedificatoria* already referred to: "*Angularem quoque testudinem sphericam modo per eius istius crassitudinem rectam sphericam interstruas, poteris attollere nullis armamentis.*"

[2] Here, again, there is an echo in this passage in Alberti, "*Sed istic nexura potissimum opus est; qua huius imbecillae partes partibus illius firmioribus arctissime illigentur,*" though there is no direct reference to inner and outer shells as in Leoni's translation (*loc. cit.*). It is possible that, with a less rigid interconnection, the inner one would not have cracked.

[3] Since Brunelleschi left no evidence of his ideas other than his buildings, it is possible only to speculate about any such recognition. Piero Sanpaolesi, "Ipotesi sulle conoscenze matematiche statiche e meccaniche del Brunelleschi," *Belle Arti* (1951), 25–54, discusses Brunelleschi's possible contacts with current theoretic thinking and R. J. Mainstone, "Structural theory and design before 1742," *Architectural Review*, CXLIII (1968), 303–10, gives a more general discussion.

[4] Prager, pp. 495–6. I do not wholly agree with Prager's reasoning.

[5] It is interesting to note, in this connection, that telltales of bronze and marble were inserted at numerous points in 1695 when considering whether the radial cracks called for the insertion of new chains and that, when such chains were proposed, it was stipulated that the bars coming from the foundry should be rigorously tested. No strength criteria were stated, however. Nor was any analysis of the structural action given to justify the proposals. (Guasti, *Cupola*, doc. 391, pp. 177–83.)

approach which it is difficult to parallel until much more recent times.[1] Some reference has already been made to the impact made on contemporaries. Most obviously it was simply that of the demonstration of a new or reawakened human power. But Brunelleschi's achievement was so woven into the newly emerging patterns of Renaissance thought that the full impact was both wide and profound.[2]

Discussion and Correspondence

In the discussion and in correspondence subsequently arising from the paper Mr. Mainstone amplified some of the points made about the method of construction, the uses of tensile reinforcement, the structural behaviour of the dome, and Brunelleschi's possible understanding of this.

The use of weighted ropes tied back behind the forward edge of the working face and hung over the bricks on this edge as they were set in place was suggested as an alternative method that might have been adopted to retain the bricks in place while the mortar hardened. This alternative to fixed or temporary formwork had been discussed by John Fitchen in his book *The construction of Gothic Cathedrals,* Oxford (1961), 180–7. Fitchen was referring there to the relatively short courses of the webs of Gothic ribbed vaults and based his discussion on a description of the method published in the *Journal of the Royal Institution of Great Britain* and elsewhere in 1831. It was not, however, the method that Mr. Mainstone had observed with Professor Fitchen when they had together inspected the construction of the nave vaults of the neo-Gothic cathedral of St. Peter and St. Paul in Washington, D.C., by courtesy of the resident architect Mr. Howard Trevillian, in 1966. There was no evidence at all for it having been used by Brunelleschi and it would be difficult to account for the upright bricks of the herringbone bond if it had been used. He thought that one of the procedures he had suggested, or something similar, was more likely. He admitted, nevertheless, in relation to his second suggestion, that it would be stretching the meaning of *ansas* in the passage by Alberti to read it as referring directly to the upright bricks. As had been pointed out, Alberti was not even referring directly to the cathedral dome. What seemed most significant about this passage was that it did fairly clearly describe at least the principle of making a limited use of light temporary formwork tied back to the part of the structure just completed.

Nothing was known of the form of the iron chains embedded in the dome of S. Maria del Fiore. The selection of documents published by Guasti did not even include any reference to their having actually been set in place. Only a careful study of the full records (which do include sufficiently full accounts of the purchase of materials to settle such matters) could show whether they were so set and give some indication of their likely size. Such study had been rendered difficult by the damage done to the records by the flood in 1966 and by their removal from Florence for repair. Professor Saalman would, however, be publishing in due course a further selection of those that now seemed important. It seemed to Mr. Mainstone more likely that these chains, if they were set in place, followed the sides of the octagon than that they were continuously curved like the wooden chain. If they had been curved in this way, he would have expected the associated stone chains to have been similarly curved, which seemed most unlikely. Certainly they did not appear at the surface in

[1] The first parallel that comes to mind is Smeaton's reconstruction of the Eddystone Lighthouse in 1756–59. See John Smeaton, *A narrative of the building and a description of the construction of the Edystone Lighthouse with stone,* London, 1793 (second edition). Smeaton, of course, went further than Brunelleschi in conducting preliminary experiments akin to the sort of preparatory testing that might be undertaken today. Brunelleschi relied more on observation in the course of construction as a guide to future action (*perche nel murare la praticha insegnera quello che ss'ara a seguire*), but he seems to have done so in a much more deliberate manner than had been the case previously.

[2] The direct influence within the limited field of dome construction is, of course, to be seen in a long succession of double and multiple domes from that of St Peter's onwards. As a tailpiece, we may note that among the documents published by Guasti (*Cupola,* doc. 350, p. 130) is a free pass into the dome for Michelangelo and two associates, presumably in connection with his studies for the dome of St. Peter's.

the places where one would expect to see them if they were curved. The timber chain was probably envisaged more in the same way as the arch-like reinforcements of the outer shell (*c* in Fig. 1), and its more-or-less circular plan form had the practical advantages of standardising and considerably simplifying the jointing of the 24 lengths of timber.

Still less was known about iron chains or any other kind of tensile reinforcement that might perhaps exist in the drum, since there is not even, among the published documents, any specification calling for such reinforcement. He would expect to find some cramping together of the masonry but nothing further. Certainly Brunelleschi had been in no position to call for any chains there. He clearly attempted, moreover, to contain the potential outward thrusts of the dome within its own masonry rather than to rely on any tensile reinforcement of the drum. A final answer must again await a study of the full records.

With adequate tensile reinforcement of the dome itself, there would, of course, have been no need for such reinforcement of the drum. Its presence would then, he thought, have been advantageous in allowing the dome some freedom to "breathe" without the sort of restraint that could lead to cracking.

His dismissal of earthquakes as a serious threat to the stability of the dome was based on a considered judgement of a good deal of evidence accumulated in recent years about the seismicity of different areas and the nature and structural effects of earthquakes. Florence had indeed experienced earthquakes, in 1695 for instance, but never, as far as was known, of the degree of severity that had been experienced further south. Also, as had been pointed out, it was lightning rather than earthquakes that the records mentioned as a cause of damage. As far as structural effects were concerned, a dome was, in itself, one of the forms best fitted to stand up to the momentary shifts in the effective strength and direction of gravity to which an earthquake might be likened. When collapses occurred as a result of these momentary shifts, they were usually attributable to failures of the supports rather than primary failures of the domes themselves. The risk of such failure would have been enhanced by the massive proportions given to the cathedral dome. But this was the only way in which he thought earthquake loading could have been important—and he had been considering the dome only and not the whole structure in this part of his structural assessment.

In support of his belief that the cracks at present seen in the dome did not develop until well after the completion of construction, he could offer no evidence beyond that given in the paper. It might be worth commenting briefly on the positions of the cracks, however. They all occurred over (and extended into) the very substantial diagonally aligned supports (Plate 2) rather than over the principal arches spanning between these supports. Since it was these arches that were most likely to spread, this might, at first sight, appear paradoxical. The explanation, he thought, was that when an arch spreads, other than by slipping of its voussoirs, it is over the haunches that the spandrels must crack to accommodate themselves to the spread, and not over the crown. What had probably happened here was that the cracks had been somewhat further displaced from the crowns of the principal arches on account of the weakening effects of the circular windows in the drum. Each crack passed, as might be expected, through one of these windows.

Finally, his remarks about Brunelleschi's "keen awareness of the general pattern of stresses within a circular dome" were not intended to imply the sort of understanding on which the analysis given in the Paper was based. Brunelleschi almost certainly had no clear abstract concept of a "stress" at all. Even Leonardo was still groping his way towards such a concept. The "awareness" referred to was more a general awareness of the two characteristic actions of a domical structure—one of which we called "compression" and the other "tension", but which Brunelleschi would have thought of in terms of their characteristic manifestations of crushing and pulling apart or tearing. It was a matter of understanding where, and roughly in what direction, each was likely to arise.

PLATE XII

S. Maria del Fiore, Florence. (*Author*).

PLATE XIII

(*a*) Longitudinal Section.

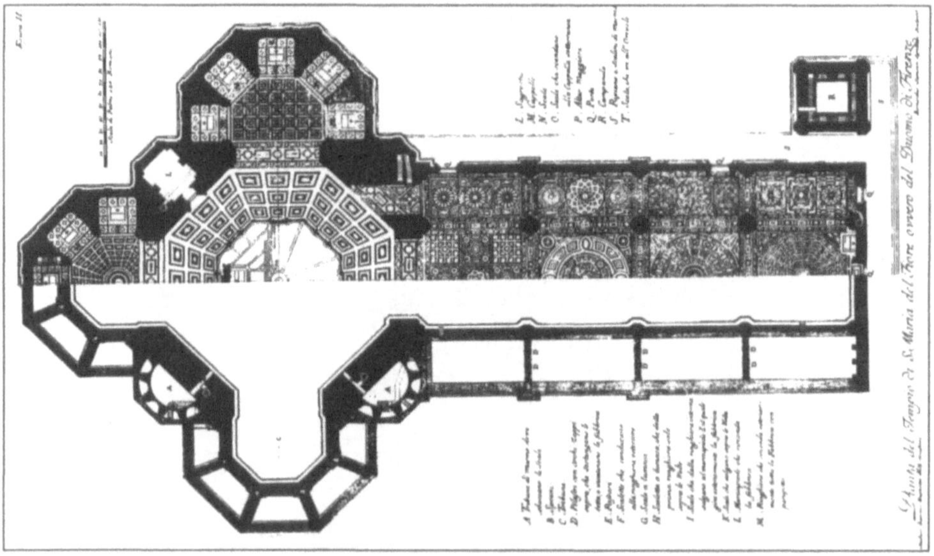

(*b*) Half plans at floor level (above) and at the level of the springings of the principal arches (below).

S. MARIA DEL FIORE (*Sgrilli*).

PLATE XIV

(*a*) View between the two shells from the point marked "b" in Fig. 1 looking into the passageway cut through one of the corner ribs.

(*b*) Detail of the crack running up one of the diagonal faces of the dome.

(*c*) View between the two shells of the dome looking upwards towards the point marked "b" in Fig. 1.

S. MARIA DEL FIORE (*Author*).

PLATE XV

Baptistery, Florence. Section (*Isabelle*).

PLATE XVI

(*a*) Baptistery, Florence: looking up between the two shells at the middle of one of the sides (*Author*).

(*b*) Temple of Minerva Medica, Rome. Detail of the dome (*Author*).

PLATE XVII

(*a*) Ramasseum, Thebes: brick vault of a storeroom (*Author*).

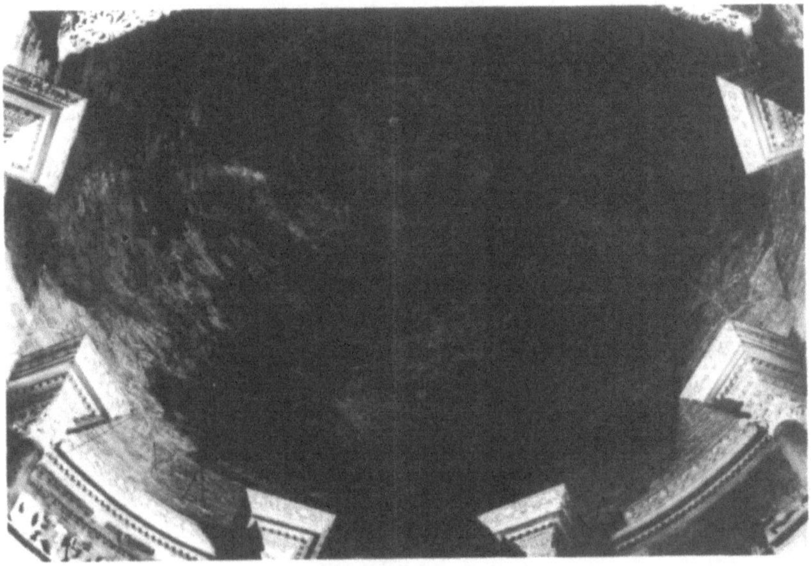

(*b*) Mausoleum of Diocletian, Spalato: detail of the dome (*Author*).

XVII

Brunelleschi's Dome

Perhaps it is in the nature of man's greatest achievements to keep something hidden from us; and there can be no doubt about the greatness of Brunelleschi's achievement in building the dome of S Maria del Fiore—the cathedral of Florence—between 1420 and 1436. After more than 500 years its towering mass still dominates the whole city and its surroundings in a unique way . When it was built its structural daring was unmatched, and there is ample evidence from places like Milan that structural expertise generally was at a fairly low ebb. To contemporary Florentines the achievement must also have been given added piquancy by a realisation that a less successful outcome could have entailed a loss of face exceeding even that suffered by the Sienese in the previous century when they were forced to abandon grandiose plans for the enlargement of their cathedral. The immediate jubilant reaction when success was finally assured was epitomised by the newly returned exile, Alberti, in the well-known words of the dedicatory prologue to the Italian version of his treatise 'On painting':

'Who is so stubborn or so jealous as not to praise Pippo the architect when he sees here so great a structure, rising above the heavens, broad enough to cover with its shade all the people of Tuscany, made without any aid of centering or other quantity of timber, an achievement which, if I am not mistaken, was believed to be impossible in our time even as, among the ancients, it may well have been unknown and unheard of'.[1]

After Brunelleschi's death in 1446 this reaction was officially endorsed in the resolution of the Cathedral Building Committee that sanctioned his burial within the church and, again, in the epitaph by Carlo Marsuppini, the State Chancellor, inscribed on a marble slab set on the south wall of the nave near his grave.[2]

Today our understanding is challenged on several levels. What sort of structure, precisely, is the dome? How was it built 'without any aid of centering'? By what steps did the design evolve, and with what objectives and intentions? What were the sources of the ideas embodied in it? And to what extent did others, besides Brunelleschi, contribute?

To help answer the first two questions and, in part, the third, we have the dome itself almost exactly as Brunelleschi left it,[3] a copious documentation of expenses and of formal resolutions by the building committee,[4] and the insights now offered by modern structural theory into such things as the problems faced and the likely consequences of proceeding in different ways. In the dome itself, much that is of interest still lies behind the visible surfaces. Until it is laid bare (or perhaps traced by electronic probes) we can only infer what is there from what we can see and from the relevant documents. The documents include, most revealingly, a long specification of what was to be built

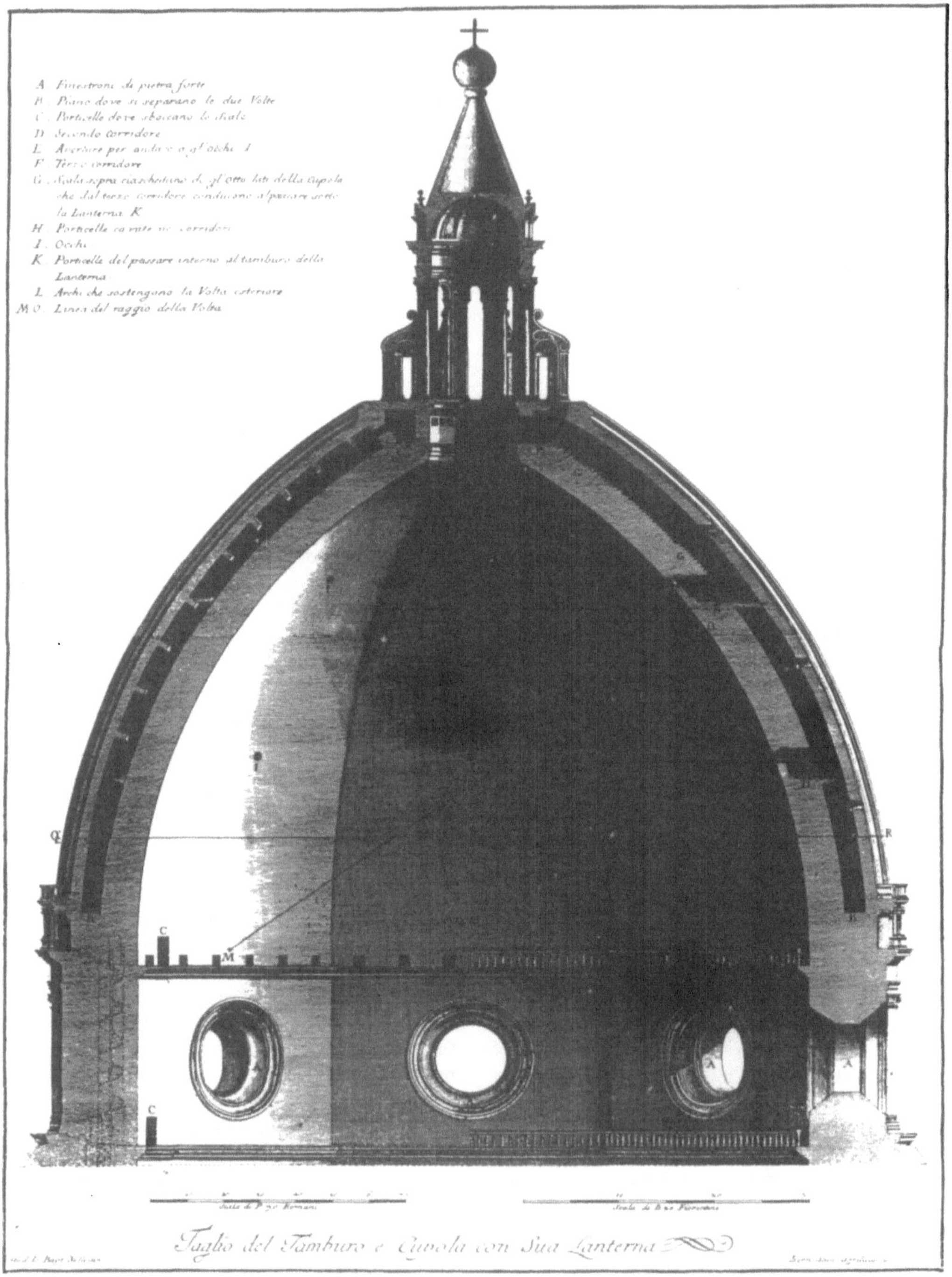

I, cross-sections of the dome near a corner rib (left) and through the centre of a side (right). (Sgrilli.)

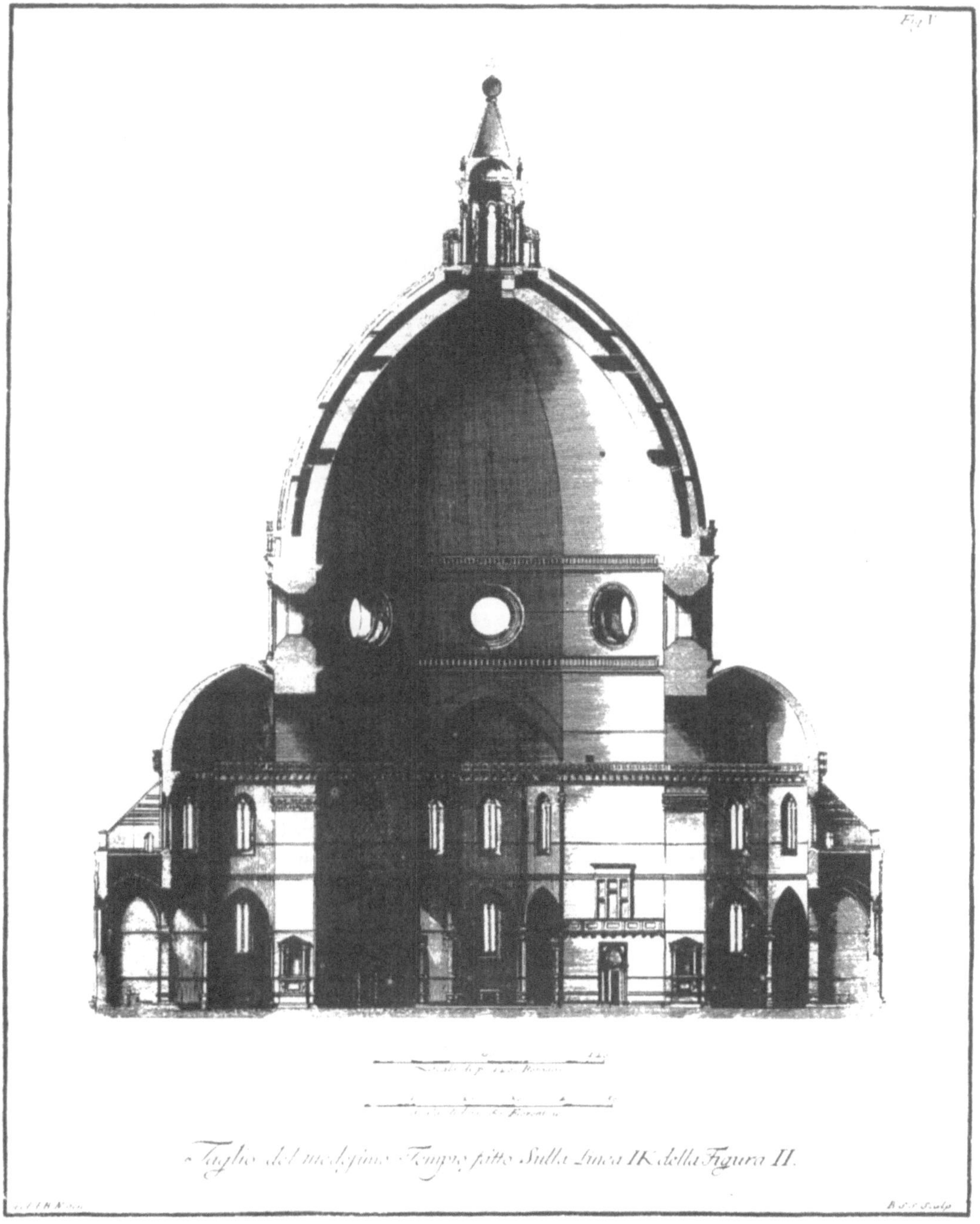

2, north-south section of the church. (Sgrilli.) Two of the sliced-off octagonal vaults over the main eastern arms can be seen below the dome at the left and (the last to be constructed) at the right.

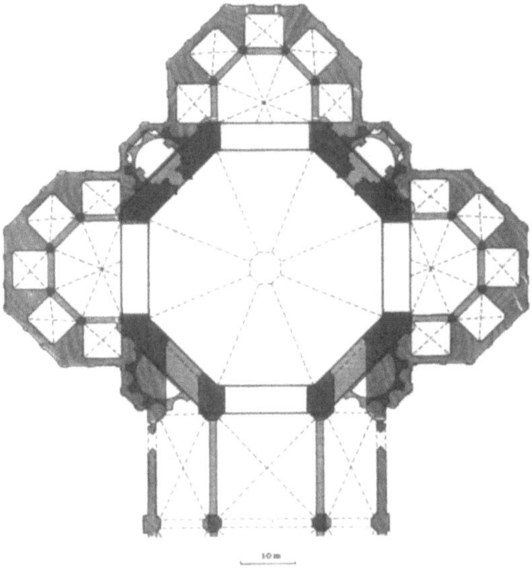

3, superimposed plans of the eastern part of the church showing the walls and piers at ground level and those at the level of the springings of the main arches and the main vaults of the eastern arms. The drum of the dome is indicated in bold outline only.
4, the dome and eastern-arm vaults as represented in a fresco by Andrea Bonaiuti in the Spanish Chapel of S Maria Novella. The campanile has been moved, for compositional reasons, to the south east of the church (from the position it already occupied at the south west), and similar artistic licence may explain the omission of the drum immediately below the dome.

4

appended to a resolution of 30 July 1420 that formally approved it, and two variations of this specification approved in 1422 and 1426.[5] But, as far as is now known, no working drawings or preparatory sketches have survived; all the models that are documented as having been made to explore different possibilities or exemplify what was to be built have been destroyed; and no detailed formal record was ever made of the construction procedure. We must therefore do our best to infer this procedure largely from the other evidence and insights, and to trace the development of the design largely through the recorded patterns of design activity leading up to the 1420 specification and the subsequent variations.

When we turn to objectives, intentions, and sources of ideas, we face more fundamental difficulties, since they were never open to direct observation and probably went completely unrecorded at the time. The difficulties are compounded by the lack of preparatory sketches and of any similarly relevant material. Indeed it seems highly probable that Brunelleschi deliberately played his hand very close to his chest to avoid having his ideas appropriated by rivals, and that it was only after the dome was completed that he discussed some of them with people like Alberti and his first biographer, Manetti. Here, therefore, we must frankly speculate or turn back. By putting all the evidence together though, I believe that we can come a little closer to the truth than hitherto.

As for contributions by others, we have to consider chiefly Ghiberti and Battista d'Antonio who were appointed with Brunelleschi in 1420 to superintend the construction. There seems to be little doubt that Battista contributed practical experience rather than ideas and that he was, in effect, the clerk of works responsible for the day-to-day running of the job. Ghiberti's role is more problematical. It cannot have been for nothing that he was, at the outset, at least *secundus inter pares*. But as construction proceeded and the real difficulties began he dropped more and more into the background, and it was to Brunelleschi that all recognition was given. It seems reasonable to conclude from this that the successful ideas were largely, if not wholly, Brunelleschi's. If so, a reluctance to disclose them fully to the building committee in advance of construction might well have been a major reason for Ghiberti's co-appointment.

As in an earlier paper,[6] I shall assume in what follows that the ideas were Brunelleschi's—a view consistent with the early biographies, though not derived from them—and shall concentrate on elucidating the ideas themselves and their likely sources. I shall, in particular, develop further the suggestion first advanced in the earlier paper that the central idea was to build the dome as if it was a circular one in spite of its actual octagonal form.[7]

The starting point: problems and possibilities

In 1417, when serious attention was first given to building the dome, only a limited freedom of choice remained.[8] The drum from which it was to rise was essentially complete and the piers and arches on which this stood had been completed many years previously, 2 and 3. Structurally, therefore, it had to accept the support that these provided without overtaxing it. Two of the three sliced-off octagonal vaults over the main eastern arms of the church had also be completed, 2.

Aesthetically, therefore, the dome had to relate both to the octagonal drum and to these vaults. This must have led to a considerable reluctance to adopt externally any form that was not octagonal in plan and that did not closely follow the low pointed profile of the arm vaults. Yet another constraint would have been the brick model of the whole church built and approved in 1367. All subsequent building committees and chief masters had sworn adherence to it, and it is highly probable—on the evidence of closely related works such as the fresco by Andrea Bonaiuti in the Spanish Chapel of S Maria Novella, 4—that it already indicated a form very similar to that finally built.

In spite of the seemingly broad scope of some of the general invitations to contribute ideas that were issued by the building committee in the second half of 1418,[9] it thus seems that what they were really seeking from the start were designs for the detailed realisation of this particular form. Proposals involving any substantial departure from it would have

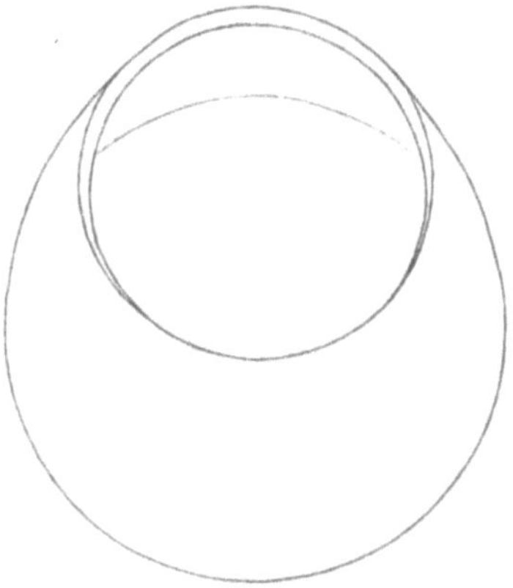

5, the true circular dome is stable without additional support at any stage of construction provided that it is completed ring-by-ring, because the uppermost complete rings act as a key.

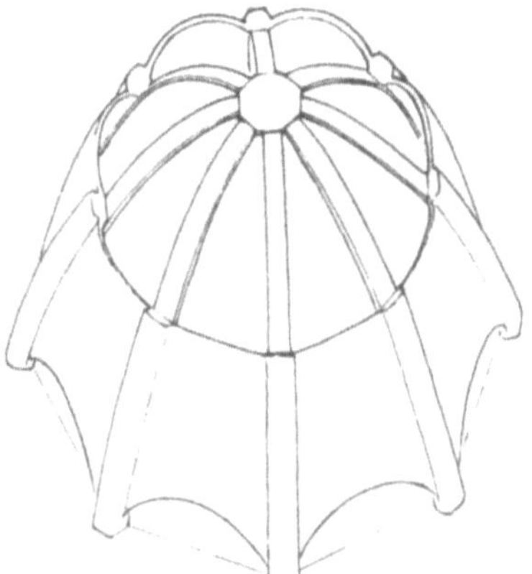

6, the ribbed melon-shaped dome needs temporary support along the lines of the ribs, at least until these are completed.

been seriously considered only as a last resort. Architecturally they would probably have allowed some freedom of choice chiefly in relation to provisions for lighting and access, provision for and detailed design of the lantern, and detailing of the junctions with the drum and of the eight corners rising externally to the lantern. But they would have been chiefly preoccupied with devising a practicable means of building an octagonal dome of unprecedented size, whether with or without centering, and with the design of any necessary scaffolding, hoists, and other machinery.

Though the assumed prior choice of a low pointed profile rather than a semicircular one probably stemmed simply from the precedent established by earlier baptistery vaults in Florence and elsewhere, it did simplify the structural and constructional problems.[10] At the time its advantage was most likely seen chiefly as a constructional one, in that it avoided the need finally to lay the masonry on almost vertical beds. Its further advantage of reducing the tendency of the dome to burst open at the base in the later stages of construction cannot then have been as clearly understood, but was no less real. Indeed, we can now calculate that (in terms of the circumferential tensile stresses necessary to resist bursting) it was roughly halved by the profile actually adopted.[11]

The octagonal form in plan, on the other hand, added to the difficulties of construction and of ensuring stability without overloading the supports. From these points of view a true dome, circular in plan, would have been the ideal choice. Any complete horizontal ring of such a dome is capable of acting as a continuous horizontal arch, 5. At any stage of construction the completed upper rings can thus, in effect, key together the incomplete radially aligned vertical arches of which the dome is also composed and thereby prevent them from falling inwards. Centering is therefore unnecessary.[12] The only temporary support that may be needed is that of an incomplete ring still under construction or of one or two rings immediately below that have not yet gained sufficient strength to act as horizontal arches—and such support can be simply tied back to what has already been built at a slightly lower level. Provided that the masonry of the lower part of the dome can develop enough tensile strength, it can also resist the bursting tendency that develops in the later stages of construction and allow the entire dome to be made much thinner (and much lighter) than if it acted finally just as a ring of vertical arches. This tensile strength is called for only gradually (as a reversal of the initial circumferential

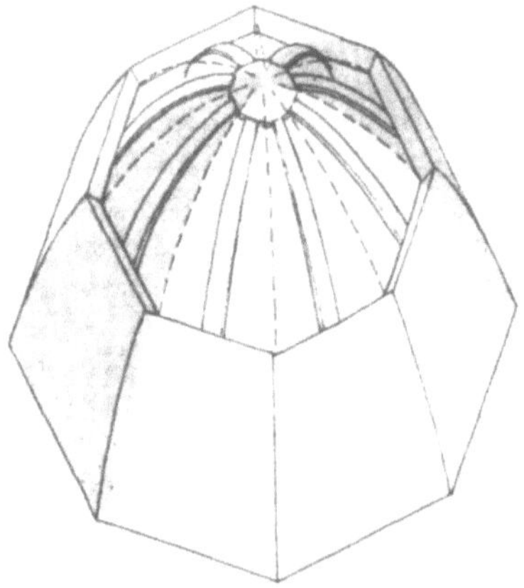

7, the octagonal form is stable, when completed, with thicknesses similar to those that would be required for continuous barrel vaults of the same profiles as each pair of opposite sides. But, like these barrel vaults, it must be supported during construction, particularly at the centres of the sides.

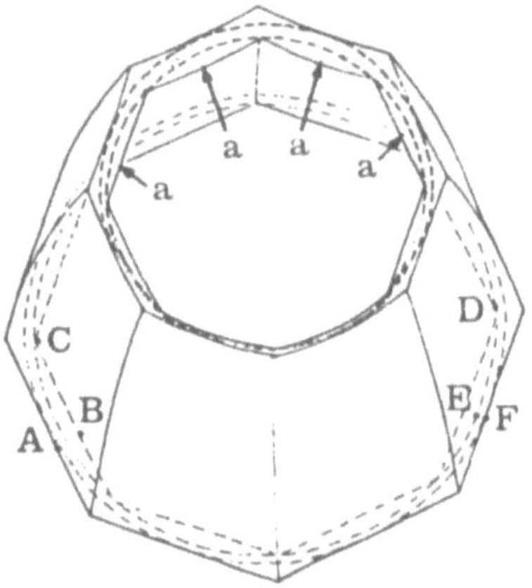

8, the octagonal form can also, however, be constructed without additional support if it is constructed in such a way as to contain within its thickness at every stage of construction similar continuous circular rings to those of the true circular dome. This calls for a somewhat greater thickness and, as soon as the bed joints become appreciably inclined towards the centre, for the whole beds to assume a conical form as in the true circular dome.

compression) as construction proceeds well above and it can, if need be, be provided by adding circumferential ties.[13]

After the true dome, the next best choice from these points of view would probably have been the ribbed form typified by the Gothic apse or the melon-shaped variant on this adopted elsewhere on a much smaller scale by Brunelleschi himself.[14] Where it is arched upwards and outwards between the ribs this can be even thinner than the true dome (on account of its sharper curvature), and it can similarly be constructed here with little or no temporary support since all the tendency to fall inwards is concentrated on the ribs, 6. The ribs must, however, be deep enough to resist this tendency as independent arches and must be supported by centering at least until they have been completed.

In the octagonal form the horizontal curvature of the surface between the corners, instead of being sharper than in the true dome, is reduced to nothing, 7. This greatly limits the possibility of circumferential arching between the corners and means that it is near the centres of the sides that the tendency to fall inwards is concentrated, particularly during construction. Corner ribs can do nothing to resist this tendency. In the completed vault it is necessary to have a thickness throughout similar to that required for independent arches or barrel vaults spanning across the octagon. (The complete form can, of course, be envisaged as being made up of sections of four such intersecting barrel vaults, the relevant section of one such vault being shaded in the drawing.) And during construction it is necessary to provide much the same temporary support as would be required for such arches or barrel vaults unless the thickness is further increased. This calls for substantial centering, particularly under the centres of the sides.

The initial intention was, almost certainly, to use centering. An earlier master, Giovanni di Lapo Ghini, had made a design and model for such centering for the projected dome as far back as 1371,[15] and the model was carefully preserved even after 1431 when the dome was nearing completion and the model in which it had been housed—the model of the whole church made in 1367—was destroyed.[16] The three similarly shaped vaults over the main eastern arms of the church were all constructed on centering, the centering for the last being actually under construction while designs for the great dome were being prepared and debated. And these designs included several new proposals for centering.[17]

The difficulty and expense of acquiring suitable timbers, erecting the centering at twice the height of the arm vaults, and subsequently dismantling it would not, however, have been the only deterrents to relying upon it. In the first place the precise requirement to be satisfied is less determinate and would have been a good deal less clear at the time than that for the support of the ribs of a ribbed vault, while experience with the vaults of the eastern arms might not have been a completely adequate guide on account of the large difference in scale. Secondly, it would have been realised in the course of constructing these vaults, if not before, that the use of centering calls for fairly rapid construction of the vault. This is on account of the changing distribution of the load imposed on the centering and its inevitable progressive yielding to this load. Ideally the vault should all be constructed at once. It would not have been too difficult to come reasonably close to this ideal with the arm vaults. But it must have seemed well-nigh impossible with the great dome. If this had simply been geometrically scaled up in all its proportions from the arm vaults it would have called for the placing of almost 20 times the quantity of material, again at twice the height and with vertical lifts greatly impeded by the presence of the centering. Construction without centering calls, on the contrary, for a much more leisurely pace so that each new ring can gain strength before another is added. It also eliminates the risk of settlement when the support is removed.

As more thought was given to the problems of construction any possibility of dispensing with centering must therefore have seemed increasingly attractive. Without departing from the surface geometry of the octagonal form—strictly speaking that of a cloister or pavilion domical vault—there was only one way in which this could be done. This was to increase the thickness beyond that needed for final stability as a set of interconnected

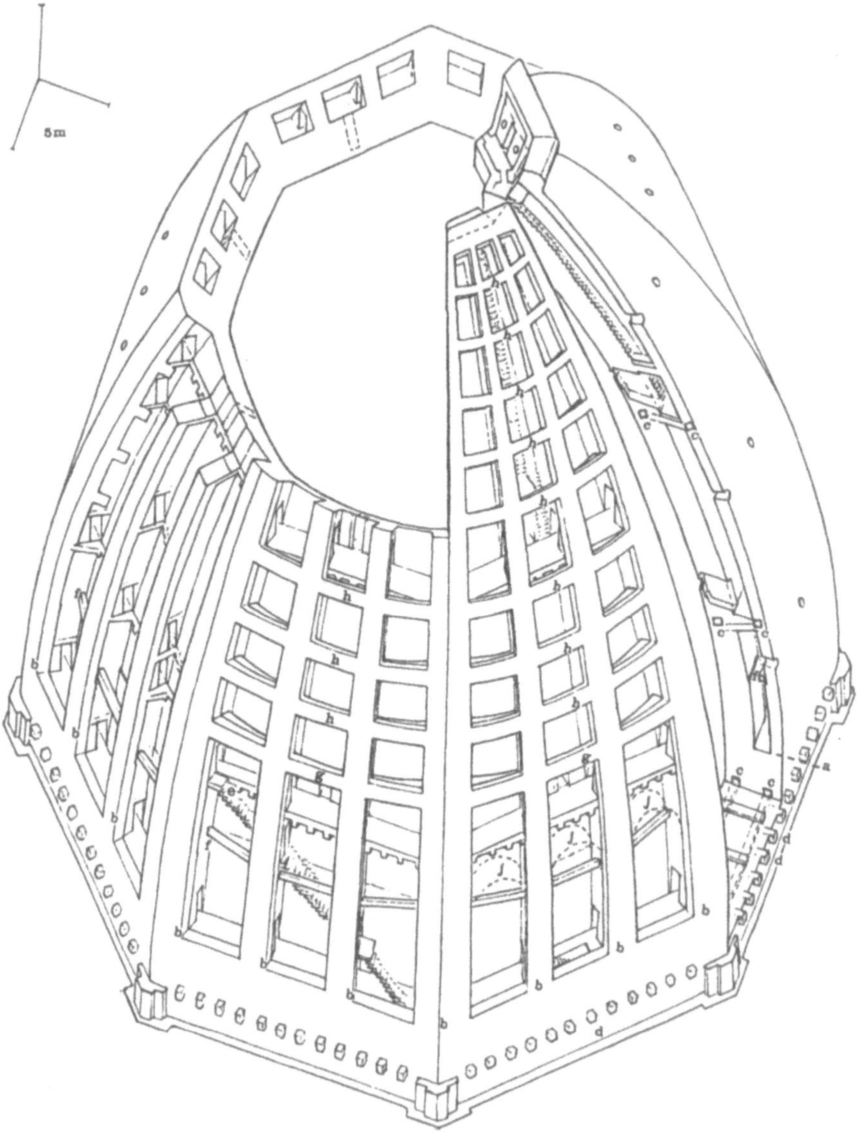

9, cut-away view of the dome to show the principal features of construction. At the right it is shown complete to the lantern base (but without the external corner ribs) and in cross section through the centre of a side. In the right central section the outer dome has been cut away between the spurs, but leaving those sections which, with the added circumferentially arched reinforcements, constitute continuous horizontal circular rings. At the left it has been shown only up to a continuous conical bed dipping to the level of the third internal walkway at the inside centre of each side. The outer dome has been cut away at the left centre as at the right centre, and the internal walkways, stairs, and circumferential rings have also been cut away in the adjacent section further left. At 'a' the solid base ends. The inner and outer domes above this level are connected by the radial spurs 'b' and reinforced circumferentially by 'rings' or 'chains' of sandstone blocks cramped together with iron, probably in the positions lettered 'c'. Each pair of rings is interconnected radially or transversely by similar blocks which project externally at 'd' (where they were intended to carry a largely unexecuted external walkway) or serve to carry the upper walkways at the higher levels. At 'e' this detail is most clearly revealed. Other circumferential reinforcements are a timber tie at 'f' and arched projections from the inside of the outer dome that help to create continuous horizontal rings of brickwork at the levels marked 'g' and 'h'. Rather than being envisaged as a reinforcement of the dome as built, each length of the timber tie was probably intended to contain the thrusts of a barrel vault spanning between adjacent spurs to carry an internal walkway. Such barrel vaults were called for in the original specification but never executed. Three are shown in dashed line at 'j'. Circumferential iron ties over the sandstone rings were also called for, but it is difficult to establish how many were set in place.

barrel vaults and up to the point where the continuous circular rings of a true circular dome were contained within it at all levels, 8. This would also introduce the possibility of containing within the lower rings the bursting tendency of the increased outward thrusts consequent on the increased thickness. Fortunately the drum was more than wide enough to carry such a dome.

The solution: the dome as built

It was the essence of Brunelleschi's design that he recognised and exploited this last possibility, 9 and 10.

Had he given the dome a solid thickness equal to the full width of the drum one might, in the absence of further evidence, question the fullness of the recognition. But he did not. He started by constructing a firm base of this thickness, but from a level at which the inward leans were still barely perceptible, 9(a), he reduced the thickness of what became the inner dome, 8(AB), to 3¾ braccia (1 braccio = 0·584 m). With an internal diameter, 8(CD), across the corners of the octagon of about 78 braccia this left a continuous circular ring, 8(EF), about ¾ braccio wide. As the sides closed in towards one another and the internal radius diminished, the continuous ring could be obtained even with a further reduction in thickness, and this also seems to have been recognised. The specification of 1420 called for a 'pyramidal' reduction to 2⅓ braccia at the open eye over which the lantern was to be built. This did not go as far as it might have done if the maintenance of the circular ring had been the only criterion. A sensible caution seems to have been exercised on account of the need to provide also a firm base for the lantern, and this caution prevailed further as construction proceeded with the result that the outer faces were stepped back only twice by a little under ½ braccio each time.

The large initial reduction in thickness was made by setting back the outer faces of this dome about 3¼ braccia behind those of the newly constructed base and still further in relation to the drum. This was too much to be left exposed to view. It did, however, provide the opportunity to add the lighter outer dome as a protection from the weather, and to provide permanent access routes between the two domes, 11. The specification called for an outer dome 1¼ braccia thick at the base reduced to ⅔ braccio at the eye, which left an internal space initially 2 braccia wide. It was to be 'bound'[18] to the inner dome by 24 spurs (or internal radial ribs). There were to be eight broad ones in the corners and two narrower ones at each side, 9(b). These also were to diminish 'pyramidally' in width to the eye. Again some changes were made in the course of construction. There was only one small reduction in thickness of the outer dome, but the width of the spurs at the sides was reduced, ostensibly 'to remove surplus weight' in the words of the 1422 variation that authorised this change.

Presumably to contain the bursting tendency and to add strength and stiffness, there was a requirement in the specification for 'six rings of long blocks of strong sandstone well cramped together with leaded iron, and over these sandstone blocks . . . iron chains'.[19] This is one of the less clear parts of the specification, probably deliberately to allow further thought to be given to the detailing of the 'rings' and 'chains' and to their incorporation in the domes. These elements—particularly the iron chains—are also the ones whose precise constructed form is most difficult to establish on account of their being largely or wholly hidden from view. As far as can be ascertained though, the 'rings' were constructed in pairs in the positions marked 'c' in 9, whether with or without the superimposed iron chains. The specification also called for the first ring, probably meaning the first pair, to be reinforced by long transverse blocks. Such blocks can be seen flanking the square holes that penetrate the base just above the inside walkway at the foot of the dome (originally to receive the cantilevered supports of the main working platform) and projecting as short stubs on the unfinished exterior, 9(d), and the presence of similar transverse blocks carrying the slabs of the second, 10(e), and third walkways between the domes can be taken as *prima facie* evidence of the presence of the upper pairs at these levels. Partial confirmation is provided in position 'e', 9, where part of the outer dome had to be cut away to accommodate the rising stair, 12, 13, and in the passageway through the adjacent spur, since something

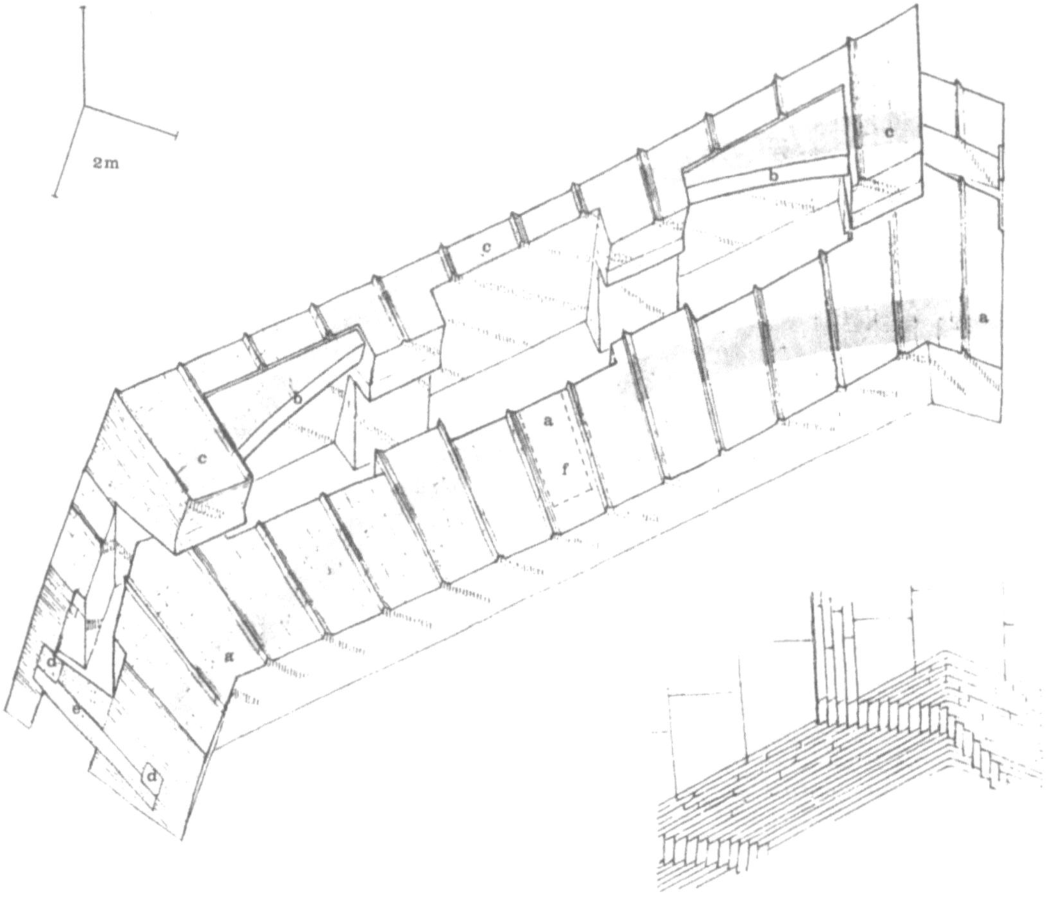

10, view of a typical side of the dome cut on two conical beds, one dipping to the level of the second internal walkway at the centre of the side and the other (in the rear) just above the top face of the first ring of arched projections from the inside of the outer dome. The circular dome contained within the inner octagonal form is indicated by the lower shaded band 'a'; two of the arched projections from the outer dome are shown at 'b'; and the continuous circular ring that they help to create is indicated by the upper shaded band 'c'. The radial alignment in the beds of the diagonally ascending bands of vertical bricks is shown and the detail of the inner right-hand corner sketched below shows also the circumferential joints in the beds. This bond commences at the level of the second pair of sandstone rings shown in section at 'd' with one radial connecting block at 'e'. At 'f' the inner dome is interrupted at the level of the walkway by a narrow passage shown by the broken lines and giving access to a small circular window on the inside face as shown in the section 9. But the 1426 variation called for the temporary filling in of the gap during construction, so the typical situation with no interruption is shown here.

can be seen of what is almost certainly the second outer ring in both places.[20]

I do not, however, now think that the one circumferential tie that is largely exposed to view (and that has therefore received most attention hitherto) can be considered in the same light.[21] This is the timber tie, 9(f), 14, that runs between the spurs some distance below the second pair of stone rings. It was also called for in the 1420 specification, though in oak rather than chestnut as actually used. But it was to have been just one of several at vertical intervals of about 12 braccia, and each was to have been set below

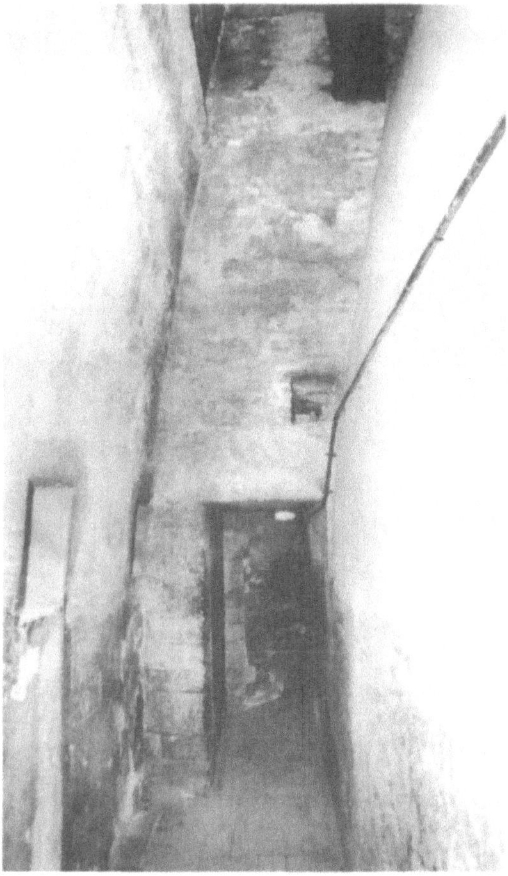

11, the space between the inner and outer domes at the level of the first internal walkway (looking towards the corner seen in the centre of 9). Part of the circumferential timber tie can be seen at the upper right.

13, detail of the ascending stair just below the viewpoint of the previous photograph showing transverse blocks and part of the second outer sandstone ring partly cut away to accommodate the stair.

12, looking towards the next corner at the level of the second internal walkway, with the ascending stair joining this walkway at the bottom left. To accommodate the stair the outer dome has been partly cut away to the left of the superimposed line, thereby exposing part of the second outer sandstone ring. Part of the lowest ring of circumferentially arched reinforcements to the outer dome (9g and 10b) can be seen at the top.

15, a corresponding view to the previous one in the Baptistery. The inner dome is again at the foot, but the spurs are now aligned at right angles to its surface and spanned by a barrel vault whose axis is similarly aligned and which carries part of the outer pyramidal roof. The timber tie seen just above the inner dome is similarly aligned in relation to the spurs.

14, looking up to a section of the circumferential timber tie. The inner dome is at the foot and a corner spur at the left. Note the radial alignment of the transverse blocks of the second pair of sandstone rings seen above the tie.

small barrel vaults spanning between the spurs to carry walkways like those that are now carried by the transverse blocks associated with the stone rings. These barrel vaults were not further described, but the implication seems to be that, possibly on the broad analogy of the vaults that carry the conical roof over the inner dome of the Baptistery,[22] they were envisaged as spanning circumferentially with their axes directed towards the centre of the dome, 9 and 15. The function of the ties would then have been to neutralise their thrusts, which would otherwise have added to the overall bursting tendency of the dome. With the decision to support the walkways in another manner both vaults and timber ties became redundant, no doubt to general relief in view of the difficulty again of finding suitable long lengths of timber. The fact that the first tie was completed provides, of course, an approximate *terminus post quem* for the decision.

The choice of thickness for the inner dome permitted its construction without centering but did not establish precisely how this was to be achieved, and neither the 1420 specification nor the later formal variations had much to say about this. The specification merely called for the use of stone up to a height of 24 braccia and thereafter of brick, tufa, or some other lighter material, and for the construction to proceed without centering (but with working platforms) for not more than 30 braccia. The situation would then be re-assessed in the light of experience. In 1422 the height at which there was to be a change of material was reduced and brick was definitely chosen for the upward continuation, but there was no further reference to the manner of construction. Only the 1426 variation was more revealing, particularly in the light of what can still be seen between the two domes, but chiefly in relation to the outer dome.

The chief problem arose with the inner dome, since it would always have been possible to support the outer one temporarily on its back. It was twofold. In the first place the bricks had to be laid in such a manner that each course did indeed contain the continuous circular ring that would ensure its stability as an open eye. Secondly it was necessary to bond in adequately the bricks that projected forward of this circular ring (in the regions lettered 'a' in 8) to prevent them from falling

16, a diagonally ascending band of vertical bricks in the inner dome partly exposed by a fall of plaster from the inside face. Three bricks of the next lower band are also visible below the right hand end of this band.

away from it, and desirable also to have some vertical bond between successive courses to reduce or eliminate the need for temporary local support while the bricks were laid and the mortar gained strength.

The first requirement led directly and necessarily to the laying of the bricks on conical beds[23]—in other words to the construction of the inner dome as if it were a circular dome of pointed profile of greater thickness, but with parts cut away from both inside and outside to leave the octagonal cloister-vault form. The necessary and desirable bonds were then both achieved by the ingenious expedient of setting bricks vertically on edge along radial lines at regular intervals in each course. Each vertical brick projected through four of the cone-shaped courses so that, together, these bricks formed diagonally ascending bands that progressively closed in towards one another, 10.[24]

The ascending bands of vertical bricks can now again be seen on the inside face of the dome both where they have been revealed by falls of plaster, 16, and where the associated surface irregularities have collected dirt on the fresco. They are also a conspicuous feature in the passageways through the spurs at the level of the second and third internal walkways, 17 and 18, since they were carried through into the spurs and even into the outer dome, and they were specifically referred to in the 1426 variation under the name *spinapescie* or 'herring-bone'.[25] Their radial alignment in each conical bed is clear both from what can be seen of them on the radial faces of some of the spurs and from the manner in which they turn the corners of the octagon.

The conical beds are less directly apparent and even the 1426 variation passed them over in silence. Between the two domes they are visible today only in short lengths, chiefly on the radial faces of the spurs and on the sides of the passageways through them, 17 and 18. On these sides they are cut at ever varying angles, but they always drop down towards the centre of the octagon side and do so to an extent that depends on the distance from this centre and on their primary inclination towards the centre of the dome. Measurements and other observations up to the level of the third walkway in 1974 and 1975 confirmed an earlier hunch that each bed was that circular cone whose apex, C in 19, was at the intersection with the vertical axis of the dome (ZZ) of the relevant radii (AB) that define the arcs (XBB′) of the internal angles of the dome surface.[26] The lack of any reference to this in either 1422 or 1426 may have stemmed in part from the difficulty of expressing the intention clearly, but probably stemmed also from the fact that it was part of the initial concept and a necessary corollary of the basic choice of the means for dispensing with centering. Certainly it is only at the very foot, where a firm solid base was being built up almost vertically, that the masonry was laid

17, 18, views looking along the third internal walkway through the intermediate spurs and into a corner spur, showing evidence of the setting out of the brick beds and the manner of incorporation of the bands of vertical bricks. Note that the beds dip towards the centre of the dome roughly at right angles to its faces and dip towards the centre of the side to an extent dependent on the distance from this centre.

in horizontal courses with joints aligned parallel with and at right angles to the sides. There is, however, an indirect reference in a surviving document submitted to the building committee by one of the substitute superintendents, Giovanni di Gherardo da Prato, late in 1425 or early in 1426.[27] Giovanni must have observed the conical beds, which by then would have become noticeable, and probably a swinging cord device, 19, that was being used to control the setting of the bricks.[28] He concluded that the dome was being built to a semicircular profile and wrongly alleged that the specified pointed profile was not being followed. There is also a somewhat later, but surprisingly accurate, account of the procedure in Bartolomeo Scala's *Historia Florentinorum*: 'For, having found and marked the centre [of the conical bed], he stretched a cord from the centre to the circumference; and sweeping it round determined in what order and according to what circular disposition the bricks and mortar were to be laid and set in the fabric to complete the sides of the vault'.[29]

In its main reference to the outer dome the 1426 variation called for a new element—'bricks acting as an arch above the door jambs of the second walkway for the perfection of the circle around the outer dome, and

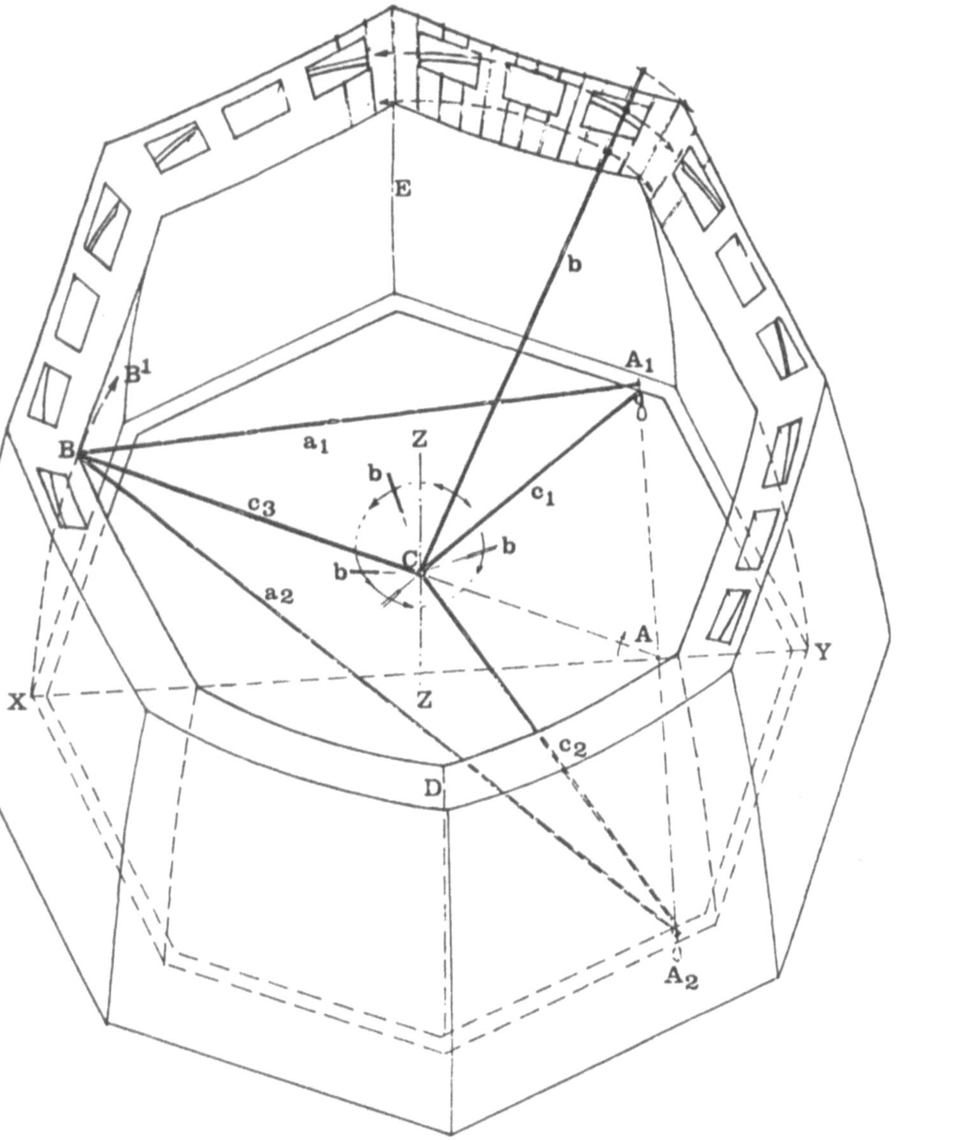

19, view of the dome cut throughout on the conical bed immediately above the top face of the first ring of arched projections from the inside face of the outer dome and showing possible arrangements for controlling both the basic geometry and the setting of the brickwork. A is the centre of the vertical arc XBB′ that defines one corner of the inside face ($AY = 0{\cdot}2\ XY$). Successive points B, B′ are established by swinging chains or cords a_1, a_2 about points of attachment A_1, A_2 on the internal gallery at the foot of the dome and on a line passing through A. Points in the other corners are similarly established periodically by attaching the same chords to corresponding points on other sides of the gallery. To guide the radial and circumferential setting of the bricks and establish the inside edges of the arched projections a cord b is required attached at the appropriate centre C and swinging about this centre. C, like A, is not directly accessible, so the point of attachment is provided by further chains or cords c_1, c_2, and c_3. Initially, at each successive working level, c_1 and c_2 pass over pulleys at A_1 and A_2 and are equally weighted to keep them under equal tension. c_3 is adjusted against their combined pull until, by sighting from D to the opposite edge E, the point where all three cords meet is seen to be in line, when it will be in the required position C on the vertical axis of the dome ZZ and on the radius arm AB. All three cords are then fixed at their ends A_1, A_2, and B under sufficient tension to minimise movement at their meeting point as the cord b is swung round. Levelling of the beds is carried out along the circular arcs swept out by markers along the length of b, and the faces are aligned either by sighting between corners or by stretching other cords between them. The system c_1, c_2, and c_3 could be stiffened to give more precise control by adding additional cords between C and other points on the gallery and at the working level, these additional cords being left slack until c_1, c_2, and c_3 have been adjusted.

20, part of the lowest ring of circumferentially arched reinforcements to the outer dome seen from the ascending stair just above the second internal walkway. The inner dome has been partly cut away to accommodate the stair as indicated by the superimposed lines.

so that this live arch shall be complete and not broken'.[30] Here is the one explicit reference to a complete circular ring, here achievable only by an addition to the thinner form originally specified. The ring referred to is that lettered 'g' in 9 and seen also in 10. It was clearly considered to have served its purpose because eight similar rings were constructed above it, 9(h). The central section of each arch on each side is an integral part of the outer dome itself, 10(c), but the outer sections project prominently below it, 10(b), 12, 20 and 21. Careful examination shows that, at all levels, their inner edges were truly circular, so that their upper and lower faces intersect the inner faces of the dome along the same rising curves as the conical brick beds: they do not run in straight lines from corner spur to corner spur as they have always hitherto been conventionally shown. In the completed structure they must have some beneficial effect in that they will help concentrate any weight transmitted from the outer dome to the inner one on the corners where it can most readily be carried. But the fact that they were seen as primarily another construction expedient is indicated by the proviso in the 1426 variation that 'if ever the addition shall appear unsightly or obstruct the walkway and stairs, it may be dismantled when the dome is completed, having made it possible to proceed to completion with greater safety'.[31]

By these and other related means (including the temporary bricking up of the small round windows of the inner dome and the passageways leading to them) construction did proceed safely to completion without ever calling for the temporary support of centering. The 16 years that it took would have allowed ample time for the brickwork to gain strength as the work proceeded and may be compared with the period of as many months that it took to complete the comparable circular dome of St Peter's using centering to support its 16 ribs. Brunelleschi's almost continuous involvement must have stemmed not only from the experimental nature of the entire procedure and his involvement with almost all aspects of the building process, but also from all the geometrical complexities of marrying the circular and octagonal forms and the virtual necessity of working out most of the implications on the job itself.

Sources of ideas

Later, or as the work was nearing completion, he must have discussed the design with the enthusiastic Alberti. This discussion is briefly recollected in the reference to the construction of polygonal domes in Book III of *De re aedificatoria* which provides one last confirmation of the basic idea: 'A polygonal dome also can be constructed without centering if a true circular one is contained within the thickness'.[32] A first consultation of the mathematician Giovanni del Abbaco in 1417 at about the same time as Brunelleschi presented his first designs suggests that the idea may already have been germinating then.[33] Setting aside for the present any further discussion of its development either during the initial planning phase or later, let

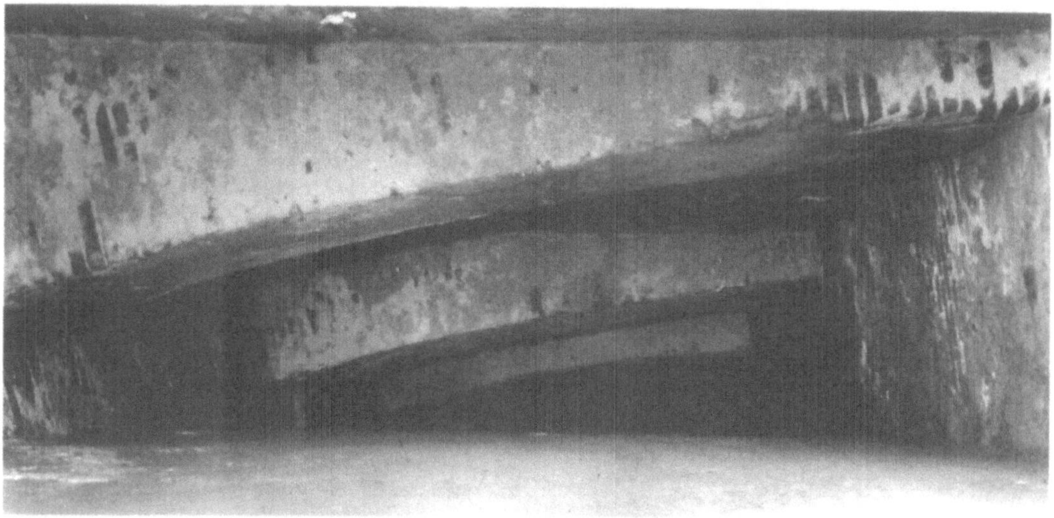

21, parts of the second, third, and fourth rings of circumferentially arched reinforcements seen from the second internal walkway. The inner dome is again at the foot and a corner spur at the left as in 14.

22, two eleventh-century (Seljuk) tomb towers with early pointed brick double domes in north-west Iran.

23, view between the two domes of the left hand tower seen in the previous photograph.

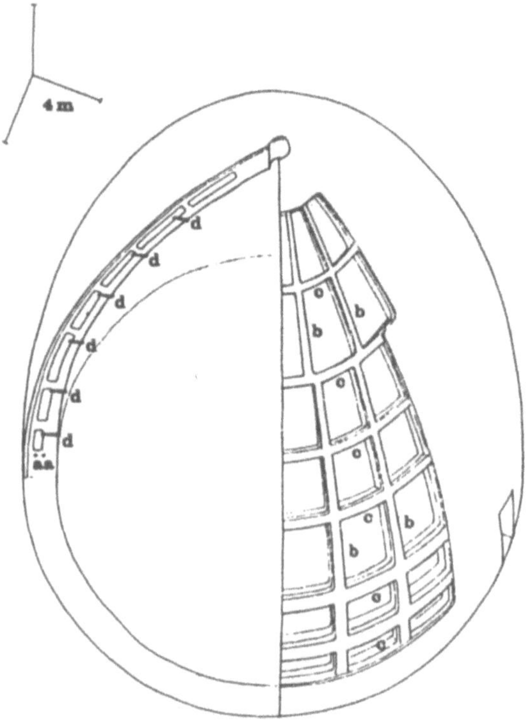

24, *cut-away view of the dome of the Mausoleum of Oljeitu at Sultaniya. The solid base is reinforced by two circular rings of large poplar trunks at 'a'. Above it the inner and outer shells of the dome are interconnected at frequent intervals by both vertical, 'b', and horizontal, 'c', bands of brickwork. These form closed compartments between the two shells only about 0·5 m wide, so that it might be more correct to describe the form as a cellular single dome than a true double one. Part of the thinner outer shell and its tile covering have been cut away at the right to reveal some of the compartments (the form of the upper ones being derived from old photographs, only some of the lower ones having been accessible to measurement). At 'd' are shown the typical positions of the surviving short beams built into the inner shell, originally to carry working platforms. The sawn-off inner ends now form a diagonal pattern on the inside face of the dome: they have been shown at all levels in a single vertical section merely for convenience of representation. This dome springs from a circular base formed over an octagon by three superimposed rows of small squinches.*

us return finally to the question of whence it came and what were the sources of the principal related ideas—notably for the bonding of the brickwork.

Manetti describes a trip to Rome with Donatello after Brunelleschi's failure in the competition for the Baptistery doors and says that Brunelleschi then applied himself to the study of Roman buildings and ruins, paying particular attention to their construction. 'He considered methods of vaulting with centering or other temporary support, and then the situations where it was possible to save the expense and forgo the convenience and the procedures that had to be followed when centering could not be used on account of the size of the vault or some other reasons.'[34] Despite the lack of contemporary documentation of any such visit, this account has naturally led scholars to seek Roman prototypes. The great difficulty is that the Romans never built large polygonal domes and always, as far as we can see from present evidence, cast their domes in some form of wet concrete in Rome itself. Experiments with centreless construction using bricks or purpose-made pots seem to have been confined to more distant parts of the empire and to have been of limited relevance to what Brunelleschi did. A drawing attributed to Guidetti (Uffizi 1330) has been held to demonstrate, on the contrary, that they even employed the same *spinapescie* bond, not just with bands of vertical bricks spiralling upwards in one direction, but with two criss-crossing sets to give an effect like that of the coffers in the semi-domes of the Temple of Venus and Rome.[35] But I find it very difficult to take this drawing at its face value and consider it most unlikely that the bond was so used in Italy before Antonio da Sangallo the Younger made such extensive use of it in the sixteenth century. The one feature of the Florentine dome for which there was an undoubted direct Roman precedent was the series of rings (or 'chains' as they are named in later documents) of cramped blocks of stone. But there was such a continuous later tradition of using these reinforcements that there is no need here to seek such a precedent.[36]

For the most direct precedents for both the *spinapescie* bond and the centreless construction of a large double dome one must look as far afield as Iran. Here there was a very long tradition of brick vaulting using a wide variety of patterns of bricklaying to reduce or eliminate the requirement for temporary support, and this included bonds virtually identical with that shown in the Guidetti drawing. True double domes were also built in brick as early as the eleventh century, 22 and 23, and an early fourteenth-century

25, detail of the construction of the double dome of the Mausoleum of Oljeitu at Sultaniya. The photograph was taken between the two shells in a position corresponding to 'd' in 24. The dome interior is on the left and it will be seen that the brick beds are much more nearly horizontal than at right angles to the dome surface.

example—that of the Mausoleum of Oljeitu at Sultaniya[37]—was the closest earlier counterpart of Brunelleschi's dome. As shown in the cross section, 24[38], it also had a thicker inner dome and a thinner outer one, both of pointed profile and springing from a solid reinforced base. They were interconnected by a grid of radial and circumferential ribs, 24(b, c), and constructed without centering —part of the evidence for this being the embedded ends, still *in situ*, of the projecting beams used to support working platforms on the inside face, 24(d). It was given a span very similar to that of the Baptistery dome. Could Brunelleschi have learnt from these sources? Distance alone would not preclude this because Sultaniya, for instance, was on a major trade route eastward from Tabriz and had undoubted Italian connections, being even the seat of a catholic bishop for about a century after the time of Oljeitu. Brunelleschi might therefore have had at least hearsay knowledge of the mausoleum. It is doubtful, nevertheless, whether he could really have learnt anything useful. All the Iranian domes, Sultaniya included, are circular in plan, whereas his main problem was the octagonal form in Florence. And closer study of these domes shows that their builders relied more on a fast-setting gypsum mortar to dispense with centering than on the bonding pattern. The inner dome at Sultaniya is, for instance, constructed throughout of rings of brickwork laid on the normal bed, but at a flattish inclination in the radial sense and virtually glued, one at a time, to the bricks below, 25.

We are thus forced to reconsider Manetti's account, bearing in mind that, where it is possible to check against surviving documentation, there is usually a factual basis for even his more garbled stories, and looking less for direct precedents than for possible stimuli to Brunelleschi's creative imagination. I suggest that there are four likely sources of such stimuli:

1 The usual Roman practice of leaving a central eye in a dome, framed by a circular ring of brick, as seen most notably in the Pantheon.

26, the so-called Temple of Minerva Medica, Rome, as seen in about 1790 by J. C. Klengel, showing some of the embedded radial bands of bricks as partly free-standing ribs. (Painting now in the Gemaldegalerie Alte Meister, Dresden.)

2 The related practice of casting the concrete in horizontal layers, each externally circular in plan but frequently scalloped, coffered, or otherwise modulated internally.

3 The further practice in some later domes—notably that of the so-called Temple of Minerva Medica, 26—of embedding radial and circumferential bands of brick on the inner surface.

4 The common use of a flat-arch form in concrete construction with about one in four of the facing bricks penetrating the full thickness and serving to divide the concrete mass into short wedge- or voussoir-shaped blocks, 27.

27, a Roman concrete flat arch in the Baths of Trajan. Where the facing has been broken away in the central section it can be seen that some full bricks penetrate the concrete mass to create wedge- or voussoir-shaped blocks in a manner closely analogous to the division of the brickwork mass in Brunelleschi's dome as shown in 10. In both cases the falling away of the part of the arch that lies below (or inside) the effective arch within the total depth (or thickness) is thereby prevented.

None of these practices is directly paralleled in Florence or could have given a direct answer to the problem there. But implicit in the first two is the possibility of building an octagonal dome as if it were a circular one and thereby ensuring its stability with an open eye of any size and without extraneous support. In the third there is a hint of the system of radial spurs and circumferential arches that helps to support and stiffen the outer dome in Florence. And, most significantly to my mind, there is implicit in the fourth the precise means adopted by Brunelleschi—using the radially aligned bands of bricks-on-edge in each conical bed—to bond in the mass lying outside the effective arch-like ring. These are just the sort of hints to which I would expect a perceptive and creative mind to react.[39]

I am therefore inclined, in essence, to believe Manetti after all.

NOTES

The substance of this paper was first presented as a public lecture at the Courtauld Institute in March 1975. It was then intended to expand it as part of a joint monograph with Howard Saalman. It is based primarily on numerous inspections of the dome and of related structures, and I am particularly grateful to the Research Fund of the Royal Institute of British Architects for their generous support of the principal journey undertaken for this purpose. I am also grateful to those in whose company some of the inspections have been made, notably Piero Sanpaolesi and Howard Saalman whose ever-questioning interest has served as a spur to the further development of my ideas.

1 *'Chi mai si duro o si invido non lodasse Pippo architecto vedendo qui structura si grande, erta sopra e cieli, ampla da coprire chon sua ombra tutti e popoli toscani, facta sanza alcuno ajuto di travamenti o di copia di legname, quale artificio certo, se io ben iudicho, come a questi tempi era incredibile potersi cosi forse appresso gli antichi fu non saputo ne conosciuto.'* (My translation.)

2 Guasti, C. *La cupola di Santa Maria del Fiore*, Florence, 1857 (reprinted 1974), documents 120, 121.

3 The best survey drawings are still those by G. B. Nelli in Sgrilli, S. B. *Descrizione e studi dell'insigne fabbrica di S Maria del Fiore*, Florence, 1733. But see also *Rilievi e studi sulla cupola del Brunelleschi eseguiti dalla Commissione nominata il 12 Gennaio* 1934, Florence, 1939, and Ferri, W. *et al. Il rilevamento fotogrammetrico della cupola di S Maria del Fiore in Firenze*, Florence, 1971.

4 With few exceptions referred to below, the documents used in this study are those published in Guasti *op cit* (referred to below as *Cupola*) and in Guasti, C. *Santa Maria del Fiore: La costruzione della chiesa e del campanile*, Florence, 1887 (also reprinted 1974 and referred to below as *SMF*). Further documents not included in Guasti's selections have been transcribed recently by Saalman. I have had the privilege of seeing these but leave discussion of those aspects of the evolution of the design on which they have a bearing until they have been published.

5 There are two versions of the specification, both published in Doren, A. 'Zum Bau der Florentiner Domkuppel', *Repertorium fur Kunstwissenschaft*, vol. XXI, 1898, 249-62. The quotations below are from the more definitive version. The other version only and the two variations are published in Guasti, *Cupola*, documents 51, 52, 75.

6 Mainstone, R. J. 'Brunelleschi's dome of S Maria del Fiore and some related structures', *Transactions of the Newcomen Society*, vol XLII, 1969-70, 107-126.

7 I did not feel able in this paper (or in my *Developments in structural form* written shortly afterwards) to put forward this suggestion as explicitly as I do here because none of the measurements referred to below had then been made and I had not yet had the opportunity to discuss with Sanpaolesi my objections to his different interpretation of the visible evidence.

8 The earlier history is discussed in Saalman, H. 'Santa Maria del Fiore: 1294-1418', *Art Bulletin*, vol XLVI, 1964, 471-500.

9 Guasti, *Cupola*, documents 11-14.

10 The profile called for in the specification was a 'pointed fifth'—*'quinto acuto'*—in the inner angles. This meant a circular arc with a radius one fifth less than the span. The profile as built is so close to this, even on the basis of the latest photogrammetric survey, that I cannot accept the conclusion of Ferri *et al* (*op cit*) that it was set out differently. It is also very close to those of the Baptistery vault and the vaults of the eastern arms of the church as far as can be seen from the best available drawings and photographs.

11 Mainstone, *op cit*, p118.

12 By centering is meant here a substantial temporary support to the structure under construction across its span. It is referred to in the documents as *'armadura'*, and is not to be confused with formwork that serves locally as a template for, and may give local support to, a surface under construction. This may be used with or without centering and is referred to as *'cientine'* and its use as *'cientinare'*. Nor is it to be confused with working platforms which serve the different function of supporting men, materials, lifting devices, and the like. These are referred to as *'ponti'*. Both templates and working platforms were used, but I largely ignore them here.

13 For a precise analysis of the requirements for a circular dome of 'quinto acuto' profile see Mainstone, *op cit* pp116-20.

14 As, for instance, in the Old Sacristy of S Lorenzo.

15 Guasti, *SMF*, document 231.

16 Guasti, *Cupola*, document 68.

17 Certainly those referred to in Guasti, *Cupola*, documents 26, 27, 39, and possibly some others that were less precisely described.

18 The requirement is that the spurs *'legano insieme'* the two domes.

19 *'sei cerchi di forti macigni e lu(n)ghi e bene sprangati di ferro stagnato; e di sopra a decti macigni . . . catene di ferro.'*

20 The visible evidence suggests that the first pair of 'rings' ran parallel to the sides of the octagon over most of their lengths, but that the second outer ring at least was curved around the corners. Contrary to my earlier opinion I now consider that the second and third inner rings are both circular in conformity with the circular disposition of the surrounding brickwork and the radial alignments of the

transverse blocks. I also now think that the prime function of these upper rings may later, if not at first, have been seen as the wider one of providing more dependable strength and stiffness at their respective levels and more secure bases for further unsupported advance into the void.

[21] This is a partial revision of my earlier view.

[22] There are now excellent drawings of the Baptistery in Pietramellara, C. *Battistero di S Giovanni a Firenze*, Florence, 1973.

[23] The necessity arises as soon as the beds become inclined downwards towards the centre of the dome: with flat inclined beds the central parts of the sides would be too high in relation to the corners.

[24] For a fuller discussion of the role of the vertical bricks in facilitating the construction of successive courses see Sanpaolesi, P. *La cupola di Santa Maria del Fiore*, Rome, 1941, p25, and Mainstone, *op cit* pp113-15 (though the former is at fault in envisaging flat inclined beds, joints aligned parallel and at right angles to the sides, and a diagonal setting back of the bands of vertical bricks on each bed; and the latter must be read in the implied context of the conical beds necessary to provide the continuous horizontal arch-rings). There is a lengthy earlier discussion by Nelli, G. B. *Ragionamento sopra la maniera di voltar le cupole senza adoperavi le centine* in *Discorsi di architettura del Senatore Giovan Batista Nelli*, Florence, 1753, pp53-74. See particularly pp65-7.

[25] The name seems a little inappropriate but is qualified by the expression 'such as shall be decided upon by those responsible for the execution'.

[26] The most comprehensive (and final) series of measurements was made (with the assistance of Howard Saalman) in the passageways through all the spurs at the level of the third walkway. For the statistically minded these showed that the mean dips of the bed joints towards the centres of the sides in the four spurs were (working clockwise) 13·3°, 4·1°, 4·1°, and 13·6° with standard deviations of 2·1°, 1·2°, 1·5°, and 2·7°, as compared with calculated dips of 15·1°, 5·6°, 5·7°, and 15·2° for the assumed conical beds. The slightly smaller measured dips may be largely accounted for by necessary adjustments of the beds to the horizontal at the tops of the passageways, which are bridged by large horizontal blocks of stone. Further confirmation has been provided very recently by measurements made independently by Salvatore di Pasquale who was able to observe large areas of the outside of the outer dome stripped bare of its tiles. See idem, *Primo rapporto sulla cupola di Santa Maria del Fiore*, Florence, 1977. Di Pasquale has also just drawn my attention to a reference to the conical beds in Choisy, A. *Histoire de l'architecture*, Paris, 1899, vol 2, p617. Sanpaolesi, in the only other relevant discussion I know (*op cit*), concluded from his careful observations that the beds were inclined planes meeting 'like an open book' in the corners, perhaps misled by analogy with the fully visible construction of Sangallo's large octagonal dome in the entrance to the Fortezza da Basso where the bricks are so laid.

[27] Frequently published. See, for instance, Saalman, H. 'Giovanni di Gherardo da Prato's designs concerning the cupola of Santa Maria del Fiore in Florence', *Journal of the Society of Architectural Historians*, vol XVIII, 1959, 11-20.

[28] There is documentary authority in the 1426 variation for at least one control device such as I have illustrated. It is described there simply as consisting of three cords: '*e murisi con gualandrino con tre corde*'. These three cords could have been $c_1 + c_2$, c_3, and b of the device for controlling the work on each bed, or $a_1 + a_2$, $c_1 + c_2$, and c_3 + b for complete control. The former seems to me the more likely since the requirement for control of the vertical arcs of the internal corners would have existed well before 1426 whereas the requirement for fairly precise control of the beds would have arisen at about this time.

[29] '*Nam centro invento signatoque filium a centro duxit ad circumferentias; quod in girum ferens quo ordine, quoque orbe lateres caementaque ducenda collocandaque essent a fabris, per parietem terminabat.*' (My translation.)

[30] '*sopra i cardinali degli usciuoli, che sono sopra 'l detto secondo andito, per perfectione del cerchio che gira intorno la cupola di fuori, accio che detto archo vivo sia intero e non rotto, si muri di mattoni in atto d'archo.*'

[31] '*se mai paresse che detta agiunta mostrasse rustica a l'occhio, o impedisse l'andito e schale si possa, fatto la cupolae disfare detta agiunta, accioche con piu sicurta si possa guidar, a murare la cupola in sino alla fine.*' There was in fact a small amount of cutting away of the third and fourth arch-rings where the ascending stair passes by.

[32] '*Angularem quoque testudinem sphericam, modo per eius istius crassitudinem rectam sphericam interstruas, poteris attollere nullis armamentis.*' (My translation.)

[33] Guasti, *Cupola*, documents 16, 23.

[34] '*e penso e modi e delle centine e delle altre armadure, e cosi doue si potessi fare sanza esse pe risparmi delle spese e delle comodita, e che modo s'auessi a tenere, e cosi doue l'armadure non possono seruire per la grandeza delle testudinj e diuerse altre cagionj.*' (Quoted from the edition by Saalman, H. *The Life of Brunelleschi by Antonio di Tuccio Manetti*, University Park, 1970, lines 341-5—my translation.)

[35] Illustrated and discussed in Sanpaolesi, *op cit*.

[36] For a brief account of this tradition with references see Mainstone, R. J. *Developments in structural form*, London and Cambridge, Mass, 1975, pp52 and 122. A further relevant reference is idem, 'The reconstruction of the tympana of St Sophia at Istanbul', *Dumbarton Oaks Papers*, vol 23/24, 1969/70, pp355-68. The commonest form of cramp was an elongated U with the ends of the U projecting downwards into the two blocks joined by it. The cross piece was also chased in and the whole cramp was usually set in place with molten lead as is still done today. In S Maria del Fiore itself there is evidence for the use of such reinforcement as far back as 1353-57 during the construction of the top storey of the campanile, presumably to resist the shaking due to ringing of the bells. See Guasti, C. *SMF*, document 70, 19 June 1353 (p76), 5 February 1355 (p86), and 31 May 1357 (p91). Continuous iron tie bars were also used across arches in numerous places, and several remain in place to this day across the nave. In this context moreover, it seems that the use of rings of cramped stones and/or continuous chains of iron tie rods embedded within the thickness of the dome was already envisaged when its basic form and size were decided upon in 1366-7. This seems to be implied by the expressed hope that it could be built without *visible* chains—'*sanza esservi chatene che si veghano*'—*op cit*, document 150, 13 August 1366 (p175). See also *ibid*, documents 195, 21 November 1367 and 272, 23 May 1377.

[37] Sanpaolesi, P. 'La cupola di Santa Maria del Fiore ed il Mausoleo di Soltanieh', *Miischengen des Kuntteilunsthistor Institutes in Florenz*, vol XVI, 1972, 221-260 gives the best recent account.

[38] Based on Sanpaolesi's cross section in the work cited above but corrected on the basis of later observations and measurements. These were made in 1974 from scaffolds by courtesy of Professor Sanpaolesi and his assistant D'Errico who have been restoring the structure. In this connection I am also grateful to David Stronach and David Morgan, then of the British Institute in Tehran, of whom the latter accompanied me to the mausoleum and the Kharraqan tomb towers (with the earliest brick double domes) and provided helpful information on contacts with Italy in Brunelleschi's time.

[39] I shall discuss the possible sources further in a paper 'The origins of the structural concept of the cupola of S Maria del Fiore' to be presented to the International Congress on Brunelleschi Studies to be held in Florence in October 1977.

Addendum (AR November 1977, p. 320).

SIR: In my article on this dome (AR September) I rejected earlier suggestions of a direct Roman prototype for the *spinapescie* bond used by Brunelleschi. These suggestions were based on a drawing (Uffizi 1330) attributed by Rivoira to Guidetti and thought to represent a lost Roman temple. I considered it most unlikely that anyone in Italy before Antonio da Sangallo the Younger used the bond in the manner shown in this drawing. I have now been able to identify the precise subject of the drawing. Through the courtesy of Ing. Francesco Vacchini and Don Cipriano Cipriani I have been able to inspect the octagonal rooms above the barrel vaults of the 'secondary naves' of St Peter's (ie above the vaults that spring from the rear faces of the four piers that carry the dome). These rooms are shown in part or full sectional elevation in Fontana, *Templum Vaticanum*, p359, and Letarouilly, *Basilique de Saint-Pierre*, pl 8, and more fully in Ferraboschi, *Architettura della Basilica di S Pietro in Vaticano*, pl IX and X. They were built by Sangallo. The drawing, whether from Sangallo's hand or not, is a precise representation of one of them, and not of any lost Roman structure.

XVIII

The dome of St Peter's: structural aspects of its design and construction and inquiries into its stability†

The challenge faced by those who designed and built the late-sixteenth-century dome of St Peter's did not quite match that faced by their predecessors in Florence almost two centuries earlier. The dome of the Cathedral of Santa Maria del Fiore already stood, demonstrating that the task was not impossible and providing a model if one was required. But this model was not slavishly followed. Design and construction were, moreover, spread over almost the whole sixteenth century. Under a succession of popes, different architects were in charge and they produced numerous different plans including almost as many different designs for the dome. These designs pose a number of questions about their structural adequacy and their designers' intentions. Further questions arise about how the dome was finally built in 1589-1591 and its subsequent stability, and about the inquiries that stemmed from renewed fears for this in the mid-eighteenth century.

Design

Views expressed in the course of the eighteenth-century inquiries were extensively published at the time. On the other hand, most details of the development of the design and the progress of construction were recorded in print only a century and more after the event, and much less fully even then.[1] To answer the earlier questions, it is therefore necessary to rely heavily on other sources, chiefly the surviving material of other kinds: a considerable number of design drawings, building accounts, artists' sketches of construction in progress, and two wooden models. For over a hundred years, these survivals have been subjected by scholars to ever closer scrutiny in order to piece together the complex sequences of events. Thanks to this, a wide consensus has been reached on most of the broad outline, leaving debate to continue on many details and on the wider questions raised. In what follows, that broad outline will be merely summarised and taken as the starting point for consideration of the questions referred to above.

1505-1514: Bramante

When Bramante assumed control in 1506, the apse of the old St Peter's had already been replaced by Rossellino with an extended vaulted choir, which it was intended to complement by vaulting the transepts and placing a dome over the crossing.[2]

Having obtained the agreement of Julius II to a more radical reconstruction of both nave and transepts of the old church, Bramante produced a number of plans centred on a larger dome.[3] These plans, incomplete though they are in the surviving drawings, were essentially expansions of the Greek cross plan adopted at the time for several fully centralised churches like S. Maria delle Carceri in Prato and S. Maria della Consolazione at Todi. They added secondary domed spaces between the arms of the main cross, and gave the four main piers chamfered inner faces to support a central circular dome with a diameter considerably greater than the breadth of these arms. This diameter was close to the diameter, measured across the corners, of the octagonal dome of S. Maria del Fiore.

The choice of a circular plan nevertheless made this Florentine dome a less directly relevant model than it would otherwise have been. Having also no Florentine connections, Bramante turned instead to its Roman counterpart, the Pantheon. Serlio, writing a few decades later, thought the design of sufficient importance to illustrate it in both plan and section alongside this and other Roman structures that are the main subject of his *Third Book on Architecture.*[4] As far as can be seen from his woodcuts and another representation on the foundation medal, Bramante intended a fairly straight copying of the form for the dome itself, apart from the addition of a lantern over the oculus (fig. 1, left). No voids are indicated in the lower part to reduce the weight, though he might have added these later if the dome had been built.

He had, of course, the assurance that the prototype had stood for many centuries. But he seems to have ignored the fact that this stability was partly dependent on the fact that it was carried by a substantial drum that rose directly from foundation level. His dome also sprang from a drum, but from a far less substantial one that consisted in the main of three concentric rings of free-standing columns. And these did not rise directly from the foundations but from four tall piers and their linking arches and pendentives.

Examination of the Pantheon dome would presumably have disclosed, then as now, that, even with its substantial, well-founded drum, it had undergone extensive radial cracking, and that there had been some outward tilting of this drum under the thrusts generated. Without circumferential ties of considerable cross section, Bramante's columnar drum would have tilted outwards much more. And, if this alone did not result in collapse of the dome, the further expansion of the base of the drum due to the tilting of the crossing piers under the thrusts passed on to them would almost certainly have done so. Even Serlio twice comments that its great mass and weight 'should have had an excellent foundation to

1. Dome designs by Serlio (left) and Sangallo (right). From Letarouilly, *Projets divers*, plates 6 and 19.

stand fast and not be made to stand on four arches of such a height'[5].

In the event it was not built. By the time of Bramante's death, only the crossing piers, the great arches that span between them, and the lowest parts of the pendentives had been completed, and all had been built with cross sections that could never have been adequate (fig. 2). They had also, as Vasari noted, been built in great haste.[6]

1541-1521:Giuliano da Sangallo, Raphael, Peruzzi

Over the next seven years a succession of fresh plans was produced that developed further some of Bramante's ideas. But the only significant undertaking from the present point of view was the first reinforcement of the piers and arches that were already standing – which, as Serlio also wrote, 'were, without any superimposed weight, already deformed and cracked in some places'.[7]

1521-1546: Antonio da Sangallo

After a virtual cessation of work between 1521 and 1534, the younger Sangallo further reinforced the main piers by increasing their cross sections, possibly in preparation for a start on the construction of the dome to his own new design (figs 2 and 3). Like others before him, he produced not one but several designs, the definitive one being shown not only in drawings but in the surviving great wooden model constructed between 1539 and 1546.[8] In his strengthening of the piers, he seems to have had a better appreciation than Bramante had of what would be required of them. Nevertheless doubts still arise about the structural adequacy of his overall proposals.

Perhaps stimulated by his Florentine background, he turned more to the Florentine dome for inspiration and adopted a steeper profile more like the one adopted there. This was better fitted to carry a heavy lantern and would have simplified the later stages of construction. But he seems, on the evidence of the drawings, to have thought again of a solid construction

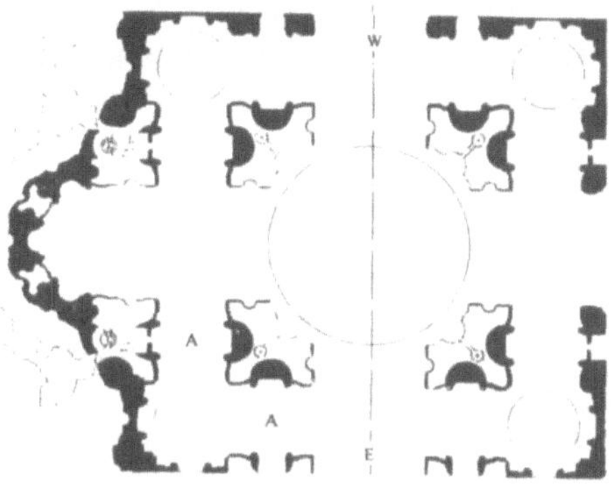

2. Partial plan showing, approximately, the principal successive stages of construction of the crossing piers and southern transept. The lightest tone denotes the initial construction by Bramante, the next darker tone denotes additions up to the time of Sangallo, and the darkest tone the final additions by Michelangelo. The unshaded outline at the left is Sangallo's outer northern hemicyle demolished by Michelangelo. Based partly (with corrections) on a drawing by Vanvitelli.

3. View from the north in about 1532. Detail of a drawing by Heemskerck.

rising from a relatively weak drum of free-standing columns around a ring of piers that were not much wider than the intervening voids (fig. 1 right). Indeed, echoing his planned adoption of multiple superimposed orders everywhere else, he even reduced the dome thickness over a substantial height at a higher level to introduce, there, a second purely columnar drum on the outside. This dome would have been about twice the weight of Bramante's dome and the lower drum little stronger than Bramante's drum. To contain the outward thrusts at the foot would have again called for very strong circumferential tying.

1546-1564: Michelangelo

It was Michelangelo who, after Bramante, made the greatest contribution to the design of the church as a whole, both in spatial planning and in plastic expression.[9] He tightened up and unified Sangallo's loosely co-ordinated aggregation of spaces and, in place of Sangallo's piling-up of elements, he showed an unsurpassed mastery in the use of a giant Corinthian order to articulate the now continuous movement of the outer walls (figs 2 and 4). This articulation culminated in the columns and ribs of a single drum and a squatter, almost hemispherical, dome.

During his lifetime construction proceeded as far as the drum as well as embracing the near completion of both transepts after the demolition of all that had yet been built of Sangallo's southern outer hemicycle (fig. 5 and fig. 2 again). According to Poleni it also included a further strengthening of the dome support by additions to the four main arches.[10] For the dome itself, the surviving evidence is limited to several preparatory drawings, engravings of what was probably the intended final design, part of the inner shell of a wooden model that is now incorporated in a later model of the

4. Exterior from the south west.

5. The state of construction in about 1570. Drawing by an unknown artist.

dome finally built, and an early copy of this model of the inner shell.

For his design, Michelangelo turned further to the Florentine model, as might be expected in view of his strong Florentine connections and his intimate knowledge of it (fig. 6 left).

Most obviously he adopted a similar double-shell construction, much lighter than the solid construction proposed previously. Saalman has even doubted whether shells as thin as those shown on the exploratory sketches now in Lille and Haarlem would have been stable.[11] There need be no such doubt if they were stiffened by intermediate ribs and if they were adequately supported at the springing level. But, if the inner shell in the Haarlem drawing was intended to stand independently and if it then cracked radially into a ring of adjacent arches with a common keystone, it would indeed be close to the theoretical limit for stability.

Whether a hemispherical form was finally intended for either or both shells is not entirely clear because some of the profiles sketched are closer to the pointed fifth adopted in Florence. It is also unclear to what extent interconnecting radial ribs were envisaged as in Florence. On a circular plan, such ribs are usually no more necessary than circumferential ones like those that project below the inner faces of the outer shell in Florence. But they could have been seen as advantageous during construction or as conferring added stiffness. The surviving copy model of the inner shell shows broad ribs rising only to the level where a crossed arrangement of stairs ascending both clockwise and anticlockwise in the lower part of the dome terminates and gives way to continuous circumferential stairs for the final ascent.[12]

The chosen profile affects both the support provided for the lantern and the outward thrust that must be contained by

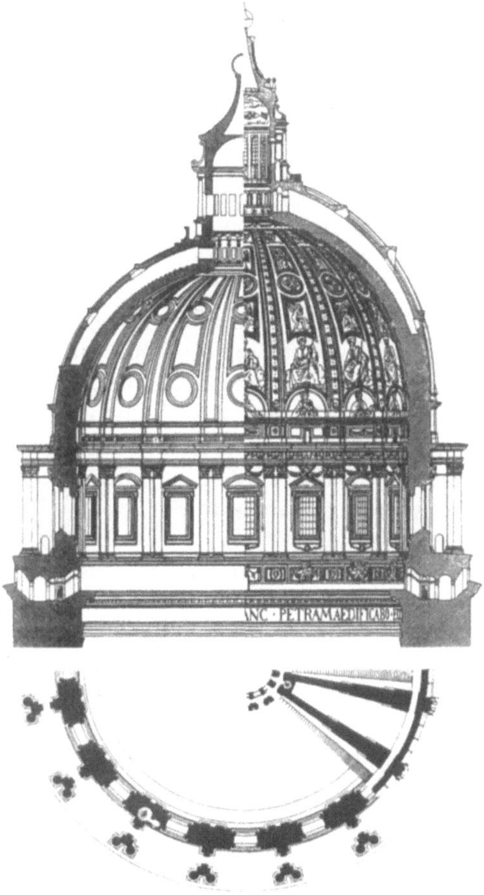

6. Dome designs by Michelangelo (left) and della Porta (right). From Letarouilly, *Projets divers*, plate 24, and *Basilique de Saint Pierre*, plates 7 and 24.

ties or resisted at a lower level by the buttressing capacity of the drum and supporting arches and main piers. For the hemispherical form it would be almost twice that for the pointed-fifth form adopted in Florence, other things being equal. Since (as will be seen later) the drum proved to be barely strong enough to carry the slightly more pointed but taller dome finally built, Michelangelo's design cannot be regarded as entirely satisfactory in this respect. Its weakness lay partly in the weak interconnection of the outer ring of coupled columns to the inner ring of piers between the tall windows. In fact this interconnection is so weak that the outer ring may have been seen as having only a visual role.

The drum was completed in the late 1550s. But, like all its predecessors, the dome remained unbuilt during Michelangelo's lifetime, leaving further doubt about what its final design might have been. For many years after his death, the drum stood open to the sky while attention was directed to such matters as the complete rebuilding of Rossellino's choir to match the transepts as Michelangelo had intended it should and the moving of the Vatican obelisk to its present position.

1588-1591: Della Porta

Construction of the dome started only in July 1588 to a modified design by Della Porta – the first for which we have fairly full details. In particular, it retained the double-shell form, the shells were given a raised profile, and they were interconnected radially over the full height by ribs that diminished twice in width as they rose, one to each pier of the drum (fig. 6 right). The precise internal profile (somewhat flatter than that of the external shell) is difficult to determine in the absence of an accurate survey and of a precise written specification such as exists for the Florentine dome. All that can be said is that the springing level was raised somewhat above the drum and that, above this level, the profile is intermediate between the hemispherical one of Michelangelo's wooden model and the pointed-fifth one adopted in Florence. Both Della Porta's shells are less pointed than those in Florence, the chief gain in height resulting from the raised springing level. As compared with what Michelangelo seems to have intended, they can be expected to thrust outwards somewhat less, are slightly better adapted to carrying the lantern, and would have made the final stages of construction a little easier, though they were not ideal for these purposes. The greater height would, of course, have made the dome more visible over the main body of the church.

Circumferential ties about 6 cm × 4 cm in cross section were incorporated at levels slightly above the drum and slightly above the level at which the two shells divide – the former tie being buried at about mid thickness and the latter one being set on the outer face of the inner shell (fig. 7; see also fig. 18, n and u). These ties should each have been able to contain thrusts of about

7. Interior of the dome at the level where the two shells divide. A section of the original upper circumferential tie is exposed at the lower right a little beneath the surface of the inner shell. The slight thickenings of the shells just beyond this point and again in the distance denote two of the ribs.

5 MN or 6 MN acting around the whole circumference if they were of sound construction and properly jointed to develop their full strengths.

The intensity of vertical load on the piers should have been comfortably within their capacity after the successive strengthening they had undergone. It was, for instance, no more than that on the piers in Florence or that subsequently imposed on Wren's piers at St Paul's.

Questions do, however, arise about how the dome was built, about the role of the ribs, and about the influence of the construction procedure on subsequent behaviour.

8. View looking from the south showing the drum under construction in about 1560. Below it is the temporary structure built by Bramante to protect the shrine, incorporating, at the west, the Constantinian apse. Drawing by Dosio.

Construction

That the pendentives were built without centring is shown by two sketches depicting the crossing while the drum was under construction. (fig. 8) The putlog holes required to support working platforms when there is no centring are clearly visible. The use of a variant of the spinapesce bond in the domes of the octagonal chambers above the vaults that spring from the outside faces of the main piers (marked A, A in fig. 2) indicates that these domes were likewise constructed without using centring.[13] A century and a half earlier, Brunelleschi had built the Florentine dome without using centring in spite of its octagonal plan form – having done so by building it as if it were of circular plan.

Della Porta had the advantage of starting with the circular drum built by Michelangelo. With this advantage as well as all these precedents, to build without centring would seem to have been his obvious choice. Yet it was never claimed that it was the one he took. On the contrary, Fontana states quite explicitly that, as well as incorporating the ribs, he used centring to support them.[14] He even reproduces a drawing of this centring (fig. 9). Rocca, writing much closer to the time, is less explicit.[15] Nevertheless, his account of the construction seems to me to be consistent with what Fontana describes and illustrates. Moreover, so far as I have discovered, neither of these accounts was ever contradicted or questioned when they appeared. Nor was their repetition by Bonani.

Saalman has however argued that construction was undertaken without centring.[16] He has done so not only because, as just noted, it would seem to have been the obvious choice and would have had all its advantages including the absence of obstruction below the working face, but also on three further grounds:

1) the fact that all Della Porta's alterations to Michelangelo's design moved in the

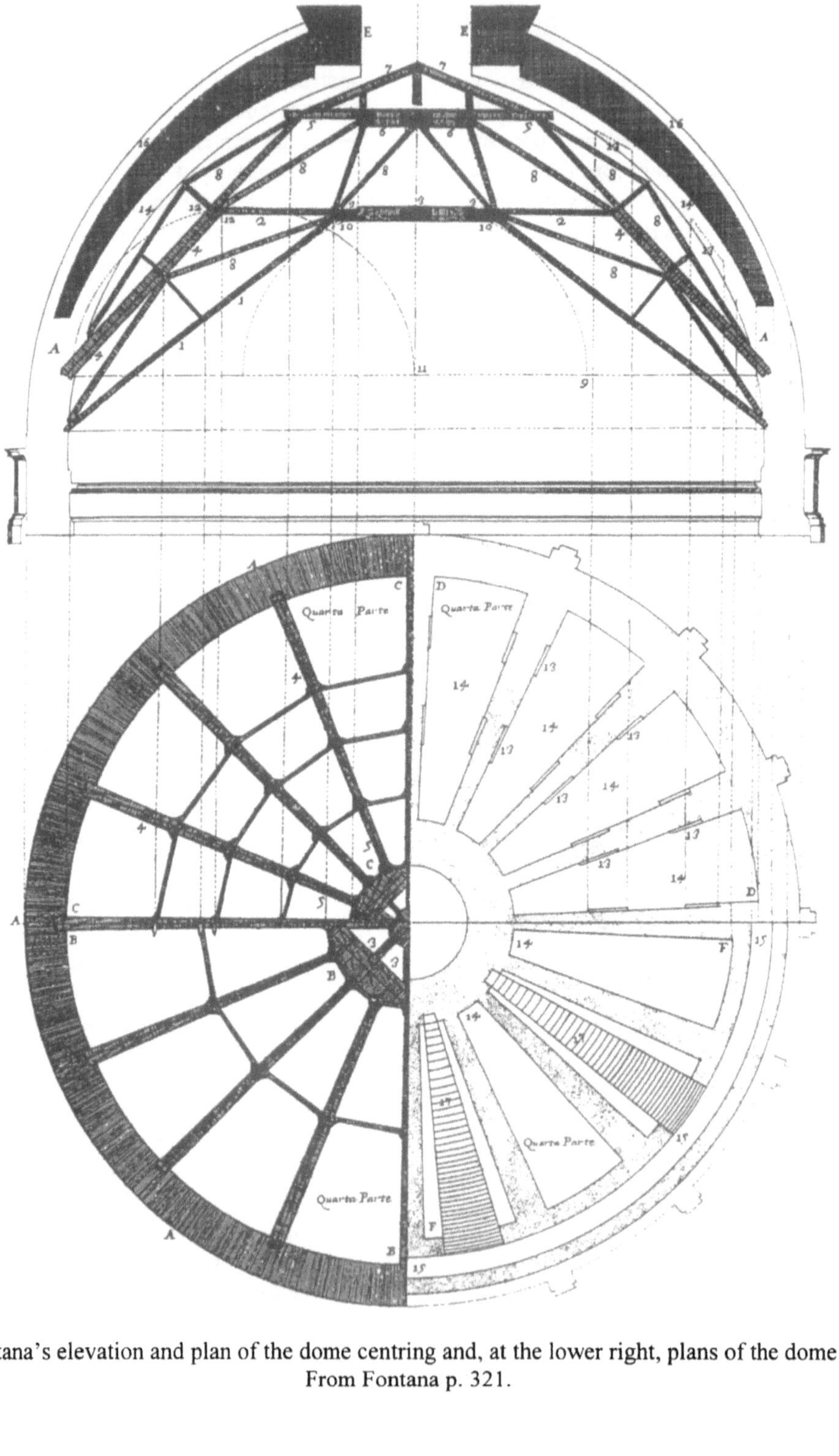

9. Fontana's elevation and plan of the dome centring and, at the lower right, plans of the dome ribs. From Fontana p. 321.

direction of a return to the Florentine model;

2) an alleged paucity of documentation at the relevant time for the acquisition of the timber needed for the centring;

3) a different interpretation of the nature and use of the timber structure that Fontana illustrated as being the centring. Having previously suggested the same interpretation for the similar (but certainly not identical) structure shown in a drawing once wrongly ascribed to Brunelleschi, he took this structure to be a scaffold erected only after the dome was finished to support working platforms for the execution of the mosaics.[17]

Taking these three arguments in turn:

1) The Della Porta design was indeed the closest of all to the Florentine one. A feature inspired by or borrowed from Florence (whether or not it was contemplated already by Michelangelo) was the incorporation of the 16 radial ribs between the two shells. But, had it been desired to follow the Florentine model also in dispensing with centring, and had the model been properly understood, it would have been seen that it was only on account of the octagonal plan form that ribs were needed there.

2) Considerable amounts of suitable timber for the main frames should have been available before work started, from the recent decentring of the new choir vaults and perhaps also from earlier arch decentring, and there should have been even larger quantities from the dismantling of the huge tower erected for the raising of the Vatican obelisk at its new site in 1586. Although there is, as yet, no complete publication of the building accounts there are also records of several fresh purchases to supplement this stock.[18] One is of a quantity of large beams and the others of much larger quantities of elm boards to be delivered between January and May 1589. The combined quantities of the heavier timbers should have been all that would be needed to allow erection of the structure depicted by Fontana to commence shortly after the level was reached at which the lowest ring of inclined struts could be set in place in prepared holes in the masonry and then to continue at intervals over the next few months in step with further advances in the level of the masonry. (It should be remembered here that, even when centring is used, it is not needed until the working face becomes appreciably inclined and that, for a dome, it does not have to be constructed all at once, as for an arch. It can be erected in stages, stabilised throughout by the rings of horizontal timbers.[19]) The lighter boards would have been used as lagging (not shown by Fontana), being nailed to the frames to provide direct support to the masonry.

3) There are several contemporary representations of the arch centring used in the church. The faithfulness of the 1561 engraving by Jacob Bos that was later reproduced by Fontana (fig. 10) and Bonani is attested by both an earlier Bramante drawing and Vasari's fresco in the Cancelleria of construction of the southern arm showing centring *in situ* under the high vaults in 1546.[20] The Bramante drawing shows also the lagging laid over the centre to support the masonry. There are also representations by Zabbaglia (figs 11 and 12) and Dumont of later working platforms in the dome.[21] These platforms were designed for use in repairs to the mosaics in the eighteenth century, but there is no reason to think that the platforms

used for the original setting of the mosaics would have differed greatly. All are very different in character from the arch centres. As well as being less obstructive and visually intrusive, they would have been more appropriate for the smaller loads to be carried, the type of access required, and the direct support available from an already existing structure. The structure which Fontana

10. (right) Elevation of the arch centring used for constructing some of the main vaults. From Fontana p. 413.

11. (left below) Platform for work on the dome mosaics. From Zabbaglia plate 26.

12. (right below) Another platform designed for work on the dome mosaics. From the reissue, c.1773, of the first edition of Zabbaglia.

illustrated as being the dome centring is definitely not of the same kind. In being three dimensional and somewhat lighter, it differs from the arch centres in precisely the ways that centring for a dome would be expected to differ.

Alongside these considerations it is also necessary to consider the implications of a factor that Saalman ignored. This is the difference between the rates of construction in Florence and Rome. As Alberti knew (at least an empirical fact), centreless construction of a dome has to proceed slowly to allow successive rings to gain sufficient strength to act as common keystones to the as-yet incomplete radial arches. Construction on centring has, on the other hand, to proceed as rapidly as possible to minimise distortion of the centring, especially while this is subjected to heavy load in the lower regions only.[22] The 16 years taken to complete the Florentine dome up to the ring on which the lantern stands was appropriate to its centreless construction. The much more rapid completion of the dome of St Peter's up to the same point would certainly have risked failure without the support of centring, at least in the later stages.

Finally there are three other documents that seem to confirm the use of centring.[23] The first two refer to the lifting in place and securing of the hub-like assemblies that key together the radial frames at their upper ends. The third refers to the dismantling, well over a year after the completion of construction, of the '*ponti*' that had been used to allow it to proceed. Although '*ponti*' might be read as 'working platforms' rather than 'centring frames' (for which '*armatura*' or '*armadura*' would be more explicit), it can hardly have had that significance here because there would have been no reason to dismantle working platforms only to have to re-erect them not much later for work on the mosaics. Heavy centring frames would, on the other hand have been an encumbrance.

For all these reasons I accept that centring of the kind illustrated by Fontana was indeed used to support the ribs. Questions nevertheless arise about how construction proceeded on it.

In the first place, there would have been a choice as to the extent to which construction of the ribs advanced ahead of that of the infills between them. Secondly there would have been a choice between constructing the ribs initially to their full depths or only to some lesser depth. Either way, some provision would have been called for to bond the infill sections to them over the depth that had already been constructed before they were added. (This would have been necessary because, in the direction of spanning, they were arched upwards far less than the severies of the typical Gothic ribbed vault or Renaissance ribbed gored dome.) Finally there would have been a choice between constructing the infills with or without temporary direct support. If the latter alternative were chosen, some temporary staging for the workmen would be required. But the ribs would be helpful in limiting the individual spans in much the same way (though not to the same extent) as did the spiralling bands of upright bricks in the spinapesce bond used in Florence.[24]

When Saalman and I inspected the dome in 1974, a limited exploratory excavation had been made into the infill alongside one of the ribs at about mid height. For what it was worth, it showed all the bricks laid on a normal conical bed and some of them apparently bonded into the rib. We found no evidence of a spinapesce-type bond. But the excavation penetrated only about half the full depth, leaving open the possibility that there was a spinapesce-type bond on the inner face. The observed bond between infill and rib could have been achieved by first constructing the rib with a vertical face, but over part only of the full final depth of the infill, then infilling to this depth and finally completing the construction of both rib and infill in a single operation. Or it could have been achieved by first constructing the rib to the full depth of a later infill, but with bricks projecting from its face to bond this infill to it.

In his description of the dome and the construction procedure, Fontana is explicit about the prior construction of the ribs and adds some details, not now open to inspection, of the seating of the infills on them. But he leaves open the likelihood that the prior construction was over less than the full depth and the likelihood that it did not involve constructing the entire ribs before starting on the infills. The relevant passage reads as follows:

> These [large walls] rise like arches up to the crowning lantern and are situated over the void in line with the sixteen spurs of the drum to carry the weight of the whole structure including that bearing on them above their closing ring. These walls rise equally spaced and each curving in the same way to define the interior and exterior profiles of the dome. As they do so, there are two set-backs in their faces to provide seatings for those parts of the vault which they support and which give the two shells their form. ... These 16 walls, known as ribs, were constructed before everything else and then left to gain strength to provide firm support for the shells.
> Between them are set 16 part-spherical sections of the vault, 6 palmi and 2 uncie [1.38 m] thick at their feet, which rise with little diminution in thickness up to the lantern with the curvature all round defined by the supporting ribs. These sections transfer their weights to the prepared seatings [in the ribs] ... and are constructed of the best brickwork laid in a spinapesce pattern to give a good bond and achieve perfect consolidation.[25]

Since this account is not inconsistent with the other early evidence and with what can be seen today, I am inclined to take it as a correct description of what was done, leaving open the precise nature of the bond within the infills, and both possibilities for achieving the observed bond between them and the ribs. The seatings referred to are illustrated, somewhat ambiguously and inconsistently, in the accompanying plate. (fig. 9 right, where they are numbered 13) On the basis of the documented dates of payments for the second chain and for the first two rings of *occhi* for the outer shell,[26] I think that it is most likely that the ribs were constructed initially over only part of their depth, and that further construction, together with that of the infill sections of the two shells, followed on, not more than a few months behind. This further construction would not impose as much additional load on the centring as might be imagined because the infills would progressively complete the horizontal rings of which they formed part, and, as their brickwork gained strength and stiffness, these rings would become self-supporting.

How much of this was understood and what part it played in the decision to introduce the ribs we cannot tell. There are several other possible reasons for the decision. Most obviously, they followed Florentine precedent and even went beyond it in channelling the weights and thrusts of the dome and lantern to the isolated piers and columns between the drum windows. They would also have further speeded the construction process by facilitating the organisation of a workforce variously numbered as 800 and 600.[27] I suspect that all these considerations played their part, with the first two probably having been the primary ones unless Della Porta was, from the start, under great pressure to complete construction in the shortest possible time.

Construction was completed up to the

base of the lantern on 12 May 1590, a mere 21 months after work started.[28] Then, on 28 September 1591, came the payment already referred to for 34 days work in removing the centring.

Work on the mosaic decoration commenced in 1592 and continued until at least 1604 or 1605.[29] It would have called for working platforms of the kind illustrated by Zabbaglia, though not necessarily over the whole dome surface at one time. The inevitable cracks in the dome brickwork associated with drying shrinkage of the mortar and adjustments to changing loads as construction proceeded do not seem to have caused enough concern for it to have been recorded, and they would have been covered by the mortar setting beds for the mosaics as work proceeded.

Growing concerns for stability and the state of the dome in 1742

Over the years cracks, once formed, always tend to grow. When they were first noticed again is not clear. In April 1680 they led to rumours that the dome was in danger.[30] Bernini, who had just died, was blamed because he had made small modifications to the main piers as part of the works he had undertaken to embellish the interior for Urban VIII. But he was justifiably exonerated after careful inquiry by Mattia de' Rossi. The fears were judged to have been exaggerated, and no action was taken.[31]

Further concerns arose after an earthquake in 1730. In 1735, swallow-tail spies, similar to those that had been used earlier in Florence, were inserted to monitor future movements.[32] Matters came to a head a few years later. On 20 September 1742 the architect Vanvitelli, having made a close inspection, presented a report recommending a number of remedial measures and thereby initiated a period of widespread consultation and intense debate about what, if anything, should be done.[33]

One legacy is a remarkably full record of the state of the dome at this time, based on a further survey next year by Vanvitelli. He produced detailed drawings, later engraved and published by Poleni (fig. 13), showing, in elevation, all visible cracks and giving measurements of the widths of many of them.[34] These records are supplemented by a plan at drum level published later by Rondelet (fig. 14). Allowing for the fact that only surface cracks could be shown, and that, where the true surface of the working masonry was covered by rendering, mosaic, or external lead weathering, the cracks were hidden or would not show at their full width, an excellent overall picture is given.

Most of the cracks were of one of five kinds:

1) Radial cracks of the kind seen in all masonry domes. These were only partly visible, being hidden on the exterior of the dome by the lead weathering and almost certainly showing less than their full width and extent internally where they ran through the mosaic.
2) Horizontal cracks through some supporting elements of the drum, most obviously through the piers between the windows.
3) Inclined or near-vertical cracks through the supporting main arches and pendentives.
4) Steeply inclined cracks through the walls linking the piers of the drum proper with the outer ring of coupled columns.
5) Vertical cracks running circumferentially around the base of the drum.

If gravity loading only is considered, the first two can have resulted only, in the usual way, from the circumferential tensions

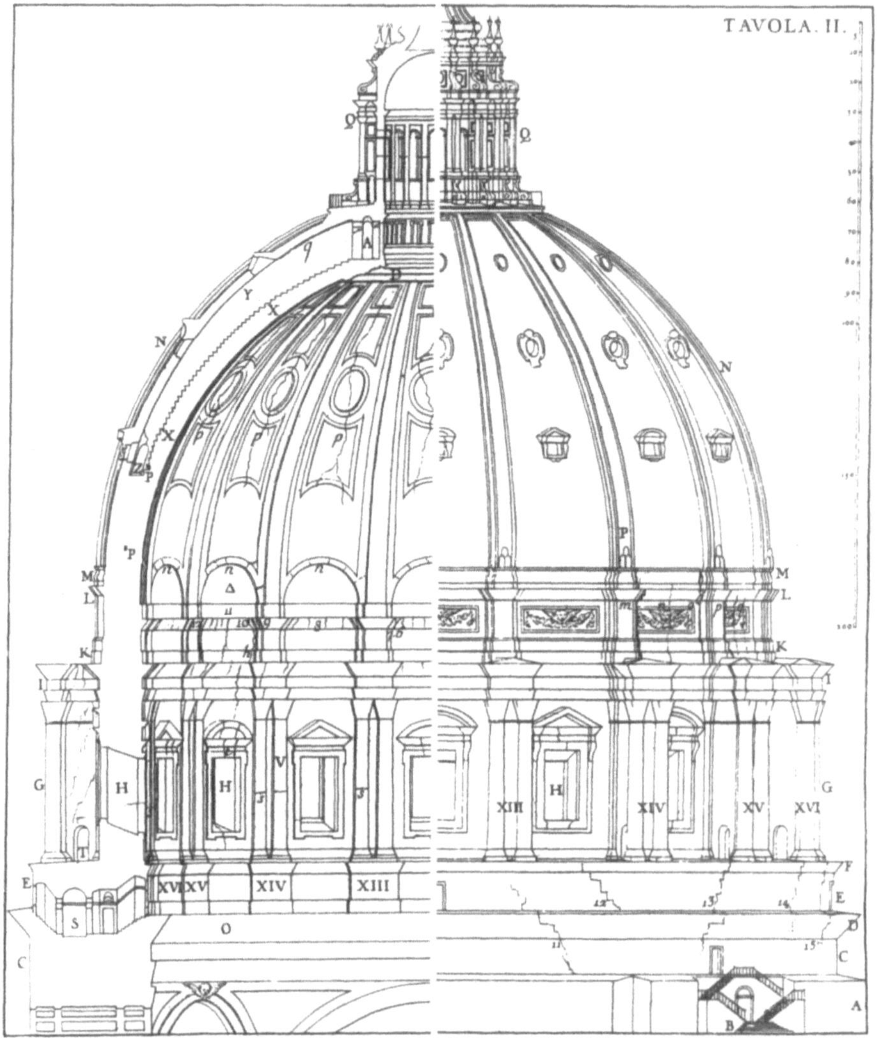

13. Plots of cracks observed by Vanvitelli in 1743 on the interior and exterior of one quadrant of the dome. From Poleni, plates XVI (left) and II (right).

that arise even in an ideally supported dome and that then extend further when the support is less than ideal and moves outward under the ensuing outward thrusts. The presence of cracks in every infill section and the fact that there were sometimes several visible in a single infill is in marked contrast to the much smaller number of larger cracks in Florence even today,[35] and is fully consistent with the manner and sequence of construction that has been postulated above.

The third, which must have been associated with some depression of the arch crowns, can have resulted only from some outward movement of the piers and spreading of the main arches. Thanks to earlier reinforcements of the piers and Michelangelo's reinforcement of the arches, the cracks were not large enough to have constituted a major risk.

For the fourth and fifth, there were, however, two possibilities. In part, at least, they must have been a further consequence of the circumferential tensions and strains already referred to – a counterpart of the first two kinds of crack. But they could also have resulted partly, or even primarily, from the greater intensity of vertical load on the piers of the drum proper than on the coupled columns around it, and the relative weakness of the masonry of the piers as compared with the near-monolithic columns if Rondelet's later depiction of their construction is correct. These disparities

14. Further records of cracks. From Rondelet, *Traité de l'art de Bâtir*, bk. IV, 1838, pl. 66.

would have caused the piers to settle more than the columns and perhaps part company with them. Only closer examination and deeper exploration of some of the cracks would have made it possible to say which of these causes was the more important.

The only other types of load that might have had an effect were those resulting from the earthquake or from a lightning strike. A lightning strike can cause masonry to burst apart by rapidly vaporising the moisture within it. But it could not have led to such a widespread and symmetrical pattern of cracks in a structure as large as the dome. The earthquake in 1730 could have led to an enlargement of cracks already existing, but the radial symmetry of those recorded a decade or so later shows it could not have been their prime cause.

Formal inquiries and strengthening measures 1742-1748

The situation was not, in one way, unprecedented: in another way it was.

Other large domes had cracked similarly and had been strengthened by encircling them with iron ties. Such ties had been applied fairly recently at San Marco in Venice and the cathedral of Montefiascone as well as to several smaller domes.[36] Following this precedent, one of the measures recommended by Vanvitelli in his 1742 report had been the addition of three or four new circumferential ties at St Peter's to help restrain the oblique thrust of the dome.[37]

The chief unprecedented aspect was that, over the previous half century, there had been a great growth in scientific interest in and understanding of the principles that governed the physical behaviour of structures like arches and domes and components like iron ties. This was all the more important because the reigning pope, Benedict XIV, was a scholar well aware of and keenly interested in developments of this kind. Very concerned as he was for the future safety of the dome, he commissioned three of the leading mathematicians of the day, Thomas Le Seur, François Jacquier, and Ruggiero Boscovich, to prepare a report for him giving their opinion. This, their *Parere*, proved to be a document of the greatest historical significance. It was presented on 8 January 1743 and published, by papal command, two days later.[38]

In a remarkably short time, the three mathematicians first familiarised themselves with the dome by making numerous inspections and itemising [39] and depicting (fig. 15) the damage they observed. Reflecting on its significance, they discounted relative settlements as a major cause of damage and saw, as the only continuing risk, the way in which the dome, weighed down also by the lantern, was thrusting out on its supports. As evidence of this thrusting action, they referred not only to the cracking in the dome but also to both the damage to the buttress-like walls connecting the coupled columns around the drum to the piers between the windows and the horizontal cracks in these piers alongside the windows.[40]

Having thus identified the risk to be guarded against, they next took the crucial step of devising a simple model of the structure that was amenable to calculation by recently developed theory. This model envisaged the actual structure as being split by the observed radial cracks into a number of radial slices, each of them consisting of a small number of rigid blocks free to rotate about effective hinges where they had seen horizontal cracks or could reasonably infer them from what they had seen. Internal resistance to rotation about these 'hinges' was ignored. But the resistance offered by the two circumferential chains was taken into account.[41]

Two limiting cases were considered. In the first it was assumed that the connections between the piers of the drum and the coupled columns were unbroken, so that they would rotate outwards as a single body. In the second it was assumed that there was no effective connection and no resistance to their rotating independently of one another. In the first case it was deduced that there would be an ample margin of safety even without the chains: in the second there was a major deficiency even if the two original chains remained unbroken.[42]

Unfortunately they did not give full details of their calculations, or even sufficient detail of some of the assumptions made to permit a precise working back from the results that they quoted. It is at least clear that, after performing tests to ascertain the densities of the materials and calculating weights, they made what would now be called virtual-work analyses for the potential collapse states that they had identified. In these, the chains would be working at their full yield strengths, and they correctly deduced from a similar analysis that each would exert an inward force around the whole circumference equal to that yield strength multiplied by the ratio of circumference to radius.

A check on the figures that they gave for the calculated overturning and restoring 'moments' shows that these are consistent with a derivation from the radial displacement of each chain around its whole circumference and the vertical displacement of the centre of gravity of each rigid block for a unit outward radial displacement at an assumed 'hinge' at the foot of the dome and with other hinges at the base of the lantern and at the foot of the drum. Because of the radial symmetry, the displacements could be determined graphically from scale drawings of a representative cross section. The calculated weights were for the whole ring of radial slices into which dome and drum were assumed to have split, so that the calculated radial force was the net total force around its circumference.

Equally unfortunately they did not fully explain the significance of the assumptions implicit in their choice and modelling of the two limiting cases and the bearing that these had on any application of the results to the real structure. Their first case was too optimistic, as they realised. Their second was too pessimistic. But it did yield

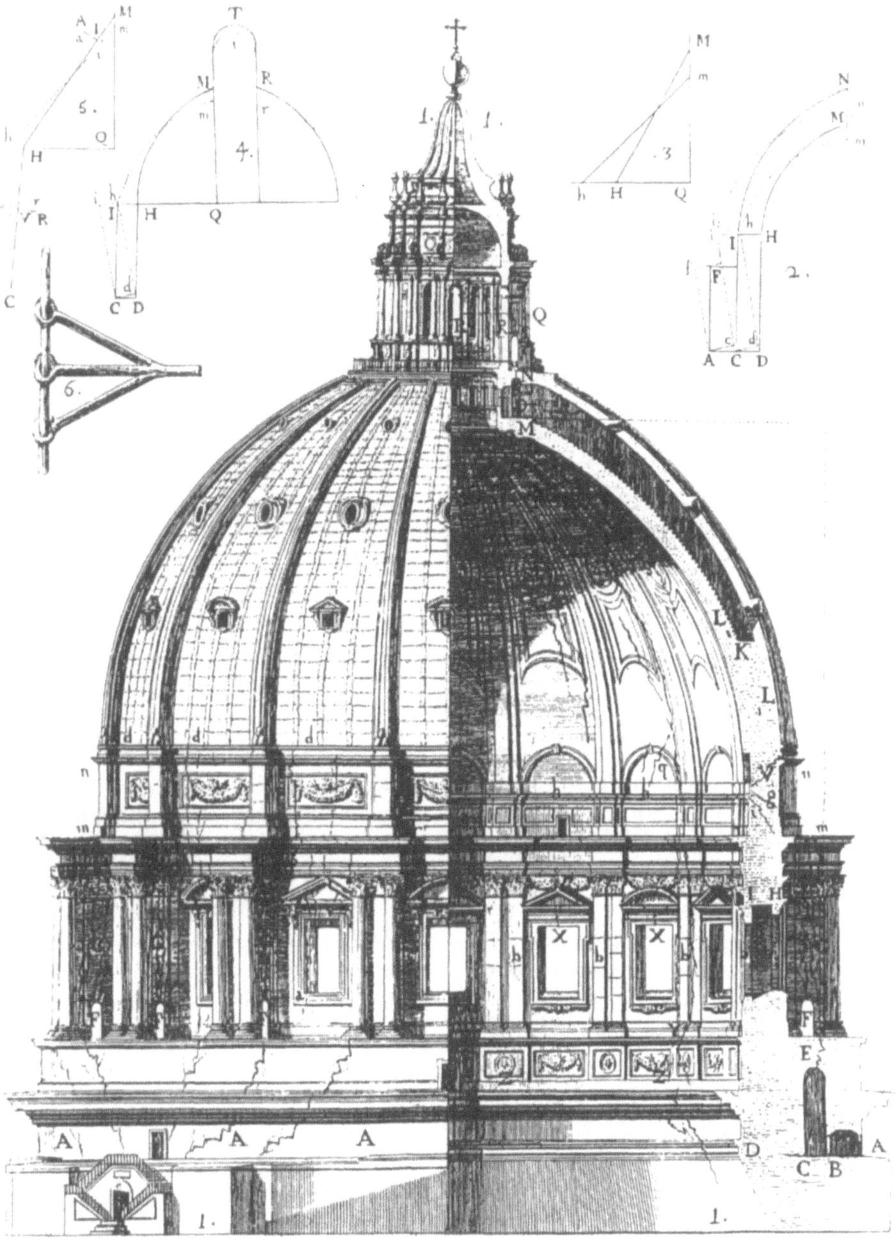

CVPOLA DI S. PIETRO

15. Internal and external part elevations of the dome showing cracks observed by the three mathematicians in late 1742 and their visualisation of possible collapse modes. From *Parere di tre mattematici*.

a figure for the strength required of an additional chain to place the future safety beyond all doubt. This had never been done before.

The novelty appears to have been too much for those who, later in the month and again next month, attended meetings to discuss the report. There is no record of any direct criticism of the calculations themselves. They seem to have been simply ignored.[43] As a result, the three mathematicians, in their *Riflessioni* published a little later, added no more detail. They contented themselves with restating their diagnosis and proposals, supporting them with further observations and arguments, and recommending further monitoring.[44]

The pope responded almost immediately, on 12 January, by seeking a second opinion from another leading mathematician and scientist, Giovanni Poleni.[45] His first report – his *Riflessioni* sent to Rome on 21 March – was based only on a study of the *Parere* of the three mathematicians and other material sent to him in Padua.[46] When it had been received he was invited to Rome to make a personal inspection and report further. This further report – his *Aggiunta* which again recommended the installation of new chains – was based also on the detailed survey by Vanvitelli that was undertaken for this purpose and has already been referred to in the discussion of the state of the dome.[47] But, unlike the reports of the three mathematicians, neither of his reports was published at the time. One reason for this was probably that the first was only an interim one, and that work on the installation of new chains was put in hand with little delay after receipt of the second. Another was that, by this time, many other views had also been expressed, and Poleni was soon afterwards commissioned to write a complete account of them and of the actions taken. All reports including his own, together with other expressions of opinion, were summarised in books III and IV of this account, which was published five years later as his *Memorie Istoriche*.[48]

Since it is Poleni's views that are of greatest interest alongside those of the three mathematicians, the views contributed by others must be passed over very briefly. They nearly all concentrated on the causes of the present damage rather than, like the three mathematicians, attempting to identify among them those aspects that might get worse and should be prevented from doing so. Most saw the great weight of the dome and lantern and poor hasty construction as the prime causes, and several specifically criticised the model of a possible collapse mode proposed by the three mathematicians, as being inconsistent with what could be observed. In spite of this, no less than nine of the fourteen proposals for strengthening favoured new chains and only two opposed them.

The *Memorie Istoriche* also contain, in book I and the latter part of book IV, much fuller accounts of Poleni's diagnoses and explorations of possible remedial works than are to be found in the summaries of his *Riflessioni* and *Aggiunta* in book III – much fuller also than those to be found in the reports of the three mathematicians (which were, for their time, models of brevity and clarity for the most part). He not only goes into far greater detail about current theories of arch behaviour and such things as the response of materials to load and temperature change than they considered necessary. He also gives far more detail of at least some calculations that he made and equally full detail of numerous experiments that he undertook.

How far these accounts present explorations undertaken to arrive at his recommendations of June 1743, and how far explorations undertaken at leisure only some time later, it is now impossible to say. Probably they were largely the latter. Even setting doubts on this score aside and taking these amplifications alongside the summaries of his *Riflessioni* and *Aggiunta*, there are serious lacunae. There is also a seeming inconsistency in his attitude towards the ideas of the three mathematicians which adds to the difficulty of fully understanding his approach.

In the summary of his *Riflessioni*, after praising their erudition, he criticises their diagnosis of the damage.[49] Like others, he prefers to interpret the cracking as chiefly a consequence of the great weight of the dome bearing down on poor hasty construction, and does not look beyond such past causes to others that pose a greater present threat because they are more likely to worsen with time. In particular he criticises their models of potential collapse modes.[50] He does this on the basis of a drawing giving his own three-dimensional interpretation of their fig. 2, which not only takes considerable liberties with perspective for greater clarity (as he admits) but seriously misrepresents it in a way that he ignores. It shows the radial cracks that divided the whole structure into separate blocks far more widely spaced than they are in reality, so that 'hinging' cannot take place as they assumed (fig. 16).

Having rejected their model, he turned, at least in his fuller exploration, to a simpler two-dimensional one of the dome alone, again considering it as cracked radially into lunes, but now supported on an unyielding drum.[51] Ignoring the internal voids between the shells and applying directly the early insights into arch behaviour of Hooke, La Hire, and Stirling (by this time well known), he modelled an inverted lune as a chain whose links carried lead weights proportionate to the weights of voussoir-like sections into which it was notionally divided. (fig. 17) These weights he calculated by dividing an arbitrary 1/50th of the three mathematician's estimate of the total weight of the dome proportionately to the horizontal distances of the centres of gravity of the sections from the vertical axis of rotation. Since, when the chain was supported to hang with its upper ends and lowest points in the centre of the cross-section of the inverted lune, it hung within the cross-section of the ribs throughout their height, he concluded that the dome was not yet in danger: '*che non è già cattiva la figura della gran Volta*'.[52] This, with minor qualifications, was perfectly valid. But it was seriously wanting in that it failed to investigate the real danger from spreading of the support provided by the drum.

Today we can easily see that he could have rectified this shortcoming in two ways. Either he could have used the model as it was to determine by calculation or further experiment the inward force or forces that would be needed to make the thrust on the drum more nearly vertical. To make it vertical at the top of the drum (going beyond what was necessary) would, for instance, have required a total inward force of about 13,000,000 libbre (45 MN) around the circumference at this level. Or he could have extended the experiment to embrace the drum as well as the lantern and dome by carrying the hanging chain

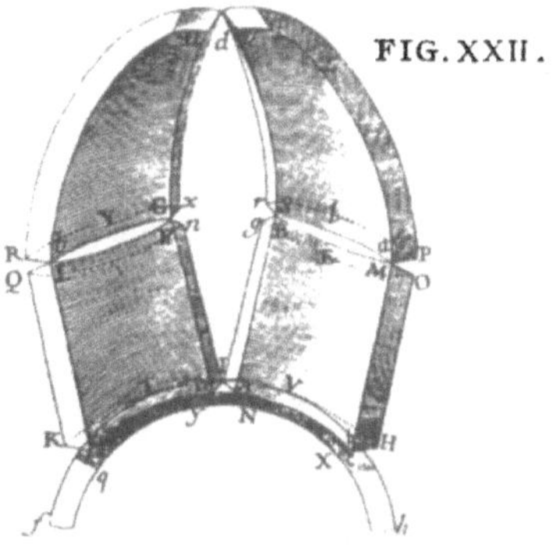

16. Poleni's representation of the collapse mode visualised by the three mathematicians. From Poleni plate H.

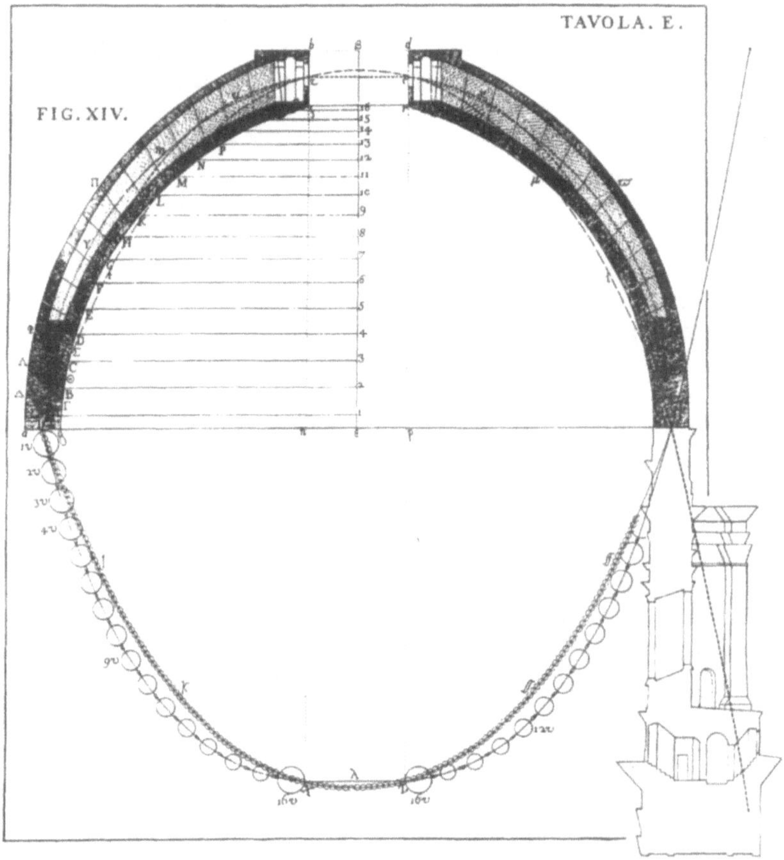

17. Poleni's determination of a thrust line in the dome and, on the right, an extension of the line to the base of the drum. From Poleni plate E, with author additions.

down to the base of the drum. This would have given the complete thrust line as shown in the addition to his plate in fig. 17. Had he wished, he could then, by similar further experiment, have explored the full range of tying possibilities. The experimental approach, which he seemed to prefer, would merely have entailed attaching horizontal weighted cords pulling outwards on the chain to represent the inward forces.

He appears to have done neither. Nor is there any other indication of how he arrived at the desirable strengths of the chains that he did recommend.

This is all the more puzzling because he wrote at great length, not only about his general reasons for recommending chains, but also about the effective strength of a chain of a particular cross-section. His argument in favour of installing them was that they would confer added strength in the same way as a tie across its feet gives added strength to an arch.[53] Though he did not regard collapse as imminent, they would similarly forestall it if the situation worsened in the future. In his explorations of strengths he conducted his own tests on sample straight bars of iron and even undertook an experimental verification of the relationship deduced by the three mathematicians between the tension in a circular tie and the inward force that it would exert around its circumference.[54] But when he refers in book IV to his recommendation that there should be four chains at different specified levels, each 5 *oncie* deep and 3 *oncie* wide (9.1 cm × 5.6 cm), he calculates their strengths and weights from his tests but says nothing about how these strengths were chosen as the appropriate ones.[55] In the absence of any word about this, it is difficult to avoid the conclusion that he simply accepted (with a slight rounding off of the height) the sizes proposed by the three mathematicians in spite of having rejected the model from which they were derived.

The overall impression that he gives is that he was not fully at ease when attempting to apply the new theories to complex practical situations to which they did not have an immediately obvious relevance. He fought shy of making the imaginative leaps to abstract models of the behaviour to be investigated that must always precede the use of abstract theory to analyse it. His preference for experiment, using simpler physical models of a kind already used or envisaged by others, is part of the same picture.

The outcome of his effective endorsement of the earlier recommendation by the three mathematicians for additional chains was, nevertheless, their installation by Vanvitelli between August 1743 and June 1744 in accordance with his instructions.[56] The added chains are shown at A, B, C, D, and E in fig. 18.

An early unexpected sequel (though something the three mathematicians had foreseen as a possibility) was the discovery in 1747, during further rehabilitation work, of two breaks in the upper original chain.[57] They did not seem to be recent and they suggested a strong likelihood that the buried lower chain was also broken since it would have been stretched more. Although no movement had been detected during and since the installation of the new chains, a further chain was nevertheless installed on the exterior slightly below the level of the upper broken one as seen at Z in fig. 18.[58]

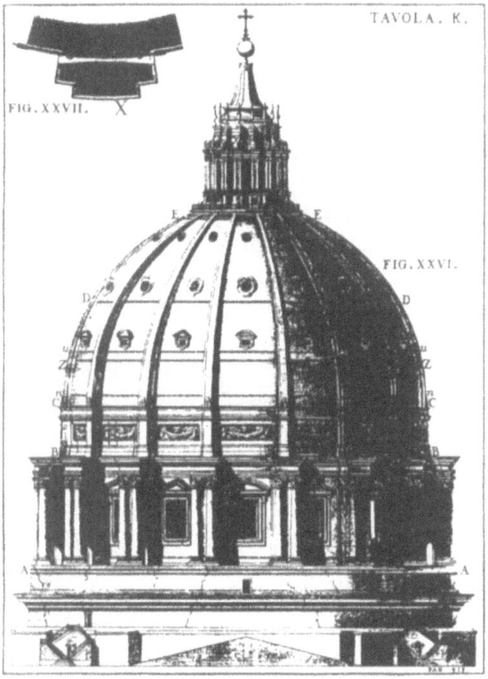

18. Positions of both the original (*n* and *u*) and added circumferential chains. The manner of installation of the added ones is shown in fig. XXVII. From Poleni plate K.

Thus reinforced, the dome has since stood secure.

The wider perspective

Both the chequered history of the dome's design and construction and its later vicissitudes contrast strongly with the history of the dome in Florence. Each history is part of the larger history of the church of which it is a major feature and stemmed in part from this. The chief underlying difference, apart from those stemming from the facts that the cathedral in Florence was built first and much earlier, was that control was in very different hands.

In Florence there was considerable continuity of control in the hands on one of the major guilds, the *Arte della Lana*. There were periods when progress was slow and others when it was faster. But there seems to have been only one major change in plan in 1366/7 when construction had advanced from the west end only part way down the nave, well short of the crossing over which the dome was to rise. After that change, the design of the dome remained essentially unchanged and construction advanced slowly and systematically towards a successful conclusion 70 years later. Concerns were expressed later about cracks. But these were far less serious than at St Peter's and they gave no cause until very recently for the sort of investigations that have just been described.

In Rome there was no continuity of control. When a new pope succeeded, work might virtually cease or there might be a fresh urgent burst of activity with a major change of plan. Many of the vicissitudes stemmed from this very different pattern of progress. Not a little of the final problem must have stemmed from the haste with which work on the crossing piers and main arches was undertaken under the ageing Bramante and the similar haste with which the dome itself was built almost a century later after long periods of inactivity and several major changes in design. Harm must also have resulted from exposure to the weather of the incomplete structure during the periods of inactivity. When concern over its extensive cracking could no longer be ignored, it was thanks as much to the pope of the day as to the new interest in the strength of materials and the statics of arches and domes that they were tackled as they were, and not (as such problems had usually been tackled previously) on the basis only of the opinions of practical men based directly on their past experience and intuitive understanding.

This change was truly epoch making, even though the application of the new theories presented problems – even greater problems than those that were to arise later in early applications to the design of new structures. The three mathematicians had the best grasp of what was needed. But they were never given the chance to develop their approach further to make proper allowance for the underlying assumptions. We have since learnt to do this better. But we still have to follow essentially the same path and still encounter similar problems, especially when assessing the safety of existing structures rather than designing new ones.

Notes

†This paper is an expanded version of the main part of a contribution to the conference on the dome of St Peter's at the Architectural Association, London on 21 March 1991. Other contributions were by Howard Saalman on 'St Peter's dome projects from Alberti to Giacomo della Porta: the Pantheon and Santa Maria del Fiore options', Henry Millon on 'The dome of Michelangelo's St Peter's: the closed oculus', and Werner Oeschlin, on 'The cupola of St Peter's in the eighteenth century context'.

1. The chief early published sources are Fontana, C., *Templum Vaticanum et ipsius origo*, Rome, 1694, especially pp. 315-323, and Bonani, F., *Numismata summorum pontificum templi Vaticani fabricam indicantia*, Rome, 1696. Letarouilly, *Le Vatican et la Basilique de St. Pierre de Rome*, Paris, 1882, sections (c) and (d), is a useful compendium of successive plans and sections (reduced facsimile edn, *Vatican 1*, Tiranti, 1953), and Ackerman, J. S., *The architecture of Michelangelo*, London, 1961, vol. 2, pp. 83-97, is the most convenient summary of the whole sixteenth-century construction process.
2. Saalman, H., 'Die Planung der neu St. Peters', *Münchner Jahrbuch der bildenden Kunst*, vol. XL, 1989, pp.103-140.
3. Discussed most recently in Metternich, F. G. W. and Thoenes, C., *Die frühen St. Peter-Entwürfe 1505-1515*, Tübingen, 1987.

4. Serlio, S., *Il terzo libro d'architettura,* Venice 1540, pp. xxxix, xl.
5. 'vorria bonissimo fondamento a farla sicura, non che a farla sopra a quattro archi di tanta altezza', Serlio, ibid., p.xxxix; echoed on p. xl.
6. Vasari, G., 'Vita di Bramante da Urbino', p. 96, in *Vite de piu' eccelenti pittori, scultori ed architetti,* vol. 3, Florence, 1771. See also pp. 92-93.
7. 'I pilastri gia fatti con i suoi archi, senza altro peso sopra, gia si risenteno, e sono crepati in alcuni luoghi', Serlio, op. cit., p. xxxix.
8. This model, normally kept in one of the domed octagonal rooms above the vaults of the church, was the centrepiece of the superb exhibition of Renaissance architectural drawings and models held in the Palazzo Grassi, Venice, a few years after this paper was presented. It is well illustrated in the exhibition catalogue, Lampugnani, V. M. and Millon, H. A. eds, *The Renaissance from Brunelleschi to Michelangelo: the representation of architecture,* Venice and London, 1994, pp. 35-47 and 632-633.
9. Michelangelo's contribution as a whole is best covered in Ackerman, J. S., op. cit, note 1, vol.1, pp. 89-102 and vol. 2, pp. 87-112. For his designs for the dome see also chiefly Millon, H. A. and Smyth, C. H., *Michelangelo Architect: the façade of San Lorenzo and the dome of St Peter's*, Milan, 1988.
10. Poleni, G., *Memorie istoriche della gran cupola del Tempio Vaticano*, Padua, 1748, art. 44 and fig. IV, pl. A.
11. Saalman, H., opening paper to the conference.
12. Illustrated, together with related drawings, by Millon and Smyth, pp. 159-167 and fig. 22.
13. The dome in the position of the lower 'A' in fig. 2 is seen on the vault behind the tip of the obelisk in fig. 5 and the interior of another is illustrated in Mainstone, R. J., 'Brunelleschi's dome', *Architectural Review*, vol. CLXII, 1977, p. 320. [Chapter XVII, p. 326, this volume]
14. Fontana, op. cit. (note 1), pp. 316, 319, 321.
15. Rocca, A, 'De tholo Basìlicae Sancti Petri', *Bibliotheca Apostolica Vaticana a Sixto V: Appendice*, Rome, 1591, pp. 417-423, especially p. 420.
16. In his opening paper to the conference and in prior correspondence with me.
17. Saalman, H., *Filippo Brunelleschi: The cupola of Santa Maria del Fiore*, London, 1980, p. 93.
18. Abstracts of the accounts have been published by Orbaan, J. A. F., in 'Zur Baugeschichte der Peterskuppel', *Jahrbuch der Königlich Preuszischen Kunstsammlungen*, vol. 38, 1917, Beiheft, pp. 189-207, and Di Stefano, R., *La cupola di San Pietro*, Edizioni Scientifiche Italiane, Naples, 1963, pp. 3-29, rev. edn, 1980, pp. 1-22. The relevant timber purchases are recorded in Orbaan, docs 47 (9 January 1589), 48 (12 January 1589), 84 (4 May 1589) and 99 (30 June 1589).
19. In this connection it is also worth noting the comments of Rondelet, J. in *Traité théorique et pratique de l'art de bâtir*, 8th edn Paris, 1838, bk V, p. 158, on the reduced centring requirements for a dome, with particular reference here to his construction of the dome of the Paris Pantheon.
20. Both drawing and fresco are reproduced in Metternich, F. G. W. and Thoenes, C., op.cit. (note 3), pp. 188-193, and the fresco in Ackerman, op. cit. (note 9), pl. 52b.
21. Zabbaglia, N., *Castelli, e ponti*, Rome, 1743, and reissue, c. 1773; Dumont, G. M., *Détails des plus intéressantes parties d'architecture de la Basilique de St Pierre de Rome*, Paris, 1763, pl. 54.
22. Alberti, L. B., *De re aedificatoria*, Florence, 1485, bk. III, ch. XIV.
23. Orbaan, docs 61 (9 March 1589) and 85 (11 May 1589); Di Stefano, p. 17 (28 Sept 1591), rev. edn, p. 13.
24. Mainstone, R.J., 'Brunelleschi's dome of S.Maria del Fiore and some related structures', *Transactions of the Newcomen Society*, vol. XLII, 1969-70, p. 114 [chapter XV, this volume]. See also idem, 'Brunelleschi's dome', (op. cit., note 13), pp. 161, 163.
25. Fontana, bk V, ch. XIV, cols 316-317 (my translation). The Italian text reads: 'S'inalzano questi [gran Muri] anco arcuatamente, fino al terminazione del Lanternino, e sono situati sopra il Viuo, & à corrispondenza delle sedici Ale del Tamburo, destinate per sostenimento della Mole, & al peso superiore nella loro unione. Si scorgono questi Muri situati in uguale distanza, e di simile piegatura, che ferue di guida a'Sesti interiori, & esteriori. Nell' elevazioni loro si ritirano con due Riseghe per banda per ritegno di quelle parti di Volte, che sopra di esse posano, e danno forma alle due Cupole. ... Questi 16. Muri, chiamati costoloni, prima d'ogni altro furono eretti e stagionati per ricevere con fermezza l'incarico del Sesto dell'edificio.

'Fra questi spiccano sfericamente 16. Falde, cioè parti di Volte, di grossezza nel suo nascimento palmi 6. un 2. quali s'inalzano con poca diminuzione di grossezza fin al Lanterno con la piegatura, che accompagna il Sesto intorno, regolato da quelli Muri, che lo sostengono. Quelle medisime vanno à scaricare il loro peso sopra quelle Intacche, ò Riseghe ...; e sono lavorate d'ottimi mattoni à guisa di Spinapesce, per ottenere l'abbracciamento della materia, e per causare una perfetta consolidazione.'
26. Orbaan, docs 76 (6 April 1589), 80 (20 April 1589), 87 (25 May 1589), 93 (31 May 1589), 94 (25 June 1589), 126 (22 September 1589).
27. Rocca, A., op. cit. (note 15) gives the number as 800. Fontana, bk V, ch. XIV, p. 316, states 600, as does Bonani, p. 93 ('sexcenti noctu diuque, Pontifice extimulante, collaborant operarii').
28. Other sources give the date as 13 May and 14 May.
29. Di Stefano, pp. 18, 19 (1592, 1604), rev. edn, pp. 13, 14.

30. Baldinucci, F., *Vita del Cavaliere Gio. Lorenzo Bernino*, Florence, 1682, p. 89.
31. Ibid., pp. 91-108.
32. Poleni, G., *Memorie istoriche della gran cupola del Tempio Vaticano*, Padua, 1748, art. 201. For the earlier use in Florence (in 1695) see Nelli, G. B., *Discorsi di Architettura*, Florence, 1753, p. 15 and pl. III, fig. 2.
33. Vanvitelli's report is summarised as 'Breve parere di N. N.' in Poleni, op. cit., arts 455-460.
34. Ibid, arts 237-238 and 243-247, and pls I-XIX. See also Di Stefano, p. 118 and figs 39-40, 42, 44-47, rev. edn, p. 89, and figs 54-58, 61-66, and idem, *Luigi Vanvitelli ingegnere e restauratore*, Edizioni Scientifiche Italiane, n.d., pp. 217-219 and figs 414-428.
35. Opera di Santa Maria del Fiore di Firenze, *Rilievi e studi sulla cupola del Brunelleschi*, Florence, 1939.
36. Poleni, arts 183-188.
37. Ibid., art. 458.
38. Le Seur, T., Jacquier, F. and Boscovich, R. G., *Parere di tre mattematici sopra i danni, che si sono trovati nella cupola di S. Pietro*, Rome, n.d. [1743]. See also Poleni, art. 214.
39. Le Seur et al., pp. 4-9.
40. Ibid., pp. 12-15.
41. Ibid., pp. 15-24.
42. Ibid., pp. 24-30.
43. Poleni, arts 218-219.
44. Le Seur, T., Jacquier, F. and Boscovich, R. G., *Riflessioni sopra alcune difficoltà spettanti i danni, e Rifarcimenti della cupola di S. Pietro*, Rome, n.d. [1743].
45. Poleni, art. 216.
46. Poleni, arts 224, 363-383.
47. Poleni, arts 239, 439-450.
48. Poleni, op. cit. The summaries of reports not already referred to appear in arts 253-271, 292-323, 354-362, 384-438, 461-517. Much briefer summaries are given in Di Stefano, pp. 36-55; rev. edn., pp. 26-41.
49. Poleni, arts 366-374.
50. Poleni, arts 369-370.
51. Poleni, arts 79-89.
52. Poleni, art. 88.
53. Poleni, arts 133, 444, 568-571.
54. Poleni, arts 141-153.
55. Poleni, art. 575 and pl. F, figs XV-XVII.
56. Poleni, arts 574, 580, 584. Details of the chains are given in arts 590-595.
57. Poleni, arts 605-611.
58. Poleni, art. 660-663.

XIX

Sinan's Suleymaniye Mosque and Justinian's Hagia Sophia

Justinian's Great Church of Hagia Sophia presented Ottoman rulers and their architects with a powerful challenge. A whole century before the fall of Constantinople its great dome must have been the inspiration for the large single dome, carried on six piers, that covered almost half of the prayer hall of the Uç Serefeli Mosque in Edirne. After the fall and the conversion of the church into the principal mosque of the new capital it served more directly as a model for the first Fatih Mosque and the Mosque of Beyazit II.

The church was, however, a masterly response to a specific requirement. Sixth-century liturgical ceremonial called for a broad central nave with a dominant longitudinal axis and for broad aisles on each side with full-width galleries above them. The vaulting system of the nave – in which a central dome was abutted at east and west by semidomes of the same diameter and these opened in turn into smaller ones set diagonally – spanned its whole length and breadth in one continuous sweep focused on the dome. The nave itself opened through alternately curved and straight colonnades into separate aisles and galleries that doubled the overall width. (Fig.1 top)

Muslim prayer and Ottoman imperial ceremonial had no such need for a strong longitudinal axis: width was more important. Nor did they call for differentiation between nave and flanking aisles or for broad galleries.

Provision of the galleries in Hagia Sophia was, moreover, one reason for its only serious structural weakness, namely the inadequate transverse stiffness of the original support system for the dome. On each side, support was provided by piers at the corners of the square bay beneath it, connected to buttress piers that faced them across the aisles. To minimise intrusion on the continuity of the aisles without excessively raising the gallery floors it became necessary to make the interconnections at gallery floor level too shallow to withstand the shearing actions soon experienced.

Thus it is hardly surprising that, in his Sehzade Mosque, Sinan adopted a fully centralised plan with large and smaller semidomes on all four sides of the dome. Thereby he not only achieved a single broad undifferentiated internal space but made it much easier to provide the requisite transverse support for the dome thrusts.

Indeed it is his turning back to Hagia Sophia as the principal model for the Suleymaniye Mosque that is the greater cause for surprise. Perhaps he found the challenge it continued to pose irresistible. Or perhaps it was Suleyman himself who did so. In the absence of any known record we can only speculate about such possibilities. For the present purpose I shall leave speculation aside and concentrate on what he took from the model and on the changes he introduced to serve better a different use.

Overall design of the mosque

The main features that he took over, at least in principle, were 1) the main vaulting system

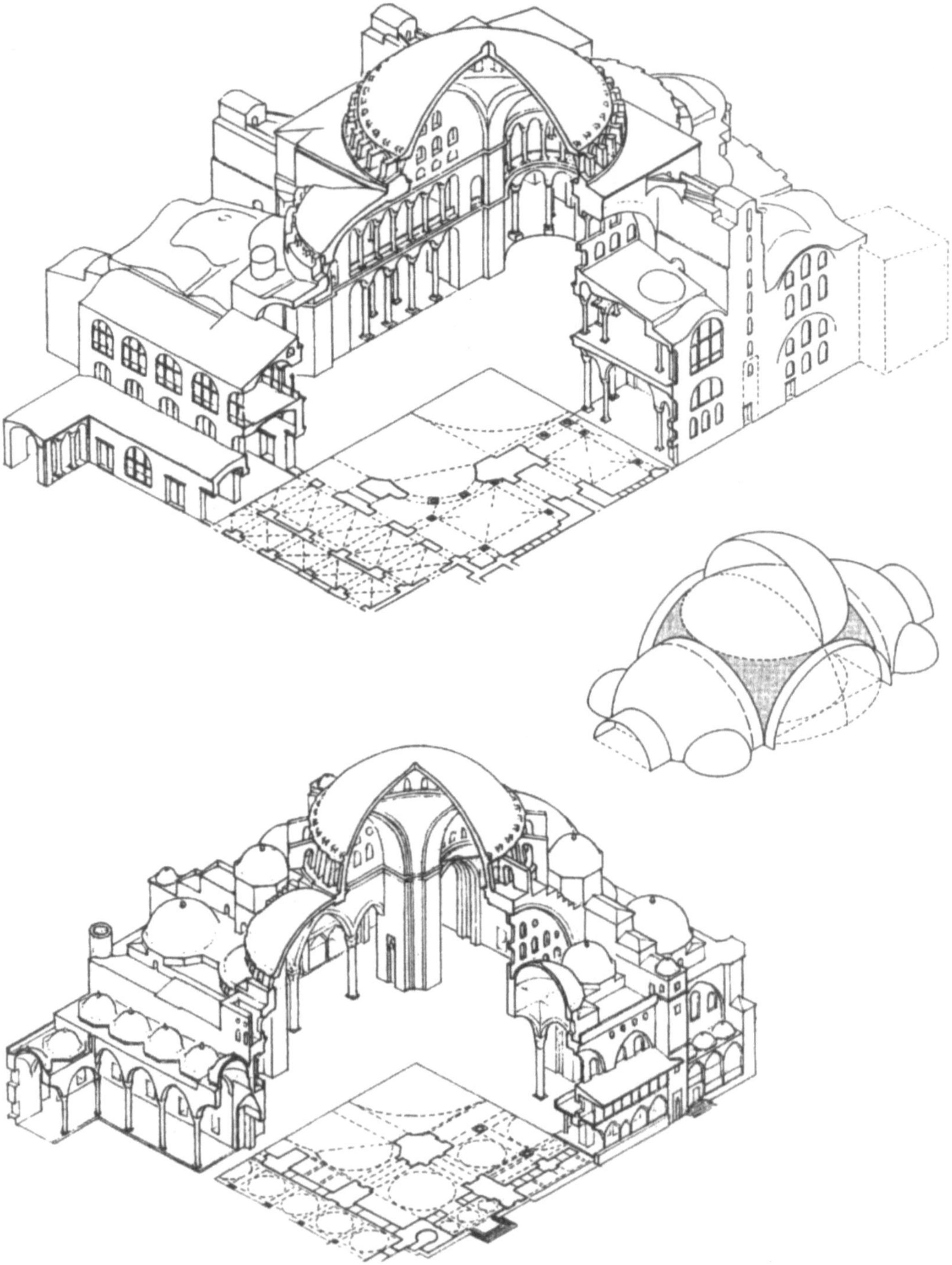

1. Partially cut away isometrics to the same scale of Hagia Sophia (top) and the Suleymaniye Mosque (bottom). Hagia Sophia is shown without its added peripheral buttresses but otherwise as Sinan would have known it. Its main vaulting system is inset diagramatically at the right.

2. Suleymaniye Mosque, details of construction

Iron tie across the north west exedra (left)

Interior view of the ashlar masonry of the north west buttress pier (below)

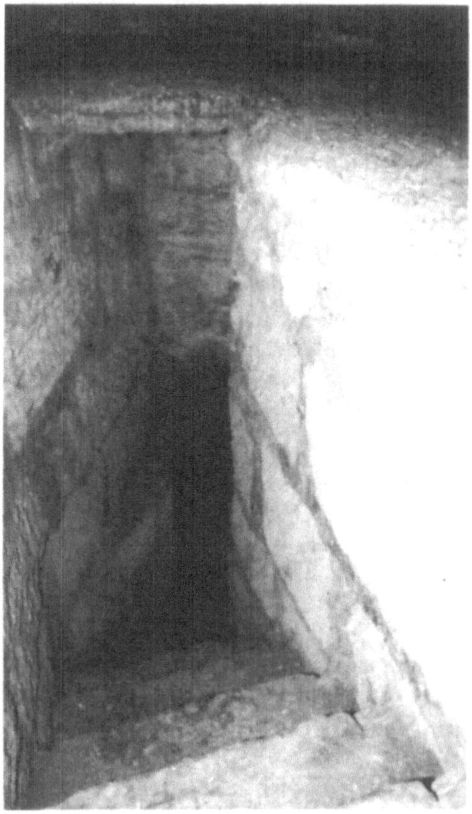

Exterior view of the ashlar masonry of the south west buttress pier (left). One of the smaller domes is seen on the left. Above it is a glimpse of the main dome.

including the broad arches on each side of the dome that were the lateral counterparts of the main semidomes, and 2) the associated system of supporting piers. (see Fig. 1 bottom)

In taking over the main vaulting system he went beyond what had been done in the mosque of Beyazit II chiefly in adding the smaller semidomes – as he had previously done in the Sehzade Mosque – in place of the pendentives that brought the main semidomes of the earlier mosque down to the square corners below. At the same time, he reduced the diameters of the main semidomes in relation to that of the dome by modifying the cross sections of the main piers to bring the springings of the transverse arches in towards one other. This also reduced the diameters of the smaller semidomes and made it possible to keep these within rectangular end bays little wider than the central bay. And instead of lifting the main semidomes well above the transverse arches – as was done in the mosque of Beyazit II – he followed Hagia Sophia more closely in making them more nearly continuous with the soffits of these arches. Whether he constructed their arches of cut stone as was done later in the Mosque of Sultan Ahmet, I have been unable to check. But this does seem likely because, unless these are later insertions, he did introduce ties across the springings of the front arches of the smaller semidomes as was done there (Fig. 2 top left).

The most significant changes to the supports were to the piers of the lateral system. Freed of the requirement for broad galleries on each side, he connected the main piers and the outer buttress piers only above the taller counterparts of Hagia Sophia's aisles, but connected them there much more substantially, stepping the connections upwards towards the dome where they rose above roof level so that they acted in much the same way as the fliers of heavy flying buttresses. Equally importantly, he corrected the two other principal weaknesses he could have observed in Hagia Sophia – the excessive reduction in cross section of the main piers at its gallery level, and the use, from too low a level in relation to the speed of construction, of brickwork with thick mortar joints instead of cut stone with closer joints. Details of the masonry at the higher levels in the mosque are seen in Fig. 2 right and bottom left.

At the two ends, the reduction in the diameter of the semidomes made it possible and desirable effectively to split into two each of the secondary piers beneath Hagia Sophia's semidomes. The outer components were aligned with the main piers and connected to them up to roof level in the same manner as were the lateral buttress piers, all these connections having one feature that can only have come directly from Hagia Sophia. This is the pair of bracing arches that spring from projections from the main mass of the piers (Figs 3 bottom and 4 bottom). The only difference is that, as far as can be seen at present, they are part of the original design rather than early additions to it as they had been in Hagia Sophia.

Had he done no more he would not, however, have wholly overcome an impression of a separate nave and aisles and a dominant longitudinal axis. To meet the different requirements of Muslim prayer, he took advantage in a masterly way of the freedom of planning of the lateral spaces that was offered by Hagia Sophia's primary structural system. In this system, both in Hagia Sophia itself and as remodelled in the Sulemaniye, all necessary support for the high vaults was provided by the main and secondary and buttressing piers. What stood between these piers was little more than an independent infill whose design could be varied at will.

In the Beyazit mosque the feeling of a central nave and flanking aisles remained. Pairs of identical tall domed bays were set on each side of the dome, separated from the central space only by a single tall column on each side, and there were further similar bays in the four corners of the overall square plan. But the columns were set, as in Hagia Sophia, flush with the inner faces of the great arches above, and the tympana that they carried were set back only slightly from these faces.

Sinan made three major changes (Figs 1 and 4). First he set back the whole infills as far as the east-west depths of the main piers and main lateral arches allowed. By so revealing most of the return faces of the piers and most of the soffits of the arches, he already went part way towards creating an east-west axis to counter the north-south one of the main vaults. Secondly he strongly reinforced this new transverse axis by dividing the spaces to each side of the dome into three unequal bays of which only the larger middle one opened to the central space through an arch that was fully visible on the face. Thirdly he differentiated between these three bays and the corner ones by placing in them balconies, set well back against the side walls and partly open to the exterior, while making the corner ones serve as vestibules to the most-used entrances.

At the same time, he gave to these lateral spaces a far more interesting character than they had had in earlier mosques and an overall A:B,A,B:A rhythm of alternating larger and smaller bays punctuated – as indicated by the colons – by the pairs of projections from the rear faces of the main piers (Fig. 4 bottom again).

Detailed design and expression

In contrast, Sinan's treatment of surfaces owes nothing to Hagia Sophia, and derives entirely from Ottoman tradition and his interpretation and development of it.

I see that tradition as being as firmly rooted in and respectful of the underlying structure as the earlier Syrian, Armenian, and Anatolian Seljuk traditions had been, and for the same reason. Anything else would have been wilfully perverse in an architecture of finely cut stone. But, like these earlier traditions, it is very different from what is seen in Western Gothic architecture because the wall was always a more important element and design tended to be more additive in character, juxtaposing forms and spaces which remained clearly defined entities. Typically, certain elements such as piers, arches, portals, and transitional vaulting elements like squinches were singled out for emphasis. Decoration elsewhere was confined to isolated panels on otherwise plain surfaces. Geometric motifs were favoured, the stalactite being the most characteristic for capitals and transitional vaulting elements.

Sinan's development of the tradition was to use the same means in a more purposefully selective manner so that they made the fullest contribution to his ultimate objective. In the Suleymaniye, he chose to emphasise the verticality of the four main piers, the balancing horizontal continuity of the springing level of the principal arches and vaults, and the transverse axis that he had created to balance the longitudinal axis of the main vaulting system. He also sought to create visual unity out of what could easily have been, as in the mosque of Beyazit II, a very restless assemblage of elements (Fig. 4).

The most obvious means he adopted were the single unbroken and boldly projecting cornice or string course at the springing level of the main arches and the expressive treatment of the faces and soffits of all arches by alternating the colours of their voussoirs or indicating in paint a similar alternation. The verticality of the piers was left largely to speak for itself by leaving their faces unadorned apart from the tall niches, sharply hooded, facing in towards the main transverse axis.

His choices of profiles for the arches also played an important role. To retain freedom of planning of the lateral spaces below the level of the main cornice, he kept to the pointed

3. Hagia Sophia, interiors from the north-west and south-east.

4. Suleymaniye Mosque, interiors from the north-east.

profile that was usual in the Ottoman tradition. To avoid a disturbing discord in relation to the necessarily semicircular profiles where the smaller semidomes met the larger ones, he also made the four great arches beneath the central dome and the smaller arches at a lower level beneath the small semidomes slightly pointed rather than semicircular. Added emphasis was given to the arches that open to the outer bays on the transverse axis by their more sharply pointed profiles. Unity was further assisted by the careful placing of circular medallions and a few circular windows.

Finally he used the typically Islamic motif of the stalactite to reconcile the semicircular springings of the small semidomes to the flat-walled square terminations of the bays below and to serve similar purposes below some of the small side domes. The real transition elements here are somewhat skew pendentives, and the covering of stalactites helped to disguise the awkwardness of their underlying geometry. The greater decorative emphasis given to the arches below also assisted, though it did not eliminate a disturbing tendency of the stalactites to cut into the faces of these arches. It is one measure of Sinan's success that such expedients are rarely obtrusive.

By all these means he created an interior that – while closely related to what he would have seen in Hagia Sophia – was just as well fitted to serve its very different purpose and gave a very different character to the space. Today at least, Hagia Sophia impresses by the vastness of a space whose precise boundaries are always somewhat elusive, enveloped as it is by surfaces that are never wholly visible and immediately comprehensible and hint always at something beyond (Fig. 3). Within it, one feels both uplifted towards the dome and impelled forwards towards the apse. In the Suleymaniye Mosque, on the contrary, everything is crystal clear (Fig. 4). The sharply expressed forms and the suggestions of movement towards the dome and towards Mecca are held in finely adjusted balance to create a sense of quiet restfulness.

Externally, even ignoring the minarets which are an integral part of the design of the Suleymaniye but merely additions to Hagia Sophia, there is also a major difference in expression. Sinan appears to have deliberately sought to create an expressive monumental form whereas there is little today to suggest that Anthemius and Isidorus had any such desire.

Comparison of the exterior treatments is, of course, difficult on account of the many changes that have taken place at Hagia Sophia. Interior and exterior are – or were – closely related in both buildings. But Sinan's exterior is certainly more clearly articulated, as is his interior, and one senses that more importance was attached to it. It better reveals the interior and more powerfully expresses it, notably in all the forms that rise above the lower roofs (Fig. 1). Whereas, for instance, the minor domes over the galleries of Hagia Sophia are covered in a way that causes them to merge almost imperceptibly with the surrounding slopes, those of the Suleymaniye are visually detached by being raised on polygonal drums and independently roofed with slightly projecting eaves in a typically Ottoman manner (Fig. 2 right). The pendentives and upward terminations of the main piers are likewise clearly expressed in contrast to the final almost cubic filling out of these features in Hagia Sophia, and the stepped cappings of the great side arches give them as much prominence as is given to those of Hagia Sophia by their more deeply recessed soffits. While on both buildings the fenestration originally introduced rhythmic interest at lower levels, there is far more such interest in the numerous recessions and projections on the flanks of the Suleymaniye Mosque and always must have been.

Assessment

Whether Sinan rated the Suleymaniye below the Selimye Mosque in Edirne simply because it was in this latter mosque only that he finally, to his satisfaction, built a dome to match that of Hagia Sophia, I do not know. Having, in the Suleymaniye, directly faced its wider challenge, he was able with less constraint to bring together in the Selimye what he saw as most valuable in the Sehzade, the Suleymaniye, Rustem Pasha, and Sokollu Mehmet Pasha and from these to create something new. But I hope that I am not alone in finding as great an achievement in the inventiveness and consistency of scale and handling of detail in the Suleymaniye.

No great work is conceived in a vacuum. It draws on what has been done before and the mark of its greatness lies in the way in which it creates a new synthesis. Anthemius and Isidorus did no more than this in designing Hagia Sophia. Sinan had the advantage of having in front of him their thousand-year-old demonstration that what he was attempting was perfectly feasible. They had no such assurance. He was also probably better acquainted with the capacities and limitations of the construction techniques of his day. And while he shared with them the benefit of an unusual client and almost unlimited resources, he probably suffered less interference from him. But none of this detracts from his achievement. The Suleymaniye is a worthy successor to Hagia Sophia.

As such, I feel that it calls for closer attention than it has yet been given. We know nothing, for instance about the way in which the design evolved. Still less can we say how, as it did so, attention shifted backwards and forwards between structural concerns and concerns of a wider nature. It is one of the merits of the final design that one is not conscious of any distinction between structure and form and such unity, though more natural then than now, was not automatic. In the light of what we know about the evolution of other designs – from that of Hagia Sophia to those of more recent structures like the present St Peter's and St Paul's and the Paris Pantheon – it is hardly conceivable that there were no developments and changes as work proceeded. The surviving accounts even hint at such changes in relation to the surrounding structures. I suspect that, if he did not leave much more to assistants than Wren did, Sinan's working method had more in common with Wren's than with that of his own contemporary, Michelangelo. But, in the absence of any preparatory drawings or other similar records, it is only through close study of visible evidences of the sequence of construction that some firmer clues may emerge.

It is hoped, therefore, that an adequate structural and architectural survey will soon be undertaken if one is not already in hand.

References and acknowledgements

For a fuller account of Hagia Sophia see Mainstone, R J, *Hagia Sophia: Architecture, structure and liturgy of Justinian's Great Church*, Thames & Hudson, London, 1988 and, for measured drawings of the structure today, Van Nice, R L, *St Sophia in Istanbul: An Architectural Survey*, Dumbarton Oaks, Washington, DC, 1965, 1988. I am aware of no comparable study or survey of the Suleymaniye Mosque. Like a paper presented to the International Symposium on Sinan the Architect (Ankara, 1988), the present paper and the isometric drawing of the mosque in Fig. 1 are therefore based largely on personal inspections, regrettably more limited than those I have made previously of Hagia Sophia. For assistance in connection with the most recent inspection I am much indebted to Professor Mufit Yorulmaz and Mr Aras Neftci.

XX

Designs for ribbed vaults by Guarini and Leonardo

Fig. 1 The dome of S Lorenzo

Fig. 2 Dome alongside the mihrab at Cordoba

In a recent review of the Polifilo edition of Guarini's *Architettura Civile*,[1] Sir Anthony Blunt drew attention to a little-known drawing by Leonardo at Windsor that may have inspired Guarini's design for the dome of S. Lorenzo (Fig. 1). The special significance of this drawing is its indication of openings between some of the characteristically interlaced ribs. It seems to me more likely, though, that Professor Wittkower was nearer the truth in seeing the principal source of Guarini's idea of a diaphanous ribbed structure and superstructure in his study and analysis of Gothic architecture.[2] Indeed, if a more direct source for the opening in the webs in the dome of S. Lorenzo existed, it is difficult to escape the conclusion that it was the dome to the right of the mihrab at Cordoba (Fig. 2), since the openings there correspond so closely to those in S. Lorenzo.

I am aware of no Gothic ribbed dome that is fully comparable with that of S. Lorenzo; but this is merely tantamount to saying that Guarini did invent a new Baroque form. Structurally this form is implicit

Fig. 3 Vault of the south tower of the Stephansdom in Vienna

rather than fully realized in many of the reticulated vaults of central European High Gothic and in the diaphanous structures of towers at Freiburg im Breisgau, Vienna, Ulm and Strassburg. The closest parallel is possibly to be found in the vault of the third (octagonal) stage of the south tower of the Stephansdom in Vienna.[3] This vault (Fig. 3) has both an open central eye and a rib system identical with that in S. Lorenzo. Other variants are grouped together in a series of sketches by Dürer in the British Museum (MS II, 167v).

That Leonardo also drew inspiration from such Gothic vaults is clear from a rather similar series of sketches in the Institut de France (MS B 10v). Immediately above the plan of a vault that is very much like those of the Stephansdom and S. Lorenzo are the words *'del tedesco in domo'* and a little below it a plan of a Gothic reticulated vault like that of the choir at Freiburg im Breisgau (Fig. 4 top and centre). The obvious inference from the reference to the unnamed 'tedesco' is that Leonardo learnt of these vaults from one of

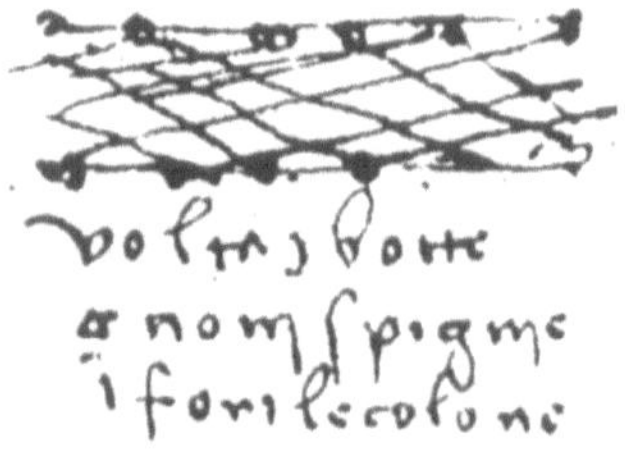

Fig.4 Sketches of ribbed vaults by Leonardo. The upper sketches are reproduced reversed to facilitate reading the captions. That at the top reads *'la ragion d'una volta, cioè il terzo del diametro de la sua camera'*. That on the left reads *'volta in botte e non ispingie in fori le colonne'*.

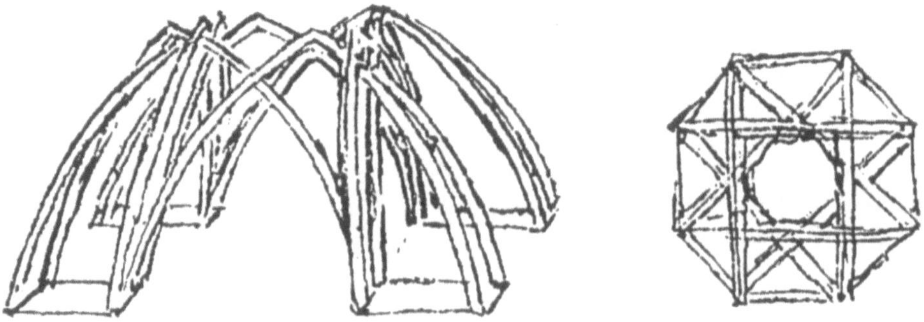

the German masters called in, as he also was, to advise on the design of the tiburio for Milan cathedral. This seems to be confirmed by the most fully worked out of Leonardo's drawings of a S. Lorenzo type vault (MS B, 27r). Alongside a drawing similar to the upper plan on the Windsor sheet, but without its suggestions of openings between the ribs, is a perspective view of a somewhat more complex rib system almost certainly related to the Milan tiburio since the ribs spring from all four corners of four squares disposed around a single square twice as large (Fig. 4 bottom). These four must represent the four aisle bays adjacent to the four piers of the crossing.

The possibility that Guarini's interest in the Gothic prototypes was, in turn, awakened by Leonardo's is less demonstrable, but it is suggested to me chiefly by Leonardo's own subsequent explorations of the potentialities of such vaults for the central domes of centralized Renaissance structures. These also appear in MS B of the Institut de France and in MS 2037 of the Bibliothèque Nationale (originally part of MS B and now also in the Institut). Leonardo's intentions are not always very clear since it is not always easy to see how the plan was to be realized in terms of the form usually shown alongside as a bird-eye perspective. Particularly relevant to S. Lorenzo, however, seems to be the plan in MS B, 24r, since this could readily be interpreted as a dome almost identical to that of S. Lorenzo set within a larger square similarly opened out into shallow niches on its main and diagonal axes (Fig. 5). It is, moreover, much more likely that Guarini knew these drawings than that he knew the Windsor sheet, because MS B was, in his day, in the Ambrosiana Library in Milan. A comparison between the subtly complex way in which Guarini supported his dome on columns set well back from its perimeter (Fig. 6) and what seems to have been

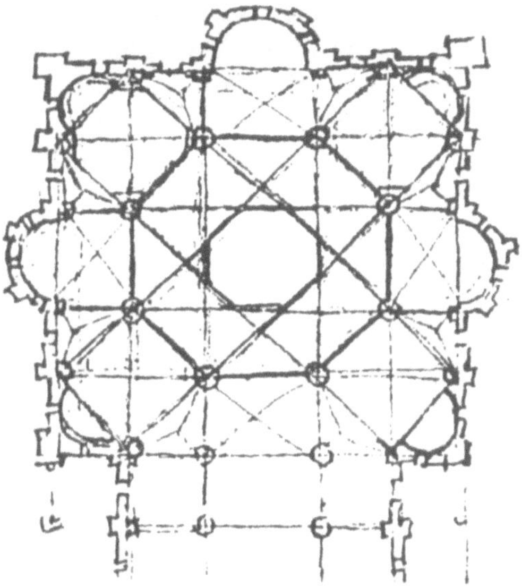

Fig.5 Design for a church by Leonardo

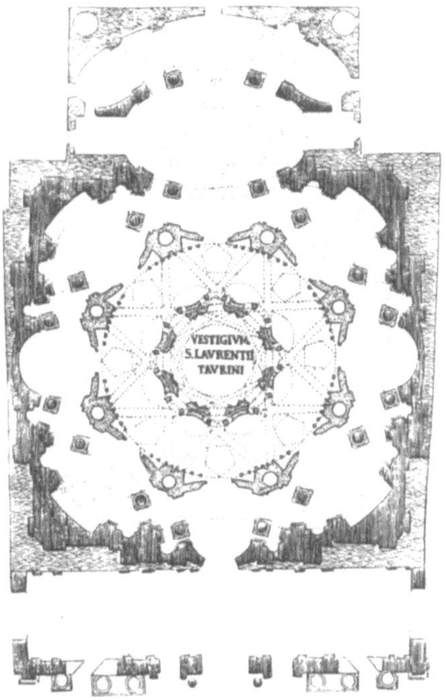

Fig.6 Guarini's plan of S Lorenzo in his *Architettura Civile*

Leonardo's intention serves, nevertheless, to emphasise the extent of Guarini's own invention and the likelihood that, here as elsewhere, he found a further stimulus in a more direct acquaintance with some of the Gothic sources.

1. *Architectural Review*, vol. CXLVII, 1970, pp. 164, 166.
2. Wittkower, R, *Architecture in Italy, 1600 to 1750*, Penguin Books, 1958, p. 274.
3. Earlier but remoter similar domes such as that over the monks' kitchen in Durham, are unlikely to have had any direct relevance.

XXI

The reinforced flat arches of the east colonnade of the Louvre

Robert W. Berger and Rowland J. Mainstone

[Before Louis XIV turned his main attention to Versailles he embarked on the partial remodelling and completion of the old Palace of the Louvre to proclaim better his glory as Sun King. In 1667, after several earlier designs had been passed over, the design of a new east front was placed by Colbert in the hands of a committee known as the Petit Conseil. This consisted of the first architect, Louis Le Vau, the first painter, Le Brun, and the architecturally inclined physician and scientist, Claude Perault. The distinctive feature became two colonnades of tall coupled Corinthian columns between central and end pavilions (Figs A, B). The columns carried straight entablatures as in a classical temple front. These had to be constructed as plates-bandes or flat arches because the available stone was incapable of spanning the requisite distances as a flat lintel. Similar flat arches were common already from Roman times onwards (Figs C, D). But they had always hitherto relied on firm abutments to resist their large thrusts. Since there was no such abutment for the arches spanning transversely to the colonnade (Fig. B), iron reinforcement was introduced, thereby creating a new structural form, precursor of the reinforced concrete beam.

Figure A. (previous page) East façade of the Louvre from the northeast.

Figure B. (left) Interior view of the east colonnade showing the transverse flat arches.

Figure C. Roman flat arch of joggled voussoirs over the west gateway of Diocletian's Palace at Spalato.

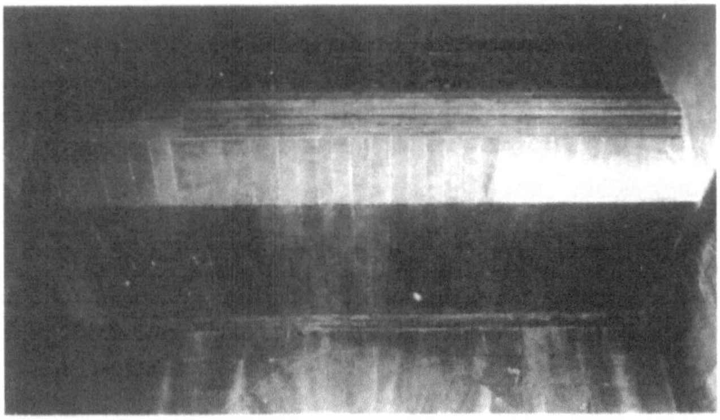

Figure D. Early Renaissance flat arch carrying the singers' gallery within the south-east pier of the crossing of Florence Cathedral.

In his book, *The Palace of the Sun*, Robert Berger discusses the whole history of Louis XIV's palace, reproduces the surviving drawings, and (in Sources I and II) appends surviving building accounts and passages in contemporary writings. The chapter entitled 'Materials and Structure', written jointly with Berger, deals chiefly with the plates-bandes and is reproduced here with the principal accompanying Figures including unique photographs of the ties. The book itself should be consulted for the Sources and for Figures not reproduced to which reference is made.

We are particularly grateful to M. Tachat of the office of the Architect-en-chef du Louvre and to Stephen Rustow of Pei, Cobb, Freed & Partners for making arrangements to allow access to the normally inaccessible "tunnel" over the entablature of the Colonnade.]

In the footnotes the following abbreviations are used:

AB	*Art Bulletin*
BN, Estampes	Bibliothèque Nationale, Paris, Cabinet des Estampes
BN, Manuscrits	Bibliothèque Nationale, Paris, Département des Manuscrits
BSHAF	*Bulletin de la Société de l'histoire de l'art français*
CBR	Jules Guiffrey, ed., *Comptes des bâtiments du roi sous le règne de Louis XIV*, Paris, 1881-1901, 5 vols.
MHF	*Monuments historiques de la France*
PV	Henri Lemonnier, ed., *Procès-verbaux de l'Académie royale d'architecture*, Paris, 1911-29, 10 vols.
ZfK	*Zeitschrift für Kunstgeschichte*

The stone masonry of the new Louvre of Louis XIV was inspected in 1678 by the Royal Academy of Architecture on orders of Colbert as part of a broader survey of the building stones of Paris and its environs. The Academy submitted a report[1] and made several revealing drawings (Figs. 89, 99, 102, 112, 114, 115)[2] that bear detailed annotations of the types of stones used

1. The report, written by André Félibien (secretary of the Academy), bears the title *Rapport de l'Académie d'Architecture sur la provenance et la qualité des pierres employées dans les anciens édifices de Paris et des environs* (BN, Manuscrits, MS fr. 12341). The only specific comment on the new Louvre is found under the date July 18: "On a aussi observé au Batiment neuf du Louvre qu'il y a de la pierre de S. Cloud qu'on a employé au Socle attenant le Portail, tant du coté du levant, qu'à celui du Septentrion, laquelle se gate et devient farineuse" (7). The report has been published in L. de Laborde, ed., "Rapport demandé par Colbert en 1678 à l'Académie d'architecture," *Revue générale de l'architecture et des travaux publics*, X, 1852, cols. 194–242, 273–293, 321–344, and in idem, *Mémoires et dissertations*, Paris, 1852, 151–290. A version of the report was entered into the minutes of the Academy (see *PV*, I, 168–249 [July 12–September 22, 1678]; notes by Lemonnier on the report are in I, 325–335). There are some differences in wording between the two published versions. In 1699 the Academy again reviewed the variety of French stones and studied the 1678 report (*PV*, III, 66f. [June 1, 1699]).

2. The drawings were first published by G. Monnier, "Dessins inédits pour la Colonnade du Louvre," *MHF*, n.s., 1972, no. 3–4, 130–137. In the space of this short article, Monnier proposes two different dates for the drawings: 1667–68 (130) and before 1674 (135). But she herself points out (133–134) the close connections between the drawings and the report of the Academy. There can be little doubt that the drawings were made at the same time as the report and were not "modèles destinés au chantier," as described by Monnier on 130. See also [G. Monnier], *Dessins d'architecture du XVe au*

and can be supplemented by another set, probably made ca. 1755 by Pierre Patte.[3] These depictions, as well as payments recorded in the *Comptes* (Sources I), indicate that the new Louvre, like the old, was constructed of a number of esteemed limestones of the Parisian region.[4] Blocks of these stones were laid up in the traditional manner to form the bearing walls of the south and north façades as well as the courtyard elevations (Figs. 95, 101, 105). The Colonnade (Figs. 43, 44), however, was a structural *tour de force*, embodying a revolutionary system of hidden iron reinforcement.

The structural problems of the Colonnade are the result of the large dimensions of the intercolumniations and the weight of the flat ceilings covering the bays between the columns and the rear wall (Figs. 51–53). The distance of the large intercolumniation measured from the side of one column to the other is 12 *pieds*, 2½ *pouces*; the distance between the side of a column and the surface of the corresponding pilaster on the wall behind is 12 *pieds*. If an attempt were made to span these intervals with monolithic stones, serious cracking would occur at the centers of the wide intercolumniations, aggravated by the weight of the flat ceilings resting on the architraves of each bay. For this reason, the architraves and friezes were constructed of a number of voussoirs to form flat arches (*platebandes*) (Fig. 79, top left). The voussoirs, carefully fitted together, inevitably produce outward thrusts. Thrusts generated parallel to the façade are sufficiently absorbed, of course, by the great masses of the central and end pavilions, but thrusts generated at right angles to the façade threatened to push over the freestanding columns with their entablatures in an outward direction, which would be a catastrophic event.

To solve this problem, the masonry was reinforced by embedding within it a network of iron tie rods and cramps completely hidden from view (Figs. 78, 79).[5] In this system, the tensile strength of the iron is used to resist displacements of the columns and slips of the voussoirs, leaving the masonry to act as it is best fitted to do: purely in compression.

Columns. The drums were bored through their centers to receive an iron rod about 2 to 3

XIXe siècle dans les collections du Musée du Louvre, Paris, 1972, 25–27. These drawings were perhaps among the ones referred to in the minutes of the Royal Academy of Architecture (March 15, 1756) as showing "le détail des différentes natures de pierre" (Sources II, no. 87).

3. BN, Estampes, Va 217a.

4. These included several *pierres dures* (Arcueil, Liais, Saint-Cloud, Bonbanc, and Vernon) and *pierres tendres* (Saint-Leu and Trossy). On these stones and their properties, see mainly A. C. d'Aviler, *Cours d'architecture . . .*, Paris, 1694, I, 202–208. See also A. Félibien, *Des principes de l'architecture, de la sculpture, de la peinture, et des autres arts qui en dependent*, Paris, 1676, 65f.; *PV*, I, 186 (July 27, 1678; Liais), 200–201 (August 5, 1678; Saint-Leu), 246 (September 14, 1678; Saint-Cloud); A. Félibien, *Mémoires pour servir à l'histoire des maisons royalles et bastimens de France*, ed. A. de Montaiglon, Paris, 1874 [1681], 76–81; L. Savot, *L'architecture françoise des bastimens particuliers*, ed. F. Blondel, Paris, 1685, 271f.; P. Bullet, *Architecture pratique . . .*, new ed., Paris, 1768, 282–289 [1st ed. 1691]; N. de Blegny (A. du Pradel), *Le livre commode des adresses de Paris pour 1692*, ed. E. Fournier, Paris, 1878, II, 111–115; J. F. Blondel, *Architecture françoise*, Paris, 1752, I, 129f. Félibien (*Des principes de l'architecture*, 65) boasted that French stones were finer than Italian stones. The marble columns reported in the *CBR* (Sources I, nos. 17, 35, 59, 65) may have been intended for interiors.

5. This discussion of the structure of the Colonnade is based on Patte (Sources II, no. 94 [1769]); the fine accuracy of his engraved analytical renderings (Fig. 79) was confirmed by the authors of this chapter in a site visit in 1990. The analysis of the central pediment is dependent on Sources II, no. 100 (Patte as continuator of Blondel, 1777). Patte's analyses form the basis for the discussions in J. B. Rondelet, *Traité théorique et pratique de l'art de bâtir* (1st ed., Paris, 1802–17; modified discussions in later editions). Two recent studies bearing on the Louvre structure are P. Saddy, "A 'Construct' of Modernity: The Reinforced Lintel," *Daidalos*, no. 8, June 15, 1983, 54–65, and R. Middleton, "Architects as Engineers: The Iron Reinforcement of Entablatures in Eighteenth-Century France," *AA Files*, no. 9, Summer 1985, 54–64.

pouces in diameter; this rod (in three segments joined together) reaches from the base of the column, through the capital, and into the entablature as high as the cornice (Fig. 79, top row, left and right). Tradition has it that between each pair of column drums there is also a flat, cross-shaped piece (Fig. 79, lower row, right, marked "A"). The vertical rod passes through the center of the column drum, and the four ends of the piece have vertical flanges; two flanges were set into the drum below, two into the drum above (Fig. 79, top row, left, showing only the downward flanges, marked "A"). This system of vertical reinforcement is connected to a horizontal system to form a network as follows:

Architraves. Above each column was placed a large stone block (*sommier;* Fig. 79, top row, left, marked "M"). The large intercolumniations are spanned by flat arches composed of voussoirs (*claveaux*) cut *à crossettes* (). It is especially in these voussoirs that much thrust is created. Each voussoir is secured to its neighbor by a Z-shaped iron joggle cramp (about 15 *pouces* long; Fig. 79, top row, left, marked "L" and "K"). Then a groove was cut into the tops of the *sommiers* and voussoirs to receive horizontal iron rods (Fig. 79, top row, left, marked "B" and "C"; bottom row, center, marked "H" and "I"), each rod having an eye at each end threaded over the projecting end of one of the vertical rods on the axes of the columns. These horizontal rods thus form a chain along the entire length of the façade from one end pavilion through the central pavilion to the other end. Other flat arches span at right angles between the *sommiers* and the rear walls and have iron rods (Fig. 79, top row, left, marked "E," right, marked "b"; bottom row, center, marked "K") set in similar grooves cut in the tops of the voussoirs. These rods are likewise attached to the vertical rods on the axes of the columns by means of an eye at one end and are held at the other end by a vertical rod set against the inside face of the rear wall and passing through a second eye. Additional rods with an eye at each end (Fig. 79, top row, left, marked "D"; bottom row, center, marked "L") run in the same direction from the centers of the small intercolumniations to the inside faces of the rear walls. They are similarly anchored at these inside faces and are secured over the intercolumniations by threading the rods marked "I" (Fig. 79, bottom row, center) and "C" (Fig. 79, top row, left) through their outer eyes. But they differ from the rods previously mentioned in consisting of two approximately equal shorter lengths joined together by a special connector or *moufle* (Fig. 79, bottom row, far left; Fig. 5 and Fig. 81) with double wedge-shaped keys to permit tensioning.

Friezes. These were constructed just like the architraves, except the voussoirs were not cut *à crossettes* (Fig. 79, top row, left). Horizontal rods were again placed in grooves cut into the tops of the stones forming the friezes. Running back to the anchorages already referred to inside the rear walls, however, are two sets of rods differently aligned. One set, identical to those immediately below, runs at right angles to the Colonnade from the centers of the small intercolumniations (Fig. 79, top row, left, marked "H," right, marked "f"; lower row, left, marked "S"). The other set (Fig. 79, lower row, left, marked "R") runs back diagonally from the centers of the columns, forming X-shaped patterns above the ceilings. Each rod of both these sets consists again of two lengths joined together by *moufles* (Fig. 81).

Ceilings. The flat ceilings over each bay of the Colonnade consist of a jigsaw-puzzle pattern of stones set around a large circular central stone (top diameter 6 *pieds,* bottom diameter $5\frac{1}{2}$ *pieds;*

Fig. 79, bottom row, left and right). Between this circular stone and its neighbors, irons shaped as upside-down T's were placed (Fig. 79, bottom row, far left, marked "Fig. 4") to help distribute the weight of the central stone more evenly to the surrounding stones. The ceiling stones thrust outward in all directions onto the friezes of the bay, but the ceiling is in thickness only about one-half the height of the frieze (Fig. 79, top row, right). Patte's cross-section also reveals the large space above the ceilings (about 6 *pieds* in height) that greatly lightens the structure and provides a tunnel in which a person can walk to perform inspection and maintenance (Fig. 80). Accessible within the tunnel are the special connectors (*moufles;* Fig. 81).[6] Thus it is possible to drive in further or loosen their wedge-shaped keys, thereby tightening or slackening, as may be necessary, the rods running from above the Colonnade to the rear wall. The tunnels (still extant and accessible) are within the entablatures above the north and south ranges of the paired columns.

Cornices. The cornices are set in three conventional horizontal courses corbeled over the interior passageway (as is the masonry opposite) to enhance stability. A single flat stone over the "tunnel" in turn forms a flat terrace with balustrade (Fig. 79, top row, right).

Pediment and Entablature of the Central Pavilion (Fig. 82). This entablature is constructed in a manner similar to those of the colonnades, but it is curved slightly upward so that at its center it is about $1\frac{1}{2}$ *pouces* above the horizontal. The entablature was built this way to help support the weight of the pediment.

The simae of the raking cornices comprise two great monolithic slabs (see Chapter V). The lower courses of these cornices consist of stone blocks with vertical joints instead of joints perpendicular to the slope (the latter system was usual). The downward pressures at the corners of the pediment, and the threat of a seesaw movement of the horizontal cornice, are counteracted by large blocks at the level of the balustrade, embedded in the wall (Fig. 48).

The tympanum of the pediment contains three hidden discharging arches, the central one ogival, the flanking ones rampant. These relieve the weights over the intercolumniations and distribute them to the columns. A network of iron rods and chains brace and interconnect the raking cornices.

The brilliance of this structural system is threefold: (1) the coupling of the columns permits increased main spans without reducing the number of points of support; (2) if the reinforcement behaves as intended, thrusts generated by the masonry are absorbed by the iron that acts in tension, so that vertical loads only are imposed on the columns and walls as in genuine trabeated construction; and (3) provision is made for varying the tension in the iron rods that restrain the tendencies of the transverse flat arches and the ceilings to thrust outward against the Colonnade and overturn it. The scheme was a notable innovation in the history of building technology. In Patte's words:

6. The rods between the architrave and the frieze marked "T" in Figure 79 (top row, center) and "L" (bottom row, center) are placed in open troughs so their *moufles* are accessible from within the tunnel. The adjacent buried rods marked "S" and "K" are not shown as having *moufles* and probably had none.

> The iron carries nothing, and its only role is precisely that of exerting [inward] pulls to counter the [outward] thrusts of the architraves, thereby stabilizing the axes of the columns, a procedure that necessarily produces the greatest resistance that can be expected from iron. [Sources II, no. 94(b), last par.]

Although the system used for the Louvre Colonnade was dramatically new, it did not spring forth into being from a technological vacuum. Various forms of iron reinforcement in masonry structures had been commonly employed in Europe since medieval times.[7] Exposed iron tie-rods were used in Early Christian and Byzantine architecture from at least the early sixth century and entered Western European architecture ca. A.D. 1000; they became quite common during the Gothic period. Hidden or partially concealed iron reinforcement (aside from cramps connecting adjacent stones, used from classical Greek times on) appears in France beginning in the twelfth century[8] and was often used in window frames during the Gothic period.[9] Because great medieval buildings had to be maintained and repaired, the "secrets" of their construction were passed on, however imperfectly, to postmedieval architects and builders. Hence, it is fascinating to note that the connecting piece (*moufle*) between the iron rods of the Colonnade (Fig. 81) is virtually identical to that used in France in the fifteenth century for "free chains" (not encased in masonry; Fig. 83), even to the detail of the double wedge-shaped keys (to permit tensioning). The cramps connecting the adjacent voussoirs at the Louvre (Fig. 79, top row, left) are similar to those used in the exterior cornice of the choir of Notre-Dame, Paris, dating from the late twelfth century (Fig. 84). Medieval builders discovered from unhappy experience that iron embedded in stone masonry was likely to rust and thereby swell, and in so doing crack the masonry or even lift whole courses of it. If iron was in direct contact with stone, there was also a risk that highly uneven bearing could lead directly to local splitting or crushing. In order to prevent these mishaps, medieval masons sometimes coated the grooves in which iron rods were placed with molten lead or mortar; even better was a type of thick mastic apparently composed of powdered sandstone, minium, litharge, and oil.[10] Charles Perrault, in a document (Sources II, no. 46) recommending the reinforcement system of the Louvre Colonnade for his brother's proposed church of Sainte-Geneviève, prescribed that the horizontal iron bars of the entablature be coated with "two or three coats of oil to protect them against rust." Perrault's recommenda-

7. On iron reinforcement in medieval masonry buildings, see mainly R. P. Wilcox, *Timber and Iron Reinforcement in Early Buildings* (*The Society of Antiquaries of London, Occasional Paper*, n.s., II), London, 1981, and W. Haas, "Hölzerne und eiserne Anker an mittelalterlichen Kirchenbauten," *Architectura*, XIII, 1983, 136–151, esp. 147f. See also E. E. Viollet-le-Duc, *Dictionnaire raisonné de l'architecture française du XIe au XVIe siècle*, Paris, 1875, II, 396–404 (s.v. "Chaînage"), and J. Fitchen, *The Construction of Gothic Cathedrals*, Oxford, 1961, 275–279 (appendix J, "Arch and Vault Ties in Medieval Construction").

8. Perhaps the earliest examples are in Vézelay (nave, ca. 1104–32, composite iron and wood reinforcement; see Viollet-le-Duc, *Dictionnaire*, II, 397–398, 398, fig. 1; 399, fig. 2) and Bourges Cathedral (iron chain in floor of high triforium of the *chevet*, 1195–1214; see R. Branner, *La cathédrale de Bourges et sa place dans l'architecture gothique*, Paris and Bourges, 1962, 82, fig. 67a; 83 and n.). At Vézelay, the exposed hooked ends of the embedded rods may have held tie-rods across the nave. As for Bourges, Branner wrote (ibid., 83): "Il est impossible de savoir si ce chaînage est rattaché aux piles par des crampons verticaux ou si sa fonction est réellement de relier les supports entre eux."

9. Viollet-le-Duc, *Dictionnaire*, II, 401; Haas, "Hölzerne und eiserne Anker," 149.

10. Viollet-le-Duc, *Dictionnaire*, II, 403.

tion indicates the procedure used at the Louvre, and this is confirmed by a payment in the *Comptes*.[11]

These instances suggest some continuity with medieval French building traditions. It is significant in this respect that in 1669 or 1670 the views of several French architects *and masons* were solicited concerning models of the Colonnade, the Arc de Triomphe du Trône (Fig. 120), and the Observatoire (the last two structures are securely by Claude Perrault) (Sources II, no. 8). Among the comments are some on the use of iron for the Louvre.

Thomas Gamard (or Jamard) condemned the iron on the grounds that it could break, adding that because the ancients never used that material their buildings were never threatened by such danger. Daniel Gittard approved of the iron, provided it was well made and well hammered, and he hoped the iron rods were composed of several flat bars beaten and joined *en liaison*. A certain Duval recommended that the rods be placed below the flat arches ("plattes bandes") for greater strength, and it could be said that this was carried out for the frieze.[12] Jean Pastel (Patel) entirely approved of the structural system but criticized the cutting of the voussoirs. Pierre Cottart agreed with Pastel's last point and did not think it was necessary to recess the flat ceilings "in order to further strengthen the cut stones" (probably meaning that the stones of the architraves could withstand the thrusts of the ceilings if they had been placed at the level of the architrave instead of the frieze; see Fig. 79, top row, right). Charles Chamois praised the Colonnade as "one of the most beautiful and daring works ever made" and recommended that the vault above the ceilings be adjusted to align with the surfaces of the pilasters of the rear wall (a suggestion given at a time when vaults with semicircular arches were still planned [as in Figure 77] instead of the eventual corbeled vaults [Fig. 79, top row, right]). He approved of the manner in which the voussoirs were to be cut and recommended that, in addition to the architrave, the frieze be built as a *platebande* and that it should be convex (this very last suggestion was not adopted).

The most extended comments were those of Pierre Thévenot, a mason whose name appears frequently from 1667 on in the royal building accounts and who later worked on Perrault's triumphal arch.[13] He also approved of the stone-cutting and structural design, noting that the

11. See Sources I, no. 103. D'Aviler reported in 1694 that it was the practice to coat iron embedded in stone not with oil but with a thin layer of lead, "ce qui à la verité le garantit un peu de l'humidité de la pierre, mais ne peut cependant empêcher qu'il ne jette sa roüille au dehors" (*Cours d'architecture*, I, 217). J. F. Blondel wrote in 1756 that only the iron rods of the central pavilion of the east façade were coated with lead; the others were coated with oil (Sources II, no. 84[b]). The use of iron in French domestic buildings was discussed as early as 1624 by Louis Savot (*L'architecture françoise des bastimens particuliers*, ed. F. Blondel, Paris, 1673, 296–297, 298, 301f.). Iron in combination with stone was criticized by Philibert de l'Orme (*Nouvelles inventions pour bien bastir et à petits fraiz . . .* , Paris, 1561, fols. 5v–6r) and by the Royal Academy of Architecture in 1677 (citing de l'Orme) and in 1683, with Blondel, the director, present (*PV*, I, 154 [November 29, 1677]; II, 30–31 [May 24, 1683]); for Blondel's criticism of the iron of the Colonnade, see Appendix B. There are discussions of iron in P. Bullet, *L'architecture pratique . . .* , Paris, 1691, 273, 390, and in d'Aviler, *Cours d'architecture*, I, 216–219.

12. Because, both between architrave and frieze and between frieze and cornice, the iron rods were set in the upper surface of the course below and anchored to the vertical rods passing through the column axes, it seems likely that the primary role of the rods was to prevent the column heads from being pushed aside by the thrusts generated—the role Patte ascribed to them (Sources II, no. 94[b]) and the role this manner of incorporation best fitted the rods to play.

13. His name first appears in *CBR*, I, col. 218. For his work on the arch, see M. Petzet, "Das Triumphbogenmonument für Ludwig XIV. auf der Place du Trône," *ZfK*, XLV, 1982, 179f.

end pavilions would absorb the lateral pressures, but he cautioned that the iron rods to be set perpendicularly to the façade should be without flaw (which suggests Thévenot was speaking from experience). He went on to criticize the circular stones in the centers of the ceilings (Fig. 79, bottom row, left and right); he preferred square ones on structural grounds. Then Thévenot advised that the arches above the ceilings be pointed (*en tiers point*), not semicircular (as they appear in Figure 77; cf. Chamois's comment, above); pointed arches will produce less thrust and will better load the large stones (*sommiers*) above the capitals. Although the latter are of Saint-Leu stone (a *pierre tendre*), they are more than sufficient in strength to support their loads, and the mason cited pillars left standing in the quarries of Saint-Leu-d'Esserent. Thévenot's last recommendation is the following:

> To be more assured that the *sommiers* will not break, it will be best to leave the part of the joint which overhangs [*qui porte à faux*] directly over the empty [narrow] intercolumniation.

If Patte's diagrams are to be relied on (Fig. 79, top row, left and center), the Colonnade was indeed originally built with vertical joints between the large stones over the capitals, with small overhangs over the narrow intercolumniations.[14] This was apparently Thévenot's idea, and Claude Perrault later elaborated its rationale in his second Vitruvius edition (1684), where, in replying to a point made by François Blondel about architraves resting on capitals, Perrault wrote:

> I also do not understand why one affirms that the end of an architrave that rests entirely on a column is not more secure than when it rests only on half the column, and that it does not bend more easily when it is only supported by its end than when that end passes beyond the column supporting it, because I have always believed that the end that passes beyond the column directly over the small intercolumniation has a weight that resists the bending of the opposite part, the one directly over the large intercolumniation. [Sources II, no. 26(a); see Appendix B]

Perrault's analysis applies to monolithic stone lintels but not to the flat-arch *platebandes* of the Colonnade. However, Perrault probably wrote this passage with the Louvre in mind, thus revealing a lack of clarity in his thinking about the distinctions between the two types of lintel.

14. The alteration of this arrangement to the present condition, where a single wedge-shaped stone rests on both capitals (see Fig. 54), is probably attributable to the architect Charles Percier, working in the first decade of the nineteenth century (see L. Vitet, *Le Louvre et le nouveau Louvre*, new ed., Paris, 1882, 166). Unfortunately, I have not been able to discover the documentary records of his work. In 1803 it was reported that "les fers incrustés dans les pierres, ne furent point placés assez profondément pour être à l'abri de toute humidité: ils se sont oxidés et se dissolvent toujours de plus en plus. La rouille en pénétrant dans les joints des pierres les a rongées et désunies. Une partie des plafonds de ce fameux portique est aujourd'hui menacée d'une destruction prochaine" (L. P. Baltard and A. P. Duval, *Paris et ses monumens . . .* , Paris, 1803, I, 37). See also P.F.L. Fontaine, *Texte explicatif joint au nos du Journal des monuments de Paris . . .* , Paris, [1891], 31 (November 1809).

The above summary of the opinions contributed in 1669 or 1670 by several French architects and masons reveals the important fact that at least one feature of the structural system of the Colonnade—the *porte à faux* over the narrow intercolumniations—was the idea of a mason (Thévenot) and that the final corbeled vault over the flat ceilings (Figs. 79, top right, 80) is probably the result of critiques by Chamois and Thévenot of an earlier scheme with semicircular arches (Fig. 77). But the document of 1669–70 does not suggest that the iron reinforcement system itself was devised by minor figures, such as Chamois, Thévenot, and the others. Could it be that, after all, the revolutionary system was invented by some member of the Petit Conseil?

Of the three members, we can immediately rule out Charles Le Brun. Although he was active to a limited extent in architecture,[15] nothing about his career indicates an interest in the structural use of iron. Louis Le Vau, however, was commercially involved in the production of sheet iron coated with tin (*fer-blanc*), as well as firearms and cannon made of iron.[16] In 1665, the King established a factory at Beaumont-la-Ferrière (Nièvre) in the Nivernais region, and Le Vau was placed in charge.[17] The firearms division was sold in 1668,[18] but Le Vau still headed the *fer-blanc* manufactory at the time of his death in 1670. This connection of Le Vau with iron is intriguing, but *fer-blanc* was not used at the Louvre[19] or in other buildings. Besides, the iron employed for the Louvre was made in Normandy.[20]

In contrast to Le Brun and Le Vau, Claude Perrault had a lively interest in the use of metal in architecture. In his obelisk design of 1666 (Fig. 116), he had proposed an iron cage for spectators at the top of the monument; the iron struts were to be sheathed in gilt copper and to descend the entire length of the obelisk, linking the masonry together.[21] In the Observatoire of Paris, designed and begun in 1667, Perrault may have used a discrete amount of iron for reinforcement.[22] His scientific interest in iron—its properties when heated, beaten, and combined with

15. For a checklist of his architectural activity, see R. W. Berger, "Charles Le Brun and the Louvre Colonnade," *AB*, LII, 1970, 402–403. For his competition designs for a triumphal arch (1668–69), see Petzet, "Das Triumphbogenmonument," 148f.

16. See A. Laprade, "François d'Orbay (1634–1697)," *BSHAF*, 1953, 91f. and idem, *d'Orbay*, 1960, 78f.

17. *CBR*, I, col. 96 (with the specific mention of *fer-blanc*). A letter from Le Vau to Colbert (November 5, 1665, written at Beaumont-la-Ferrière) mentions muskets, cannon, and bullets (G. B. Depping, ed., *Correspondance administrative sous le règne de Louis XIV*, Paris, 1852, III, 743 n. 1). In a letter to Le Vau (October 26, 1669), Colbert complains that the *fer-blanc* is not being produced (P. Clément, ed., *Lettres, instructions et mémoires de Colbert*, Paris, 1863, II, 2, 493–494 [doc. 67]; reprinted in Laprade, *d'Orbay*, 82). Well before this, in 1650, Le Vau had invested in a metal forge at Conches (Eure) (ibid., 79).

18. Ibid., 81.

19. Iron for the Louvre (payments began in 1669 for iron delivered in 1668; see Sources I, no. 33) is regularly called "gros fer" in *CBR*. Payments for "gros fer" for the Louvre and Tuileries are recorded as early as 1664–66 (*CBR*, I, cols. 12, 68–69, 242); the use of this iron then is unknown.

20. As mentioned by Charles Perrault in his *mémoire* on a proposed church of Sainte-Geneviève (Sources II, no. 46), where Perrault speaks of "une barre de fer de Normandie."

21. "La cage est composée de vingt montants, quatre à chaque face, sans ceux des quatre coins, qui laissent 20 ouvertures d'un pied de large pour regarder. Ces montants sont couverts de grosses lames de cuivre doré de la largeur de huit pouces qui descendent tout le long de l'obélisque, en lient la maçonnerie et représentent les rayons du soleil qui partent du globe doré qui en est comme le corps" (Cl. Perrault, "Dessin d'un obélisque" [October 20, 1666], in Ch. Perrault, *Mémoires de ma vie*, and Cl. Perrault, *Voyage à Bordeaux (1669)*, ed. P. Bonnefon, Paris, 1909, 238). The publication of Perrault's essay in A. Hallays, *Les Perrault*, Paris, 1926, 258–263 is incomplete. On the obelisk monument, see M. Petzet, "Der Obelisk des Sonnenkönigs: Ein Projekt Claude Perraults von 1666," *ZfK*, XLVII, 1984, 439–464, esp. 442.

22. C. Wolf, *Histoire de l'Observatoire de Paris de sa fondation à 1793*, Paris, 1902, 13.

other metals—is revealed in one of his scientific papers published in 1680 but perhaps reflecting experiments conducted over twenty years.[23] Claude's brother Charles—who always claimed that Claude was the real designer of the Colonnade—wrote in his autobiography (ca. 1700) about a special Louvre model and stated that some of Claude's drawings illustrated the details of the intricate reinforcement:

> In order to dispel all worries that M. Colbert could have about the construction of this building, I suggested to him that it would be a good idea to make a small model of the peristyle with small cut stones of the same shape and number that the full-size work was to have. When it was finished and secured by small bars of iron, proportional in size to those to be used in the actual building,[24] M. Colbert was entirely satisfied about the firmness and solidity of the entire work, in which the iron carries nothing and only resists the thrust of the architraves with such strength that there is no weight, whatever it might be, that can break it, only making it clench more while carrying nothing. There was made, moreover, a space between the ceiling of the peristyle and the covering above, where several men can walk and work without difficulty to correct problems that could arise over the course of time. The entire work has been so well built that nothing has given way, and it is likely that the building will stand forever. The details of this construction are in the first volume of my brother's architectural designs, which are among my books. [Sources II, no. 53(b)]

Charles here refers to the two manuscript volumes of Claude's drawings that were later destroyed in 1871. Charles's familiarity with the details of the Colonnade's structure is also revealed in his *mémoire* of 1697 on Claude's unexecuted project of the 1670s for Sainte-Geneviève, Paris (Sources II, no. 46; see above).[25] Charles's technical knowledge about the Louvre can be attributed to the fact that Claude Perrault was his brother and that Charles was intimately familiar with Claude's drawings (which he inherited in 1688). On the other hand, Charles Perrault, by virtue of his key position in the Royal Building Administration under Colbert, could have acquired his special knowledge through that post alone.

Other evidence suggests that Claude Perrault (along with Le Vau and Le Brun) may indeed have devised the iron-reinforcement system in collaboration with Parisian masons and/or *serruriers*. Leibniz, who was personally acquainted with Claude Perrault in the 1670s, wrote in 1676: "M. Le Brun believed the Louvre design of M. Perrault, although beautiful, would be

23. Cl. Perrault, "De la pesanteur des corps, de leur ressort, et de leur dureté," in his *Oeuvres diverses de physique et de mechanique*, Leiden, 1721, I, 1–51 (1st ed.: *Essais de physique . . .* , Paris, 1680–88, 4 vols.). Perrault discusses hardening by heat of a range of building materials (brick, marble, stone; 1721 ed., I, 24f.).

24. This is perhaps the first explicit reference to a model used in a test of this sort. The test described could have demonstrated the validity of the basic concept, but it would have given a totally misleading idea of the cross-sections of iron called for, both because it ignored the likely greater unit strength of thin model bars and, more serious—if "grosses proportionellement à celles qu'on employeroit dans l'ouvrage effectif" means the model bars were cut to the same linear scale as the rest of the model—because it ignored the fact that, whereas weights increase proportionately to the cube of linear dimensions, strengths increase proportionately only to the square of these dimensions.

25. See also Sources II, no. 37[a].

very difficult to execute. But M. Perrault found a very able entrepreneur, Preaux or Preat, I believe, who is admirably precise; the stones are well cut, everything is done with admirable beauty" (Sources II, no. 19). Leibniz was undoubtedly referring here to Pierre Bréau, described in documents (beginning in 1668) as a royal mason, and as a master mason and *entrepreneur des bastimens de sa Majesté* in a document of 1670 in which he was one of several masons who gave their opinions about the foundations of Perrault's Arc de Triomphe du Trône (Fig. 120).[26] Leibniz characterized Bréau only as a master stone-cutter, not as someone skilled in the use of structural iron; that person may have been Étienne Doyart, the master *serrurier* who appears regularly in the *Comptes* as supplying iron for the Louvre (see Sources I, nos. 33, 34, 57, 58, 83, 102, 125).[27] However that may be, Leibniz's account (the result of a personal conversation with Perrault) suggests the type of communication between architects and builders that is an integral part of the architectural process and that must have taken place in the planning and execution of the Colonnade. (On Claude Perrault's silence about iron reinforcement in his theoretical writings, see Appendix B.)

It was argued in Chapter III that François Le Vau's Louvre project of ca. 1662–64 (Figs. 24–32) was the immediate source for the Petit Conseil when it first met in 1667. The younger Le Vau had proposed wide intercolumniations of 12½ *pieds* (Fig. 32), slightly more than those eventually executed (see above). Hence François's design, like the Colonnade, could only have been realized with a system of iron reinforcement, but such a system is not indicated in the sectional view (Fig. 31), where the entablature over the peristyle is shown with voussoirs arranged in a flat arch.[28] It may have been that, when François Le Vau's design was taken up by the Petit Conseil in 1667, the architects consulted with masons experienced in the use of iron in architectural structure and that from these discussions the system of rods and cramps was devised.

26. In 1681 he rose to the rank of *architecte des bâtiments du roi* (see Sources II, note 106).

27. Doyart appears in *CBR* from 1664 on.

28. It is doubtful that the sketchy, double lines of the entablature in Figure 31 were meant to indicate iron rods.

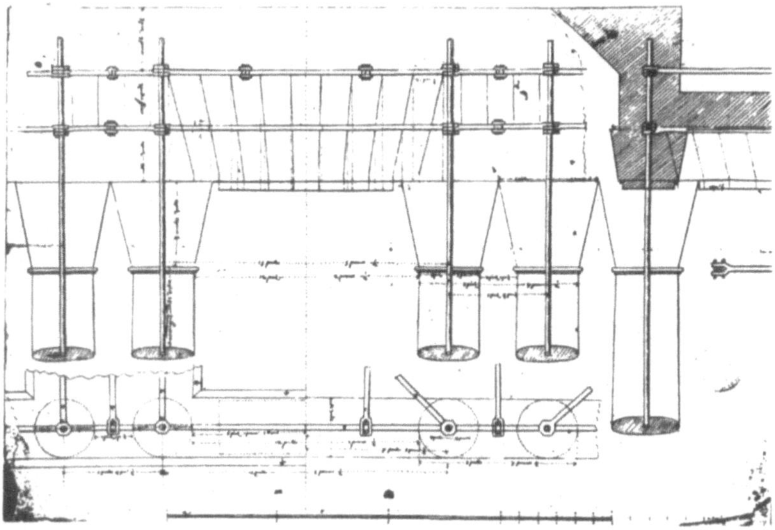

Figure 78. Anonymous draftsman, studies of the iron reinforcement of the Colonnade, 1668. Drawing, Paris, Louvre, Cabinet des Dessins, Receuil du Louvre, vol. II, fol. 74.

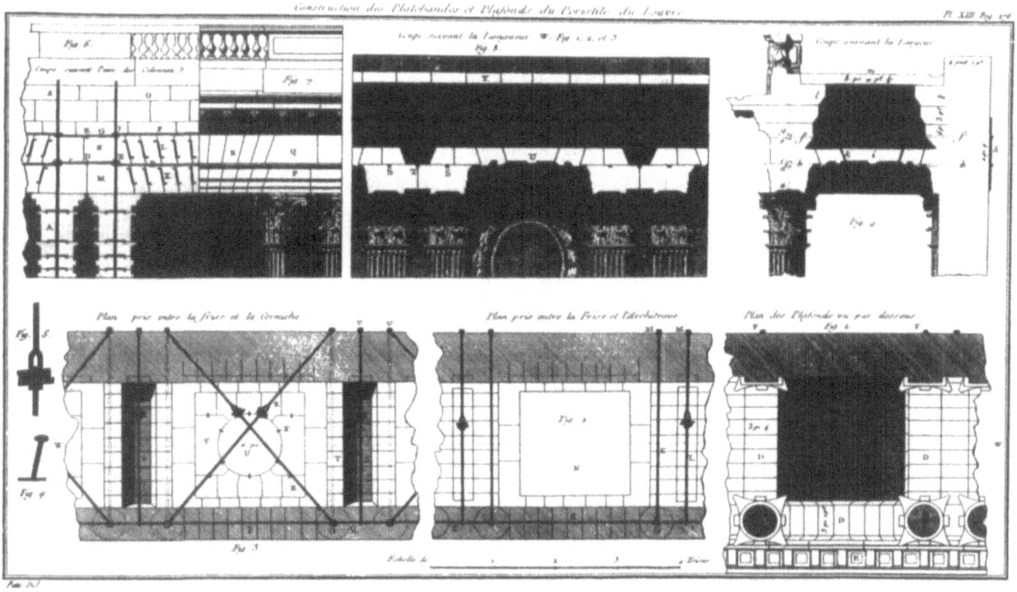

Figure 79. Louvre, Colonnade. Iron reinforcement system. Engraving from Pierre Patte, *Mémoires sur les objets les plus importans de l'architecture*, Paris, 1769.

Figure 80. Louvre, Colonnade. View of tunnel within the north entablature, with iron tie rods.

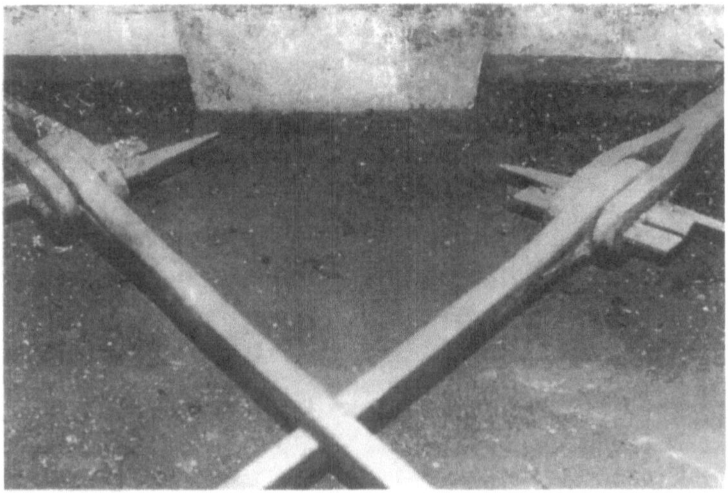

Figure 81. Louvre, Colonnade. Iron tie-rods with *moufles*.

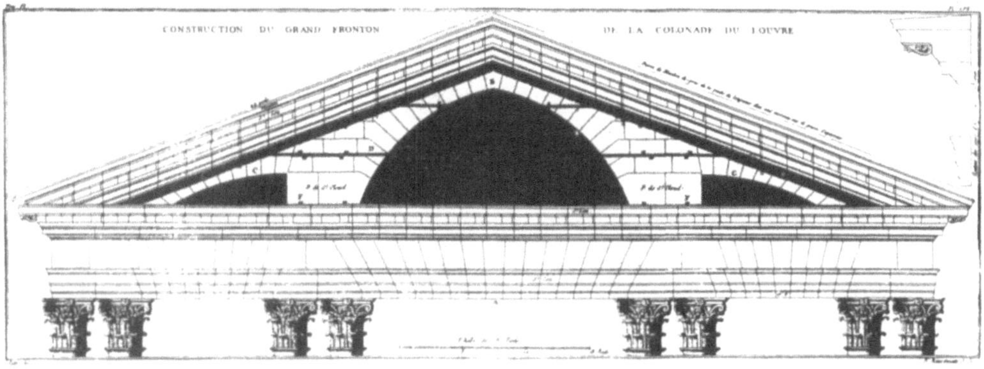

Figure 82. Louvre, Colonnade. Pediment of the central pavilion showing the iron reinforcement system and relieving arches. Engraving from Jacques François Blondel, *Cours d'architecture ...*, Paris, 1777, vol. III, Atlas.

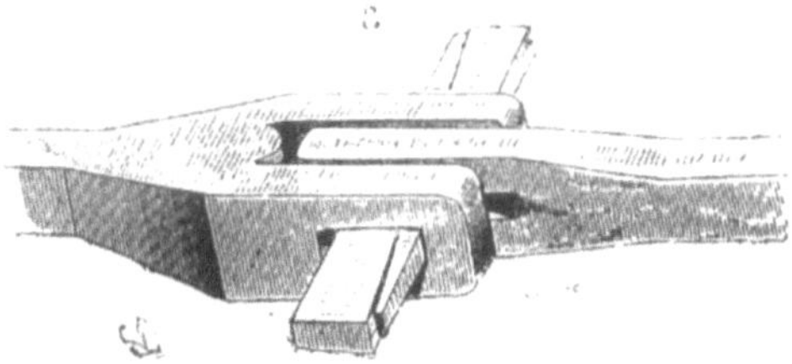

Figure 83. Iron 'free chain', French, fifteenth century. Engraving from E. E. Viollet-le-Duc, *Dictionnaire raisonné de l'architecture française du XIe au XVIe siècle*, vol. II, Paris, 1875.

Figure 84. Iron cramps, Notre-Dame, Paris, late twelfth century. Engraving from E. E. Viollet-le-Duc, *Dictionnaire raisonné de l'architecture française du XIe au XVIe siècle*, vol. II, Paris, 1875.

XXII

The Eddystone Lighthouse

Smeaton's memorial on the wall of the old parish church of Whitkirk is appropriately surmounted by a relief of the Eddystone Lighthouse. The memorial inscription refers to it as his principal work "erected on a rock in the open sea (where one had been washed away by the violence of a storm, and another had been consumed by the rage of fire) secure in its own stability and the wise precautions for its safety, seeming not unlikely to convey to distant ages, as it does to every Nation of the Globe, the name of its constructor".[1]

The invitation to rebuild it, after the destructive fire, came in January 1756 from Robert Weston, and was the turning point in Smeaton's fortunes. Weston was the principal shareholder of the small group of proprietors who held the remainder of a 99-year lease from Trinity House that enabled them to collect dues from shipping in return for maintaining a light on the Rock.[2] He was acting on behalf of them all and had first sought the advice of the Earl of Macclesfield, President of the Royal Society. Smeaton had only 3 years ago turned from making scientific instruments towards the practice of civil engineering, so he was not widely known in the field, yet the President felt able to recommend him in warm terms.

The task was a daunting one, hardly to be compared with more usual civil engineering works of the time—roads and bridges, canals and fen drainage, or even dock and harbour works. The main problems stemmed from the site and its extreme exposure to the fury of the sea.

The Eddystone Rocks, as well as being a great hazard to shipping passing up and down the Channel or entering or leaving Plymouth, present one of the most difficult sites imaginable for the construction of a lighthouse. They lie 14 miles south of Plymouth and almost midway between the Lizard and Start Point, fully exposed to gales from any direction between west and east and to the ground swells that persist for days after storms in the Bay of Biscay. At high tide they are largely covered. At low tide when the sea around them is calm, they are exposed in several roughly parallel reefs that are nowhere much above 40 ft wide. Their surfaces have a sharp dip towards the west which continues for a considerable distance below the water and leads to a massive build-up of any seas coming from this direction. The jagged rocks below water for some distance around also add greatly to the difficulty of mooring supply vessels in the vicinity. The rock itself is a hard gneiss, difficult to cut, partly on account of its tendency to split along the grain if a cut is attempted in any other direction.

As well as having to face these difficulties, Smeaton was heavily dependent on wind, tide and sea conditions in getting to and from the Rocks, and had to contend with the further hazards arising from war with France. With only sails and oars as means of propulsion the journey from Plymouth was always lengthy and sea conditions sometimes made landings impossible for weeks at a time even in the best seasons of summer and autumn. The war presented the double hazard of French privateers and the English press gang, the latter being in fact the more troublesome in taking away sailors at crucial moments. Local fishermen interfered more than once with essential moorings.

On the other hand, Smeaton was extremely fortunate in having the complete confidence and unswerving backing of the proprietors, chiefly in the person of Weston. He discussed his plans fully with them and reported regularly on progress, but was given complete authority to proceed as he saw fit.[3] In particular they sanctioned the adoption of a design completely in stone instead of a less costly one partly in timber as the previous structures had been. This certainly promised a saving in maintenance to offset partly the greater cost; but it was still a notable decision since the structure burnt down had served for almost half the term of the lease so that there was a good chance that a slightly modified one of the same type would last for the remaining term.

The successful completion of the lighthouse in 1759 in this more durable form fully vindicated the proprietors' confidence and the recommendation from which it initially stemmed. Smeaton's achievement aroused widespread interest—so much interest, indeed, that visitors wishing to see his models and talk about what had been done threatened, for a while, to take up all his time. Many years later, at the request of friends and of Trinity House, and to satisfy the continuing curiosity, he wrote his *Narrative of the Building*, which became one of the classics of the literature of civil engineering.[4] Any subsequent writer must start by acknowledging his great debt to that work.

Design

At the outset Smeaton had no first-hand knowledge of the site and the conditions he would encounter there, and he knew little of the previous lighthouses. Since a visit to Plymouth was unlikely to be profitable so early in the year, he attempted first to learn what he could about the previous structures. Apart from demonstrating that it was possible to build on the Rocks, they could provide valuable indications of the types of structure that would not, and would, survive the action of the sea. The first house, built by Winstanley between 1696 and 1698 and considerably strengthened in 1699, had been completely washed away in a storm only 4 years later. Its successor, built by Rudyerd between 1706 and 1709, had, on the other hand, survived many storms until burnt down in December 1755.

Weston was able, at his first meeting with Smeaton on 14th February, 1756, to provide a number of drawings and models, chiefly of Rudyerd's structure.[5] They left important details open to doubt, but nevertheless gave useful leads. Winstanley's house, as finally completed, had a whimsical elaboration of form and architectural detail that seemed almost to invite the sea to do its worst.[6] It appeared also to have been anchored to the sloping surface of the Rock only by twelve irons. Rudyerd simplified the form to a simple frustum of a cone devoid of projections apart from a cornice just below the lantern to throw off water rising up to this level, and an open iron stair to give access from the Rock.[7] This cone had a base diameter about one-third of the height up to the lantern, and a diameter at that level about one-fifth of the height. To anchor the structure to the Rock he first cut the sloping surface into steps and then inserted a larger number of ingeniously designed iron keys. The other most notable feature was the way in which timber and stone were combined. Externally the whole structure might easily have been taken for a purely timber one. It consisted here of 71 tapered uprights butted side by side. These were secured at intervals throughout their height to timbers running circumferentially inside. Internally the construction was entirely of timber in the upper half, where a store-room and living quarters were provided. In the lower half, which was solid apart from the necessary provisions for access, it was also mostly of timber, here in the form of superimposed horizontal beds of squared oak logs laid parallel to the steps in the rock and at right angles to them in alternate beds. Four or five courses of large blocks of Cornish moorstone were, however, interposed at intervals. A timber mast rose vertically in the centre and continued through the first two rooms above. Structural integrity was ensured largely by bolting the timbers together and otherwise interconnecting them, and by tying down the horizontal logs at the base to the Rock by means of the upward projections of the inset iron keys. In Smeaton's words the structure was "a piece of shipwrightry", merely ballasted by the courses of moorstone.[8]

In considering why Rudyerd's house had stood so much longer than Winstanley's, Smeaton concluded that it had done so partly because of its simpler overall form and partly because, by ballasting it, Rudyerd had prevented the layers of timber from being lifted if water seeped between them. Rudyerd had here "laid hold of the great principle of Engineery that WEIGHT is the most naturally and effectually resisted by WEIGHT".[9] Smeaton was less impressed by the widely held view that the "elasticity" of the timber structure was its chief merit, enabling it to give way to the sea without being overthrown. Of its considerable movement in a storm there seemed to be no doubt; but Smeaton attributed this chiefly to a want of sufficient diameter at the base, so that it was able to rock slightly. He preferred to place his trust in adequate weight, adequate strength, and a broader base.[10]

Rudyerd's use of timber also had two clear drawbacks. It had not only permitted the final destruction by fire; it had also been the cause of repeated trouble from rot and worm attack. These necessitated difficult piecemeal renewals, few of which were completely satisfactory, so that there was a gradual weakening of the whole structure that might soon have led to its collapse even without the fire.[11] A house made completely of stone would undoubtedly last longer with less maintenance if it could be given a similar strength. Against this it would cost more and take longer to build unless time could be saved by careful planning of all the operations.

Reflecting along these lines, Smeaton produced his first schematic design for a structure entirely of stone.[12] He kept essentially to the simple form and internal planning of Rudyerd's house; but, to obtain a broader base without a proportionate increase in diameter throughout the height, he adopted a curved profile suggested by the trunk of a growing oak firmly rooted in the ground. Bearing in mind the sloping surface of the Rock, he tested this profile by turning a piece of wood to it and then fitting this to the sloping surface of another block.

Without an enclosing timber structure, completely new means were devised to hold the masonry together. No mortar could be relied upon to do this during the crucial construction period; it would always be liable to be washed out before it had gained strength. The common expedient of using iron cramps—both vertically and horizontally between blocks—was ruled out chiefly because of the great number that would be needed and the great expenditure of time in fixing them, out on the Rock. This left only the alternative of shaping the blocks themselves so that they interlocked, like dovetail joints in timber. With some further cutting back of the steps in the rock surface to form dovetail recesses, the lowest courses of stone could similarly be engrafted into the rock.

This dovetailing could not be worked out in detail until the Rock surface had been examined and precisely surveyed, and other details of the design had been considered further. Smeaton drew an idealised plan of two typical successive courses showing, in each, six outer rings of dovetailed blocks around a single central block, with all joints breaking from course to course.[13] He also considered means of preventing whole courses above the level of the Rock from sliding on one another if hit by a heavy sea before the mortar was set between them, and of temporarily securing individual blocks of an incomplete course. Final choice could again be made later.

At this point he presented his proposals to the proprietors. They agreed, in principle, that the house should be rebuilt in stone.[14] He therefore felt free to travel to Plymouth, inspect the Rock, develop his ideas further, and make other necessary plans and arrangements. To assist in making an accurate survey of the Rock surface, he made an instrument with a long horizontal arm swinging about the centre of the plate of a large theodolite mounted on a tripod.[15]

Smeaton was in Plymouth from 27th March to 15th May, 1756. Soon after his arrival he was visited by Josias Jessop, a foreman shipwright in the dockyard, who had been in charge of repairs to Rudyerd's lighthouse for some time before the fire.[16] Though he found Jessop wanting in powers of invention, he formed a high opinion of his judgement and executive skills. Jessop also knew a great deal about the Eddystone Rocks, about conditions there, and about possibilities of approach and landing. During Smeaton's stay the two men set sail for the Rocks whenever the weather seemed suitable. There were only ten such occasions and only on four of them could a landing be made. With perseverance, however, it was possible eventually to make the necessary survey as well as to test the rock itself to see how it could be cut or drilled and how long these operations were likely to take.

The remaining time was not wasted. When forced to lie off the Rocks, Smeaton observed the action of the sea around them and considered means whereby the approach to them and the landing of materials could be facilitated, including the possible mooring of a supply vessel in the vicinity. When unable even to approach closely, he twice changed course for other harbours to see how they might serve as refuges in case of need, and to enable him to visit nearby quarries. While waiting on shore for better weather, and not engaged in setting his observations on paper, he inspected other sources of stone nearer Plymouth and looked for a suitable site for a shore base and workyard. He was also shown a piece of Portland stone from one of the docks that had been badly holed by a small shellfish, and subsequently saw marble rocks on the shore similarly eaten away.

He concluded from the last observations that the more easily worked Portland stone should not be used in positions exposed to the sea if the new structure were to be made as permanent as possible. He now saw no real obstacle to the carrying out of his initial plan for a stone structure if one of the Cornish granites were used everywhere externally. To proceed in this way would require much more work than a simple rebuilding to a slightly improved version of Rudyerd's design. In particular there would be a far greater quantity of stone to be taken to the Rock and set in place. He nevertheless thought that, with proper planning, he should be able to complete the construction in no more than four seasons. To achieve this it would be necessary to limit the initial cutting of the rock surface to the essential minimum, to do everything possible on shore to reduce what had to be done on the Rock, and to plan the whole operation so that progress was dependent as little as possible on the weather. He regarded the last two as fundamental. What he had already seen convinced him that it would be worth doing a week's work on shore to save an hour on the Rock.

Immediately on his return to London (having visited the Portland quarries on the way) Smeaton again met the proprietors, who confirmed their previous agreement to the use of stone.[17] The most important aspect of the design still to be considered was the precise way of fitting the structure to the Rock. The actual work here—both in cutting the rock and setting the first courses of stone—would be the most difficult that had to be done. It was therefore essential to plan it

Fig. 16. Eddystone Lighthouse, 1759 (engraving by Edward Rooker from a drawing by Smeaton—Narrative, *pl. 9)*

with the greatest care and to leave nothing to chance. To assist in planning, and to take full advantage of the measurements he had taken, he first made with his own hands a precise three-dimensional model of the rock surface as he had found it. He then made a second model showing how he intended to cut the level steps and dovetail recesses for those courses that were to be engrafted into the Rock. These models were supplemented by further models of the lower courses themselves and of the whole house and by a paper cross-section of this to show the internal arrangements proposed. Some details were still tentative, but the design was now essentially complete, so it will be convenient at this point to describe the structure as built (Fig. 16), before turning to its construction.

The structure as built

Six steps were cut into the Rock, each with two dovetail recesses and with its top surface slightly below the lowest adjacent rock surface so that the blocks laid on it would be further prevented from sliding. These steps were filled in with dovetail blocks breaking joint in successive courses—granite blocks on the outside and Portland stone on the inside (Fig. 17). In line with the objective of reducing work on the Rock to a minimum, the size of the blocks was as large as it was judged could be safely landed and placed in position with suitable tackle. A reasonable limit for most blocks was considered to be about 1 ton, with a few of up to 2 tons. To secure each block as it was set in place, and before a course was complete or the mortar hardened, it was wedged in place on each side of the engaged dovetail and treenailed down to the rock or blocks beneath to prevent lifting.

Course VI provided the first complete circular level platform. Above it the anchorage directly to the Rock was lost but it became possible to adopt the simple symmetrical arrangement of blocks in concentric rings envisaged in the first idealised plan. To prevent sliding of whole courses on one another an additional means of interconnection was introduced in the form of marble joggles. These were cubes set in matching recesses on the top and bottom faces of successive courses, shown shaded in the cross-sectional elevation, up to course XIV. From course XV to course XXIV, some further modification was needed on account of the entrance passage and central stairway. There was no longer a single central block. To give cohesion to the course, the four blocks of the inside ring were now designed to interlock with hook-scarf joints. The number of marble joggles was also increased and their size reduced. The use of treenails

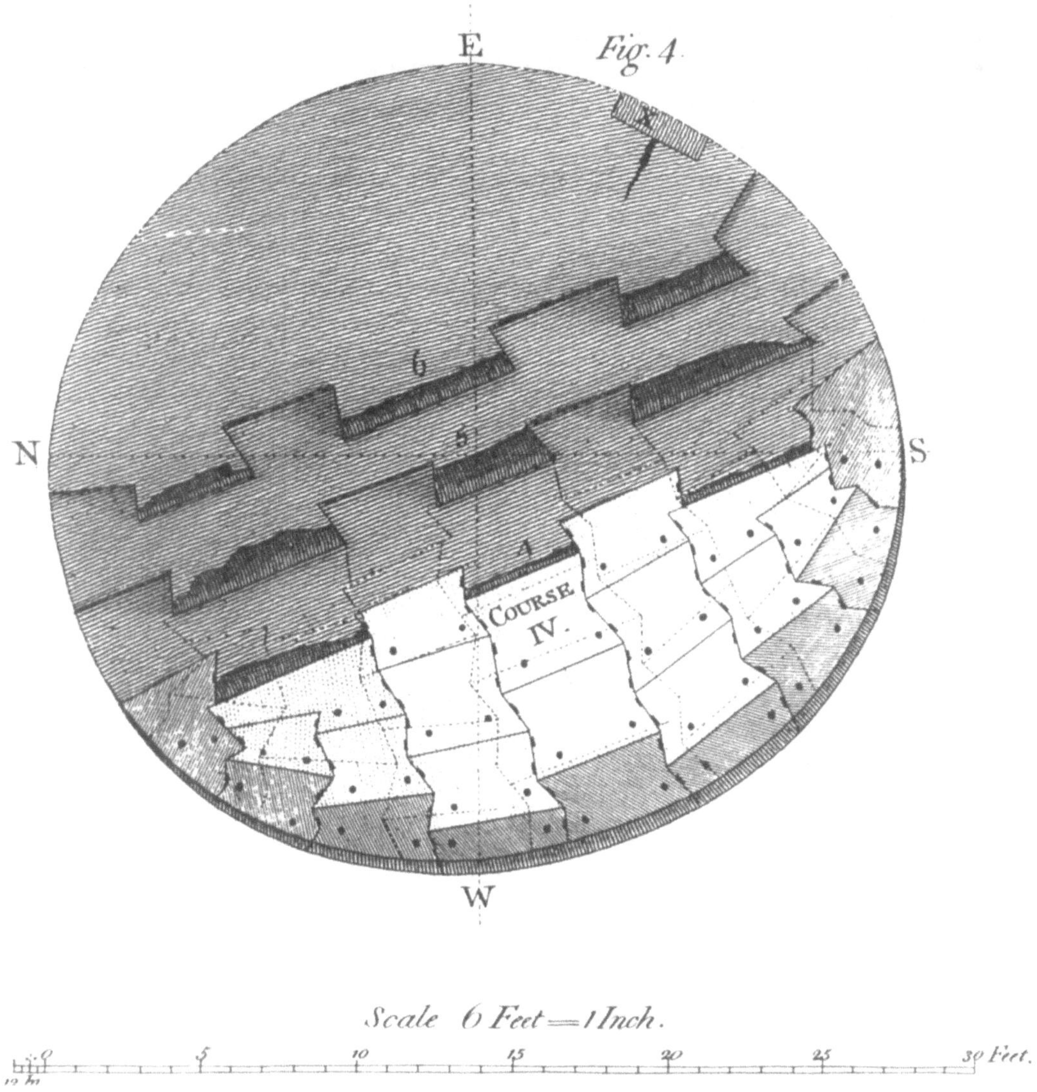

*Fig. 17. Plan of course IV (*Narrative, *pl. 10)*

and wedges continued; so did the use of Portland stone for the interior masonry.

Above course XXIV the structure changed to a circular wall, one block of granite thick, spanned at intervals by vaulted floors each with a central opening for a removable ladder. Dovetailing of the blocks was now abandoned, except in the vaulted floors. Iron cramps on the tops of the blocks were substituted, or pairs of continuous chains level with each floor. Smeaton felt that the weight of the masonry above would keep the horizontal joints watertight in the finished structure;

but he feared possible leaks through the vertical joints. Each vertical edge was therefore grooved and a flat lath-like piece of marble was inserted and mortared into the adjoining grooves of each pair of blocks as a water bar.

Though he still kept essentially to the same internal arrangements as had been previously adopted by Rudyerd, his careful consideration of every aspect of design led him to interchange the uses of the two habitable rooms just below the lantern. Rudyerd had made the upper room the kitchen/living room to minimise the necessary length of flue from the stove. Smeaton made the bedroom the upper room, so that it would be warmed by the rising warm air from below and the heat radiated by the flue pipe passing through it, and so that it would be free from the dampness from which Rudyerd's bedroom had suffered.

The lantern was framed in iron, with iron ties extended down to the main masonry wall at the level of the cornice. The copper flue pipe rose through it into a copper ball at the top. Warned by the damage he had observed early in 1757 to the steeple of the church at Lostwithiel, Smeaton also decided to provide a lightning conductor and considered that this flue could serve as its upper part.[18] It was continued downward, via the lead kitchen sink and drain pipe and an external strip of lead, to an iron chain bolted to the rock and long enough to have its foot at all times in the sea. The light was provided by two concentric rings of candles, suspended over pulleys so that they counterbalanced one another.

Construction

Smeaton fully realised the crucial importance of proper planning of the whole construction process. This was not merely a matter of planning the necessary operations with respect for the conditions under which they would have to be carried out. Success would also depend on ensuring that those working on the rock, or responsible for getting them to it and supplying them with what they needed, were able to take full advantage of all opportunities that the weather offered, and given sufficient incentive to do so. Planning of the operations was closely linked with the detailed design of the structure, and we have already seen instances of ways in which the design was influenced. It also entailed the design of suitable lifting shears and the design or selection of much more besides, from boats to mooring arrangements. Obtaining the best output from the workforce was partly a matter of choice and disposition of men, partly a matter of conditions

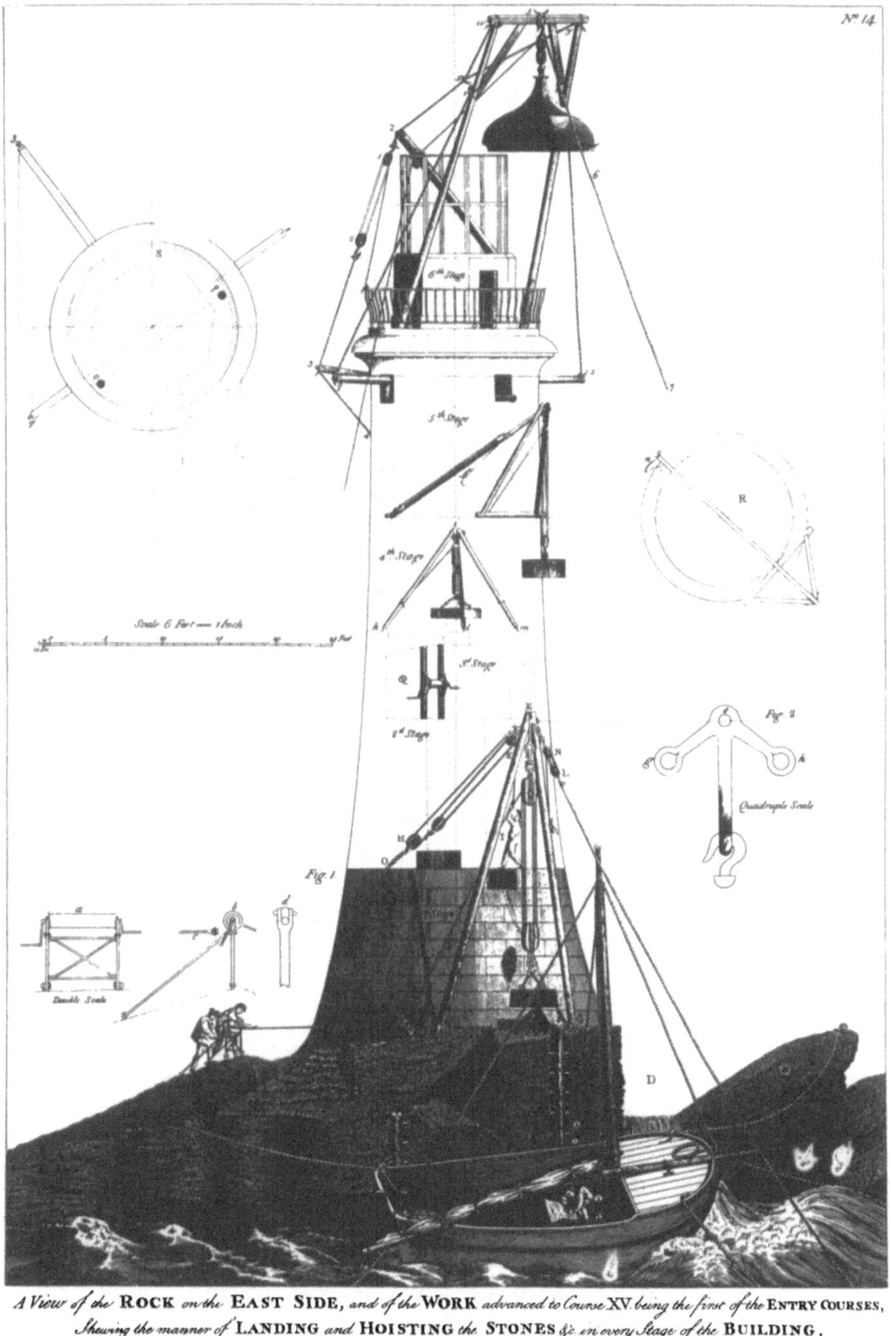

*Fig. 18. Construction of Eddystone Lighthouse (engraving by James Record from a drawing by Smeaton—*Narrative*, pl. 14)*

of pay and engagement, and partly a matter of supervision and encouragement.

For the shore base and work-yard a site was secured on Mill Bay, just to the west of Plymouth Hoe. Here there was provision not only for handling shipments of stone and cutting the blocks exactly to size but also for their trial pre-assembly and marking.[19] At about ½ mile north-east of the Rock a vessel was moored to serve as a store and provide accommodation for those working on the Rock. To facilitate construction there, several different lifting devices were used, each appropriate to a particular phase of the work. These are well illustrated in one plate of Smeaton's *Narrative* (see Fig. 18).

In addition to the necessary seamen, two companies of masons were established. It was arranged that they should work alternate weekly turns on shore and on the Rock. Terms of engagement and rates of pay were devised to encourage both seamen and masons to play their full part in achieving the maximum hours of work on the Rock that weather and tides permitted.[20] They were both fair and, in relation to other available employment at the time, generous. Smeaton therefore felt fully justified in dismissing anyone who did not readily accept them.

In reading the *Narrative* one is, however, struck most forcibly by the part that Smeaton himself played throughout the construction. At the outset he appointed Jessop his "general Assistant" or, as we should now say, his resident engineer.[21] This did allow him, from time to time, to attend to other business. The Eddystone Lighthouse, however, remained his main interest—even his consuming passion—until it had been completed and lit. He personally set out work on the Rock, made moulds for cutting the stones, supervised every new or critical operation, stepped in when men fell sick, shared all hazards, and even took the "post of honour" when this entailed a risk that others were reluctant to take.[22] Once, when the work was nearing an end, his zeal led to his collapse, probably from carbon monoxide poisoning, as he watched too attentively the heating of iron bars for the lantern in a charcoal fire inside one of the rooms of the lighthouse.[23] Much earlier he almost suffered shipwreck, which was averted by the presence of mind of one of the sailors.[24] It was perhaps partly this that, shortly afterwards, drew from Weston the admonition "venture not too rashly", to which Smeaton replied:

> "With respect to any rashness that may be in my constitution, I hope it is counterbalanced by Mr Jessop's calmness, whom I think Providence has

put along with me to keep me out of dangers that I might sometimes run into."[25]

Though he confessed, in connection with his having to dismiss one of the two foremen first appointed, that he found the management of men even more difficult than control of the elements, his sustained personal involvement must have contributed greatly to the successful completion of construction within the estimated four seasons.

The remainder of the first season, after the design had been completed and approved, was devoted to the necessary cutting of the rock surface and other preparation for the start of building.[26] The first landing was made on 3rd August, 1756, and the setting out of the steps was soon completed. Cutting and levelling then commenced, the first step for the base course being at a height of about 7 ft above LWST and therefore 9 ft below high water springs. With some interruptions from bad weather, this work was nearly finished in mid-September. Yet it took until early November to do what remained, and the store vessel was not finally brought back to Plymouth for the winter until 26th November, having been almost wrecked and then blown nearly to the Scilly Islands on the way.

There was much to be done on shore during the winter if everything was to be in readiness for building to start early next summer—from the cutting of stone to the building of additional vessels to carry it out and the selection of a suitable mortar for the very testing conditions to which it would be exposed. Smeaton therefore decided not to return to London. He remained in Plymouth, usually visiting the work-yard twice a day and devoting his evenings and other free time to the experiments on hydraulic limes that have already been described in Chapter II.

It was also in the winter of 1756–57 that Jessop, under Smeaton's direction, made drawings of the masonry courses and several models for Mr Weston. Most of these have survived and are now in the Royal Scottish Museum, Edinburgh (see Chapter XI). They include a model of the lighthouse itself, to the scale of 1 in. to 4 ft, and of Course VII (Fig. 19). The latter we know was sent to London in March 1757.[27]

The first stone, weighing 2½ tons, was landed and set on 12th June, 1757, and the remaining three stones of this course were set the next day. They were bedded in the lime/pozzolana mortar selected as a result of the experiments, wedged and treenailed down as described above, and then grouted with a more liquid mortar to fill the vertical

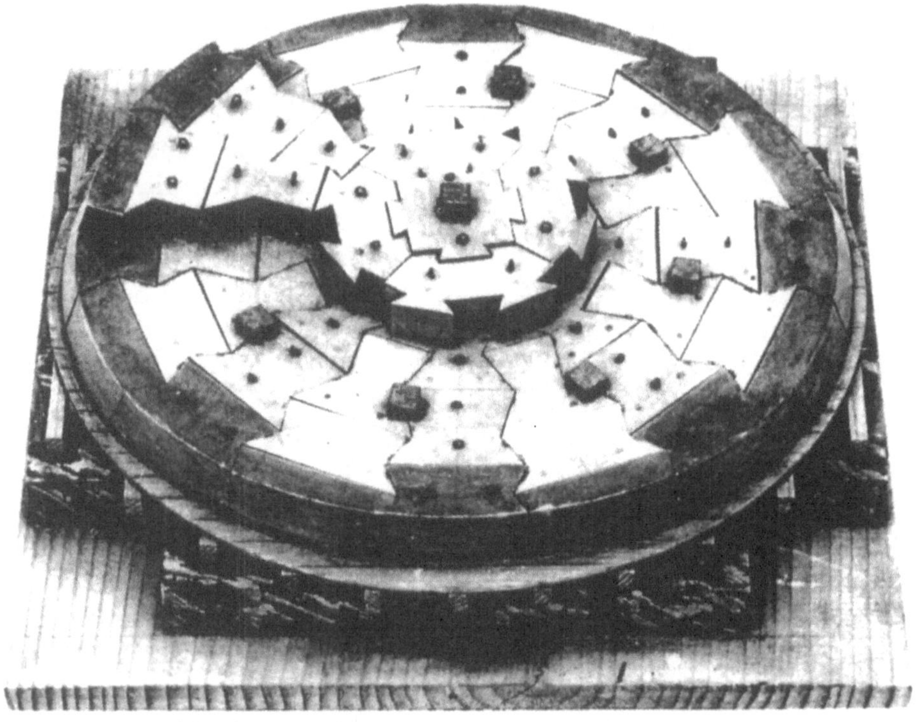

Fig. 19. Model of course VII made by Josias Jessop, 1757 (Royal Scottish Museum)

joints. Finally all the exposed ends of the joints were given a temporary protection of plaster of Paris to prevent the mortar from being immediately washed out by the next high tide.

The first course represents the beginning of a careful and often difficult series of operations in which 1,493 pieces of stone weighing nearly 1,000 tons were cut to shape, transported to the Rock, placed in position, and fixed together into a practically monolithic whole. Setting of the second course, consisting of thirteen stones, proceeded less smoothly. It was twice interrupted by gales and, on the second occasion, five stones that had not been fully set and fixed were washed away. Smeaton himself immediately returned to Mill Bay to obtain replacements when this was discovered, satisfied, however, that all his permanent fixings had held. The course was completed on 30th June. After this, work proceeded without major incident until the worst storm yet experienced interrupted the completion of course VII—the first full circular one rising above the top of the Rock. This storm washed away the lifting shears and two stones that were awaiting setting, and it badly damaged the windlass. Work resumed on 3rd

September after an enforced break of 15 days, and rapid progress was made until 20th September. There was then a further interruption which delayed the completion of course IX until 30th September. In view of the length of time that this course had taken, and the poor progress at the end of the first season, Smeaton decided to leave course X until next year.

He was pleased, on making a first inspection on 12th May, 1758, to find that all this masonry had become a single solid mass, the mortar having now become as hard as the Portland stone itself. A long series of delays nevertheless ensued, due to trouble with the moorings—including the loss of the buoy. Thus it was not until 2nd July that construction could be resumed. However, as the height above the water increased, there was less interruption from bad weather, and by 2nd October the centring was erected for the vaulted floor over the lowest room. Smeaton's hope was to complete this floor, so as to make the room habitable and make it possible to exhibit a temporary light from the house during the winter—much as Winstanley and Rudyerd had each done at the end of their third seasons. He was disappointed, however, by the onset of bad weather before the floor was complete and by the refusal of Trinity House to allow a light to be exhibited in this way. No more was done after 7th October.

Work on the Rock was resumed on 5th July, 1759. It was now becoming more akin to normal building construction on shore, apart from occasional difficulties in getting to the Rock or landing materials, and the much greater inconvenience in finding anything missing or having a man fall ill. Smeaton records one instance of improvisation when a quantity of block tin that had been ordered to harden the lead for setting door hooks was forgotten. He wished to see the hooks set before returning to Plymouth and accordingly melted down all the pewter plates and dishes that could be found on the store vessel. The main structure up to the cornice was completed on 17th August. It then rose about 50 ft above the bottom of the well hole and was found to lean eastward a mere ⅛ in. at this height. Since it was not now likely to be wet internally by the sea, the seamen were directed to wet the insides of all rooms to prevent the mortar from drying out too rapidly and assist its proper curing. The cupola of the lantern was lifted in one piece and fixed in place on 17th September, and on the following day Smeaton personally screwed in place the gilt ball on its summit. Glazing and all other preparations for exhibiting a light were completed by 9th October, when Smeaton and all others apart from the light-keepers left.

Smeaton's achievement

The light was first exhibited, as previously arranged with Trinity House, on the evening of 16th October. Stormy weather prevented Smeaton from being present, as he had hoped to be; but he went out again as soon as possible and was told that all was well in spite of the sea having washed up to the cornice and, being broken up there, had cast spray over the top of the lantern. Some motion had been felt, but not enough to cause any fear or surprise. Smeaton had, indeed, observed a similar motion on 5th October and had concluded, from observing that it was almost as sensible when the sea rolled over the Rock without touching the house, that it was essentially a motion of the rock imparted to the house.[28]

This weakness of the rock itself, coupled with a desire that the light should have a greater range and should not be obscured periodically in bad weather by being enveloped in a column of water or spray, led to the construction of a new lighthouse in a slightly more sheltered position on the parallel reef to the south-east between 1878 and 1882. After more than 120 years the upper part of Smeaton's tower was then dismantled and re-erected on a new base on Plymouth Hoe, the original base being left to stand on the Rock. The only weakness of the tower had proved to be in the upper horizontal joints of the masonry. These had opened under the action of the upward force exerted by waves hitting the underside of the cornice, necessitating the addition of internal vertical ties and, in 1865, a slight reduction in the projection of the cornice. Dismantling showed the mortar to be still in excellent condition.[29]

The best tribute to Smeaton's achievement was the extent to which his ideas inspired the designs of all later masonry lighthouses built in similarly difficult situations. The next, that of John Rennie and Robert Stevenson on the Bell Rock, built between 1806 and 1811, followed Smeaton's design in almost every respect.[30] In later structures the manner of interlocking was usually simplified, and advantage was taken of the availability of new cements and of new sources of power to expedite construction. The Fastnet Rock lighthouse, completed in 1904, still closely followed Smeaton's profile and general arrangement.

The most revealing comparison is with the new Eddystone structure—for this might be regarded as a direct constructive criticism of Smeaton's tower, just as his was of Rudyerd's.[31] The different choice of site was necessitated by the two new factors of the old site being still occupied and the now proven weakness of the rock itself at this point. But it involved founding the new structure below low-water level and, hence, the construction of a cofferdam and repeated pumping to allow work to commence. With the only means at his disposal this would probably have taken Smeaton another year. Sir James Douglass's simplification of the manner of interlocking the blocks of stone allowed the elimination of Smeaton's marble joggles by introducing interlocking bed joints as well as interlocking vertical joints. An increased wall thickness in the upper part of the structure (necessitated by the increased height) allowed the same manner of interlocking to be adopted throughout. There was therefore no need either for Smeaton's separate water bars in the vertical joints of the wall. Finally a vertical-sided cylindrical base was introduced below the inward curving profile of the upper part of the tower. Construction was very greatly facilitated and speeded by mooring a steam tender about 120 ft away to the south-east and transferring all stone and other materials directly from the tender to the place where they were needed on the structure. Each transfer, powered by a steam winch in the stern of the vessel, was completed in about 3 min.

The cylindrical base of the new structure was wide enough, in relation to the foot of the tower rising from it, to leave a narrow platform all round on which it was possible to land in the same way as it had been possible to land on the Rock itself in the case of the earlier structures. But its primary purpose was to divide an approaching wave and divert it to each side rather than, as Smeaton's sloping base had done, diverting much of it upwards.

Smeaton did foresee that, in a heavy sea, waves would rise towards the lantern, and he provided the cornice to protect it. However, he was unable to foresee the extent to which his choice of a continuously curved profile would lead to large quantities of water being thrown up vertically in some conditions of the sea. His choice of this profile was governed only by considerations of stability and strength under the assumed horizontal pressure of an advancing wave. It was guided by the analogy of the oak tree, already referred to. This, he assumed, was loaded horizontally by the wind and, stripped of its branches and foliage, should be just as well fitted to resist the pressure of a rapid current of water.[32] In adopting the simplest appropriate form that he could conceive for the purpose, he was also following that "first principle of taste, *Judgement*" that he had recognised in Rudyerd's eschewal of all the architectural elaborations of Winstanley's design.[33]

Arguing more generally he wrote:

"That the building should be a column of *equal strength*, proportionate in

> every part to the stress it was likely to bear, . . . was a view of the subject I was naturally and forcibly led to. . . . I therefore endeavoured to form it, and put it together so, that . . . no man should be able to tell me at *what* joint it would overset; for, if at any height the uppermost course was, when compleated, safe, it became more safe by another course being laid upon it; and that upper course, though somewhat less in weight, and in the total cohesion of its parts, than the former; yet every course, from the first foundation, was less and less subject to the heavy stroke of the sea."

At the same time he candidly admitted that he had followed his "*feelings*, rather than *calculations*".[34] He gave no mathematical construction for the profile, and seems to have drawn it by eye.

This recourse to his "feelings" was equally relevant to the choice of the other means for ensuring cohesion of the masonry. His approach here was a wisely cautious one: "yet, when we have to do with, and to endeavour to controul, those powers of nature that are subject to *no calculation*, I trust it will be deemed prudent not to omit . . . any thing that can without difficulty be applied, and that would be likely to add to the security."[35] Thus he used marble joggles as well as oak treenails to hold successive courses together in the circular solid above the Rock surface. He had originally considered using interlocking bed joints rather than joggles, but he decided against them because, in his view, they would introduce "unnecessary trouble and intricacy" as compared with "the much more plain, easy, and simple method of joggles".[36] He realised that, until the mortar hardened, everything would depend on the treenails, and the joggles would be of little value; also that, once it had hardened and there was more weight above, there should still be little need for them. Even in retrospect many years later, he felt that they contributed something to the consolidation of the whole mass, and that he could rely on them to continue to do so more confidently than he could rely on any wooden elements.

The dovetailed interlocking of the blocks that formed the vaulted floors in the upper part of the structure was more questionable. The intention was to unite the blocks so that each floor became, in effect, a non-thrusting monolith that would simply sit on the horizontal ledges cut to support it in the outer wall.[37] In fact these dovetails could never have achieved the necessary cohesion, though hook-scarf joints just might have done. The encircling chains—modelled on those that Wren placed around the dome of St Paul's and seen merely as a possibly redundant "extreme precaution"—were really the essential means for preventing the floors from pressing outwards.

In any critical assessment it should, however, be remembered that hardly any of the analytical tools we now have at our disposal were in existence in 1756.[38] Structural theory was very much in its infancy and, even today, it is still necessary to mount very costly research programmes to learn more about wave actions and wave loads on static structures in the open sea. Smeaton could, indeed, learn little of value from what he described as "*systematical* writings".[39] He was dependent on his own observations and experiments, the lessons he could learn from the available information about Winstanley's and Rudyerd's lighthouses, and the ideas he could pick up from accounts of other works, notably from those of Wren in the *Parentalia*, from Price's account of Salisbury Cathedral, and from Belidor's *Architecture Hydraulique* (the main source for the dovetail jointing of the masonry).[40] Simply by exercising "a turn for contrivance in the mechanical branches of science", he was required "to take into consideration the peculiarities, the advantages, and disadvantages of the situation, as well as other circumstances" and "solely from the nature of the thing . . . find out the proper methods of executing" the new structure.[41]

His achievement was also at least as much one of organisation and practical execution as of design. Though successors like the Stevensons and Douglasses improved on it, they learnt much from Smeaton's approach and followed it in all essentials, merely taking advantage of new means that were not available to him. Perhaps Rudyerd's earlier achievement should be counted a partial precedent; but, since he never recorded what he did, its only value to Smeaton was as a demonstration that he was not attempting the impossible. Of earlier achievements of which more is known, there was only one that was of such far-reaching importance, and called for such radical innovations in design and such close and continued involvement in all aspects of execution by the designer. This was Brunelleschi's construction of the dome of Florence Cathedral in the early fifteenth century—an achievement without a similar prior assurance that it was not impossible but without the problems that arose from a tiny exposed site far at sea.[42]

Finally we give a few facts and figures relating to Smeaton's lighthouse taken, except where noted, from the text and drawings in the *Narrative*:

General Assistant	Josias Jessop
Foreman	Thomas Richardson

Clerk	John Harrison
Work on the rock began	5th August 1756
Light first exhibited	16th October 1759
Light last exhibited[43]	3rd February 1882
Volume of masonry[43]	13,340 cu. ft
Height of light above mean sea level	80 ft
Construction cost[44]	£16,000

Notes

1. Smiles, S. *Lives of the Engineers*, Vol. 1 (London, 1861) p. 14.
2. Weston, R. H. *Letters and important Documents relative to the Edystone Lighthouse* (London, 1811). Hereafter referenced as *Letters*. The lease is given on pp. 257–262.
3. *Letters*. Weston to Smeaton, 6 Apr., 1756.
4. Smeaton, J. *A Narrative of the Building and a Description of the Construction of the Edystone Lighthouse* (London, 1791; 2nd edn, 1793; 3rd edn, 1813). Hereafter referenced as *Narrative* and by paragraph numbers in the (corrected) 2nd and 3rd editions.
5. Models of Rudyerd's lighthouse and its masonry courses are in the Weston collection at the Royal Scottish Museum, Edinburgh, together with some drawings; but most of the drawings (some possibly originals, others by Jessop and his predecessor John Holland) appear to have been retained by Smeaton as they are in the Royal Society (*Designs*, **5**, f. 193v, 193v 1.s. (i)–(vii), and f. 194, the latter being incorrectly catalogued).
6. *Narrative*, paras 17–21 and pls 4, 5.
7. *Ibid.*, paras 34–54 and pls 6, 7.
8. *Ibid.*, para. 50.

9. *Ibid.*, para. 40.
10. *Ibid.*, paras 79–80.
11. *Ibid.*, para. 85.
12. *Ibid.*, pl. 13, fig. 6.
13. *Ibid.*, pl. 13, fig. 5.
14. *Ibid.*, para. 84.
15. *Ibid.*, para. 97 and pl. 7.
16. Josias Jessop (*c.* 1710–60) was the father of William Jessop, the eminent civil engineer.
17. *Narrative*, paras 117–118.
18. *Ibid.*, para. 174. See also Smeaton, J. "An Account of the effects of lightning upon the steeple and church of Lostwithiel". *Phil. Trans.*, **50** (1757) 198–204.
19. *Narrative*, paras 90, 130, 154, 167 and pl. 17.
20. *Ibid.*, para. 101.
21. *Ibid.*, para. 130.
22. *Ibid.*, para. 223.
23. *Ibid.*, para. 305.
24. *Ibid.*, para. 159.
25. *Letters*, Smeaton to Weston, 18 Jan., 1757.
26. *Narrative*, Books III and IV describe the progress of the works in considerable detail. There is a tabular abstract after para. 331.
27. *Letters*, Smeaton to Weston, 22 Mar., 1757.
28. *Narrative*, paras 313, 309.
29. Douglass, J. N. "Note on the Eddystone lighthouse". *Min. Proc. Inst. Civ. Eng.*, **53** (1878) 247–248, and Douglass, W. T., *Min. Proc. Inst. Civ. Eng.*, **75** (1884) 54–55.
30. Stevenson, R. *An Account of the Bell Rock Lighthouse* (Edinburgh, 1824).
31. Douglass, W. T. "The new Eddystone lighthouse". *Min. Proc. Inst. Civ. Eng.*, **75** (1884) 20–36.
32. *Narrative*, para. 81.
33. *Ibid.*, para. 30.
34. *Ibid.*, para. 246.
35. *Ibid.*, para. 244.
36. *Ibid.*, para. 245.
37. *Ibid.*, para. 274.
38. It was possible, for the first time, to carry out a valid *post facto* statistical analysis in 1742 in relation to observed cracking of the dome of St Peter's, Rome. No similar analysis of a structure yet to be built is recorded until 1771 in relation to what is now the Pantheon, Paris. See Mainstone, R. J. *Developments in Structural Form.* (London, 1975) pp. 283–292.
39. *Narrative*, Introduction.
40. Wren, C. *Parentalia* (1750); Price, F. *A Series of Particular and useful Observations . . . upon . . . the Cathedral-Church of Salisbury* (1753); Belidor,

B. *Architecture Hydraulique* (1737–1753). The source of the dovetail jointing is Vol. 3 (1750) pp. 200–201 and pl. 26, fig. 4.
41. *Narrative*, para. 75.
42. Mainstone, R. J. "Brunelleschi's dome". *Architect. Rev.*, **162** (1977) 156–166.
43. Douglass, W. T. "The new Eddystone lighthouse". *Min. Proc. Inst. Civ. Eng.*, **75** (1884), pp. 33, 36.
44. Report by Trinity House (1809) and reply by R. H. Weston, in *Letters*, pp. 282, 294. The usually quoted figure of about £40,000 seems to stem from an erroneous interpretation of this Report.

Additional notes and comments

Chapters I to V

Looking back, I see little need to revise the ideas central to Chapter I. Some further thoughts and clarifications were presented (this time to an audience and readership composed largely of engineers) in the paper reproduced in Chapter II. But I still see as crucial the kinds of intuitive understanding identified in the earlier paper, both in the past and when they are now given greater depth and precision by the insights that stem from modern theory and a vastly wider range of experience.

On the other hand, I now think that, in Chapter IV, I may have overestimated the extent to which, in the eyes of designers, the geometric discipline of the *scientia geometriae* could have acquired structural authority from its association with Platonic ideas about the nature of the physical world, and thus have been thought capable, by itself, of guaranteeing structural stability. Faced with the need to make choices, they would have been too much aware of the inherent ambiguities referred to on page 85, and the consequent need for operational rules to remove them. They were probably less aware of the limitations of the more prescriptive rules referred to on pages 90-91. But, because of the way in which these were derived directly from experience in particular building traditions, even they had little in common with more recent geometric criteria derived from statical analysis such as those about which Heyman has frequently written.[1] I do not, however, agree that structural design underwent a long series of 'revolutions' akin to those detected by Kuhn in approaches to scientific explanation. For reasons stated elsewhere, I think that, in advancing this thesis, Addis pushes much too far the analogy that he sees between design and the very different activity of scientific explanation.[2]

In relation to current design practice, the chief shortcoming of these chapters is the limited attention they pay to the role of high speed computers – a role that can no longer be ignored as it was in a sequel to Chapter IV limited to vaulted masonry structures and assessments of their stability,[3] and that has increased considerably since the later paper reproduced as Chapter II was written. The concluding section of the chapter on design in the revised second edition of my *Developments in structural form* gives a more recent brief discussion of this role, with additional selected references.[4]

The other main aspect that receives less attention than it deserves is the emphasis that, throughout the design process, must have been placed on construction procedures at times when the possibilities were more limited and the stability of the incomplete structure must often have been a major preoccupation. Two papers not reproduced here presented a comparison between construction and natural growth and a more detailed study of past construction procedures for masonry vaults and domes.[5]

Chapters VII to XXII

Since many of these chapters were written (or first delivered in lecture form) in response to specific invitations or as comments on other papers, their emphasis varies. There are also some overlaps where they were addressed to different readerships or audiences. A more balanced review of the whole history, with much more emphasis on the last two centuries but less detail on the topics covered here, will be found in the book referred to above, especially in chapters 6 to 15.[4]

Chapters VII to XII

As mentioned in the Preface, the paper reproduced as Chapter X largely summarises the main content of my book on the church.[6] It is included chiefly to give context to the other chapters. The chief supplements to these chapters are the survey drawings of Van Nice[7] and the book itself, although the survey drawings depict it only as it is today (or rather as it was several decades ago) and the book concentrates very largely on the sixth-century structure. Because of the great complexity of this structure, neither can cover every aspect. Numerous details have never been accessible to inspection since they were first hidden when construction advanced beyond them, and much that has been observed remains recorded only in field notes. Likewise only a few of the results of numerous analyses made in the course of the investigation – numerical, graphical, physical, and chemical – have been published. For reference, the results of the chemical analyses of samples of the brickwork mortars are now given below. Since all samples had to be taken from near the surface, they may have shown more carbonation of the lime than would be found in the interiors of the larger masses. But the pozzolanic character, referred to on page 70 of the book, was clearly established by the significant presence of soluble silica in all samples, including the sixth-century ones.

Sample Source	piers, buttresses, and main arch springs						dome		
	SE	SE	SW	NE	NE	mean	rib 4	rib 18	rib 3
Sample Date	early 6th century						6th C	10th C	14th C
Insoluble Residue %	58.9	53.9	53.4	60.9	59.5	57.3	59.3	29.8	51.6
Soluble Silica %	5.6	6.2	8.5	7.6	5.7	6.7	6.3	10.8	14.8
Carbonate %	11.8	13.8	12.1	10.5	11.9	12.0	9.4	17.5	7.8
Total Lime %	15.4	17.9	17.0	13.7	15.9	16.0	13.8	25.2	14.4
Free lime %	0.07	0.15	0.02	0.07	0.03	0.07	0.05	0.08	0.06
Free lime/Total lime %	*0.4*	*0.9*	*0.1*	*0.5*	*0.2*	*0.4*	*0.3*	*0.3*	*0.4*

More recent tests of considerable interest have included a series of loading tests on brickwork prisms with wide mortar joints, as in the church.[8] These confirmed the relatively slow gain in stiffness that, with hasty construction, was assumed to have been responsible for the large early deformations.

The profile of the collapsed first dome is discussed further in Chapter XII, and certain passages in the book that had been misread in a review by Thomas Mathews were clarified in my reply.[9]

Publication of the book also stimulated a number of numerical analyses of the possible structural behaviour of the church, using finite-element models. But even the most ambitious of these models, contrived to take advantage of the most powerful computer then available, bore far too little relationship to the real structure.[10] Moreover, none of these analyses has taken properly into account the effects of large inelastic deformations that occurred during construction or the effects of the numerous built-in discontinuities and subsequent extensive cracking. Consequently, the results have borne little relationship to the demonstrable actual behaviour. Of greater value, in relation to structural behaviour, have been the records obtained of accelerations at the springing levels of the main arches and the dome during the Karacabey earthquake in 1992.[11] They showed very clearly a fall in the frequency of response when the amplitudes reached about 1 mm. Such a fall in frequency with increase in amplitude is characteristic of rocking as distinct from bending. It thus confirmed the major significance of cracking and other discontinuities.

Subsequent conservation work from scaffolds under the eastern semidome and dome has partly exposed mosaics already known to exist on the soffit of the main eastern arch, but previously hidden by plaster, and has permitted a more precise dating of some of those in the dome. It has also disclosed a lower ring of holes penetrating the dome brickwork, similar to those higher up through which lighting hoops have been suspended. These lower holes are likewise lined with clay tubes and can have served only to tie back circumferential supports for the dome and associated working platforms during construction, thus throwing further light on the probable construction procedure.

Other subsequent publications have included a detailed study of the main cornices (based partly on Van Nice's records), and an excellent purely historical account of the church during the Ottoman period.[12]

Finally, the conservation of the structure, the principal subject of Chapter XI, was discussed in greater detail and with specific recommendations in a report to the Turkish authorities prepared, at their request, by a UNESCO commission.[13]

Chapter XIII

The paper reproduced here was intended only as an interim discussion of the novel thesis by John Warren referred to at its outset. It was hoped that more definitive conclusions would be possible when the photogrammetric survey then in progress was complete, although it was feared that less precise setting out and less precise construction would make it more difficult than it had been in Hagia Sophia to determine the tell-tale deformations that had occurred during construction. Unfortunately this past lack of precision, coupled with certain limitations of the data furnished by the survey, did indeed leave uncertainties which could be resolved, if at all, only by considerable further study and measurement. The investigation undertaken after the paper reprinted here was written is described in two papers presented to conferences marking the ninth centenary of the consecration of the present church.[14]

Chapters XVI and XVII

The paper reproduced in the first of these chapters was written while the detailed study of the manner of construction of the dome was still in progress. Thus, although more obvious features like the *spinapesce* bond of the brickwork are mentioned, and although the statical analysis remains valid, this paper does not discuss the means whereby the shells were constructed to obtain the circular rings at successive levels assumed in the analysis of the inner one. Nor does Fig. 1 attempt a precise representation of the complex three-dimensional geometry, but instead shows the horizontal arches that project inwards from the outer shell conventionally meeting it along horizontal lines as in all previous representations.

In the paper reproduced as Chapter XVII, these shortcomings are made good. The figure just referred to is superseded by the new Figs 9 and 10, and the brief tentative reference on pages 297-298 to the likely alignment of the stone chains is superseded by note 20 on page 324.

The Brunelleschi centenary that stimulated the preparation of this second paper also led (together with concerns for the stability of the dome) to numerous other papers. Notable among them was an independent static analysis of the construction procedure by Salvatore di Pasquale.[15] This reached similar conclusions to those on page 292 of Chapter XVI. Several publications by Rossi presented a series of seductive perspective visualisations of the procedure that I described in Chapter XVII and put a slightly different gloss on it.[16] But these visualisations seriously misrepresented the circumferential arches, and the accompanying texts betrayed a limited understanding of some features of the design. More recently an analysis of the construction procedure followed at San Vitale in Ravenna has usefully explored the consequences, for a dome constructed without centring, of the time taken for each successive ring to stiffen and gain strength.[17]

Of the papers anticipated in notes 4 and 39 of Chapter XVII, only the second was published and this only in the Italian version presented at the conference.[18] It chiefly amplifies what is suggested on pages 164-166 about the likely origins of ideas. The first would have been based largely on my alternative draft of chapter III of what, finally, became Saalman's single-author monograph.[19] It would have concentrated on those aspects of the development of the design on which we were unable to agree and that were not covered in the present chapter XVII. I abandoned it because almost all of that draft on which we did agree was included in the monograph and too much repetition would have been called for to give context to what remained. For the previously unpublished documents and for a more comprehensive discussion of the whole documentation, reference must, therefore, be made to Saalman's monograph, treating it with due caution and preferably reading alongside it the excellent review by Trachtenberg.[20] This largely echoed (unwittingly) my own views on the principal matters of disagreement. Matters not covered by the present Chapter XVII were chiefly several aspects of the development of the design up to 1420. These included the parts played by some of those who contributed to it, and the nature of some of the models that were constructed in the course of this development, especially the 'new and final model' that was made shortly before the 1420 specification was agreed as a basis for starting work. This, I am sure, was a new model of both dome and drum. It was probably built to a scale similar to that of the surviving model of St Peter's dome and drum and likewise partly cut-away to reveal features that would otherwise have been hidden.

References

1. Heyman, J., 'On the rubber vaults of the Middle Ages, and other matters', *Gazette des Beaux-Arts*, vol. 71, 1968, pp. 177-188; idem, 'Beauvais Cathedral', *Transactions of the Newcomen Society*, vol. 40, 1967-1968, pp. 15-35.
2. Addis, W., 'The evolution of structural engineering design procedures', *Transactions of the Newcomen Society*, vol. 61, 1989-1990, pp. 51-62; Mainstone, R. J., comment on this, ibid, pp. 63-64.
3. Mainstone, R. J., 'Stability concepts from the Renaissance to today', *Stable-unstable?*, Leuven University Press, Leuven, 1988, pp. 65-78.
4. idem, *Developments in structural form*, second edition, Architectural Press, London, 1998. See also Groak, S., 'Representation in building', *RSA Journal*, issue 4/4, 1998, pp. 50-58.
5. Mainstone, R. J., 'On construction and form', *Program* (Columbia University School of Architecture), no. 3, 1964, pp. 51-70; idem, 'Einflusse der Konstruktion auf die Form von Kuppeln und Gewolben' ('Construction requirements as a factor in the choice of form of masonry vaults and domes'), *Zur Geschichte des Konstruierens*, DVA Stuttgart, 1989, 211-223.
6. idem, *Hagia Sophia - architecture, structure and liturgy of Justinian's Great Church*, Thames and Hudson, London and New York, 1988. (Note that, in plates A3 and A10, the earlier and later under-floor ties in the west gallery are interchanged: those in plate A3 should appear in plate A10 and vice versa.)
7. Van Nice, R. L., *St Sophia in Istanbul; an architectural survey*, Dumbarton Oaks, Washington, 2 instalments, 1965 and 1986.
8. Falter, H., et al., 'Experimental tests on the mechanical behaviour of reproduced Byzantine conglomerates', *Second RILEM/CSIRO/ACRA International Conference on Rehabilitation of Structures*, 1998, pp. 1-10.
9. Mainstone, R. J., Letter, *Journal of the Society of Architectural Historians*, vol. XLIX, 1990, pp. 230-231, replying to Mathews, T., Review, ibid, vol.XLVIII, 1989, pp.181-182.
10. Mark, R., Çakmak, A. S and Erdik, M., 'Preliminary report on an integrated study of the structure of Hagia Sophia: past, present, and future', in *Hagia Sophia from the age of Justinian to the present*, ed., Mark, R. and Çakmak, A., Cambridge University Press, 1992, pp. 120-131; Mainstone R. J., 'Questioning Hagia Sophia', ibid, pp. 158-176.
11. Çakmak, A. S., Davidson, R., Mullen, C. L. and Erdik M., 'Dynamic analysis and earthquake response of Hagia Sophia', *Structural repair and maintenance of historic buildings III*, ed., Brebbia, C. A. and Frewer, R. J. B., Computational Mechanics Publications, Southampton, 1993, pp.67-84.
12. Butler, L. E., 'Hagia Sophia's nave cornices as elements of its design and structure', in *Hagia Sophia from the age of Justinian to the present*, ed., Mark, R. and Çakmak, A., Cambridge University Press, 1992, pp. 57-77; Necipoglu, G., 'The life of an imperial monument: Hagia Sophia after Byzantium', ibid, pp. 195-225.

13. Mainstone, R. J., *Present state of the Hagia Sophia monument with recommendations for its preservation and restoration*, UNESCO, Paris, 1996 (serial no. FMR/CLT/CH/96/104).
14. idem, 'La storia della costruzione della basilica', *Storia dell'arte marciana: l' architettura*, ed. Polacco, R., Marsilio, Venice, 1997, pp. 201-206; 'The construction history of the basilica', *Convegno di scienza e technica del restauro della basilica*, Venice, 1995 (in the press).
15. Di Pasquale, S., 'Structural analysis of the dome in Florence', *IABSE symposium Venice 1983: Strengthening of Building Structures – Diagnosis and Therapy, Final Report*, IABSE, Zürich, 1983, pp. 161-168.
16. Rossi, P. A., *Le cupole del Brunelleschi*, Calderini, Bologna, 1982.
17. Roberti, G. M., Lombardini, N. and Falter, H., 'Late Roman domes in clay tubes: historical and numerical study of San Vitale in Ravenna', *Spatial structures: heritage, present and future*, International Association of Shell and Space Structures, Milan, 1995, pp. 1237-44.
18. Mainstone, R. J., 'Le origine della concezione strutturale della cupola di Santa Maria del Fiore', (The sources of structural ideas for the dome of S. Maria del Fiore), *Filippo Brunelleschi: la sua opera e il suo tempo*, Centro Di, Florence, 1980, pp. 883-892.
19. Saalman, H., *Filippo Brunelleschi: the cupola of Santa Maria del Fiore*, Zwemmer, London, 1980.
20. Trachtenberg, M., Review of the preceding, *Journal of the Society of Architectural Historians*, vol. XLII, 1983, pp. 292-297.

Index of subjects

INDEX OF SUBJECTS

INDEX OF SUBJECTS

Index of bridges, buildings, and places

Only the reference in the text is indexed where an illustration appears on another page unless the caption does more than name the structure.

INDEX OF BRIDGES, BUILDINGS, AND PLACES

INDEX OF BRIDGES, BUILDINGS, AND PLACES

INDEX OF BRIDGES, BUILDINGS, AND PLACES

Index of persons

Recent authors are referred to by name only.

INDEX OF PERSONS

INDEX OF PERSONS